BIOSCIENCE METHODOLOGIES IN PHYSICAL CHEMISTRY

An Engineering and Molecular Approach

BIOSCIENCE METHODOLOGIES IN PHYSICAL CHEMISTRY

An Engineering and Molecular Approach

Edited by

Alberto D'Amore, DSc, A. K. Haghi, PhD, and Gennady E. Zaikov, DSc

Apple Academic Press

TORONTO NEW JERSEY

Apple Academic Press Inc.
3333 Mistwell Crescent
Oakville, ON L6L 0A2
Canada

Apple Academic Press Inc.
9 Spinnaker Way
Waretown, NJ 08758
USA

Exclusive worldwide distribution by CRC Press, a member of Taylor & Francis Group

Printed and bound in Great Britain by TJ International Ltd, Padstow, Cornwall

International Standard Book Number-13: 978-1-926895-54-3 (Hardcover)

Library of Congress Control Number: 2013942331

Library and Archives Canada Cataloguing in Publication

Bioscience methodologies in physical chemistry: an engineering and molecular approach/ edited by Alberto D'Amore, DSc, A.K. Haghi, PhD, and Gennady E. Zaikov, DSc.

Includes bibliographical references and index.
ISBN 978-1-926895-54-3
1. Chemistry, Physical and theoretical. 2. Life sciences--Methodology. I. D'Amore, Alberto, editor of compilation II. Haghi, A. K., editor of compilation III. Zaikov, G. E. (Gennadiĭ Efremovich), 1935-, editor of compilation

QD461.B56 2013 541'22 C2013-903441-2

Apple Academic Press also publishes its books in a variety of electronic formats. Some content that appears in print may not be available in electronic format. For information about Apple Academic Press products, visit our website at **www.appleacademicpress.com** and the CRC Press website at **www.crcpress.com**

ABOUT THE EDITORS

Alberto D'Amore, DSc

Alberto D'Amore, DSc, is currently Associate Professor of Materials Science and Technology at Second University of Naples-SUN in Rome, Italy. He has authored more than one hundred scientific papers published in international journals and books. He is a member of the scientific committees of many international conferences and is on the editorial boards of several international journals. Currently he is the Chairman of the International-Times of Polymers (TOP) and Composites conferences.

A. K. Haghi, PhD

A. K. Haghi, PhD, holds a BSc in urban and environmental engineering from the University of North Carolina (USA); a MSc in mechanical engineering from North Carolina A&T State University (USA); a DEA in applied mechanics, acoustics and materials from the Université de Technologie de Compiègne (France); and a PhD in engineering sciences from the Université de Franche-Comté (France). He is the author and editor of 65 books as well as 1000 published papers in various journals and conference proceedings. Dr. Haghi has received several grants, consulted for a number of major corporations, and is a frequent speaker to national and international audiences. Since 1983, he served as a professor at several universities. He is currently Editor-in-Chief of the *International Journal of Chemoinformatics and Chemical Engineering* and the *Polymers Research Journal* and on the editorial boards of many international journals. He is also a faculty member of the University of Guilan (Iran) and a member of the Canadian Research and Development Center of Sciences and Cultures (CRDCSC), Montreal, Quebec, Canada.

Gennady E. Zaikov, DSc

Gennady E. Zaikov, DSc, is Head of the Polymer Division at the N. M. Emanuel Institute of Biochemical Physics, Russian Academy of Sciences,

Moscow, Russia, and Professor at Moscow State Academy of Fine Chemical Technology, Russia, as well as Professor at Kazan National Research Technological University, Kazan, Russia. He is also a prolific author, researcher, and lecturer. He has received several awards for his work, including the Russian Federation Scholarship for Outstanding Scientists. He has been a member of many professional organizations and on the editorial boards of many international science journals.

CONTENTS

LIST OF CONTRIBUTORS

Anton A. Artamonov
Scientific Company "Flamena", Moscow, Russia

Marina I. Artsis
N. M. Emanuel Institute of Biochemical Physics, Russian Academy of Sciences, 4 Kosygin St., Moscow 119334, Russia

Levan Asatiani
Ivane Javakhishvili Tbilisi State University I, Ilia Chavchavadze Ave., 0128 Tbilisi, Georgia

V. V. Belov
N. M. Emanuel Institute of Biochemical Physics, Russian Academy of Sciences, 4 Kosygin St., Moscow 119334, Russia
E-mail: emal@sky.chph.ras.ru

E.V. Belova
Plekhanov Russian University of Economics, Stremyanny per. 36, Moscow 117997, Russia

M. Biryukova
Federal State Budgetary Institution of Science, N. M. Emanuel Institute of Biochemical Physics, Russian Academy of Sciences, 4 Kosygina St., 119334 Moscow, Russia

T. I. Borodina
Joint Institute for High Temperatures Russian Academy of Science, Moscow, Russia

A.V. Bychkova
Federal State Budgetary Institution of Science, N. M. Emanuel Institute of Biochemical Physics, Russian Academy of Sciences, 4 Kosygina St., Moscow 119334, Russia
E-mail: annb0005@yandex.ru

Jan Chłopek
Department of Biomaterials, AGH University of Science and Technology, Krakow, Poland

J. A. Djamanbaev
Institute of Chemistry and Chemical Technology, Kyrgyz National Academy of Sciences, 720071, Kyrgyzstan, Bishkek, 267 Chui Prospect
E-mail: Djamanbaev-J@mail.ru

Alberto D'Amore
The Second Naples University, Dipartimento di Ingegneria Aerospaziale e Meccanica, 19 Via Roma, 8103 1Aversa (CE), Italy
E-mail: Alberto.Damore@unina2.it

E. Ya. Davydov
N. M. Emanuel Institute of Biochemical Physics, Russian Academy of Sciences, 4 Kosygin St., 119334 Moscow, Russia

Gennady Efremovich Zaikov
N. M. Emanuel Institute of Biochemical Physics, Russian Academy of Sciences, 4 Kosygin St., Moscow 119991, Russia
Tel.: +7 (495) 9397320, E-mail: chembio@chph.ras.ru

I. S. Gaponova
N. M. Emanuel Institute of Biochemical Physics, Russian Academy of Sciences, 4 Kosygin St., 119334 Moscow, Russia

Eteri Gigineishvili
Ivane Javakhishvili Tbilisi State University I, Ilia Chavchavadze Ave., 0128 Tbilisi, Georgia

O. A. Gololobova
Joint Institute for High Temperatures Russian Academy of Science, Moscow, Russia

A. K. Haghi
University of Guilan, Rasht P.O.BOX 3756, Guilan, Iran
E-mail: Haghi@Guilan.ac.ir

Victor Kablov
Volzhsky Polytechnic Institute (branch of) Volgograd State Technical University, 42-A, Engels St., Volzhsky 404120, Russia

V. T. Karpukhin
Joint Institute for High Temperatures Russian Academy of Science, Moscow, Russia
E-mail: karp@oivtran.ru

G. Kirshenbaum
Brooklyn Polytechnic University, 333 Jay St., Six Metrotech Center, Brooklyn, NYC, NY, USA
E-mail: GeraldKirshenbaum@yahoo.com

Mikhail A. Klimovich
N. M. Emanuel Institute of Biochemical Physics, Russian Academy of Sciences, 4 Kosygin St., Moscow, Russia

Dorota Klimecka-Tatar
Institute of Production Engineering, Czestochowa University of Technology, Poland

Dmitry Kondrutsky
Volzhsky Polytechnic Institute (branch of) Volgograd State Technical University, 42-A, Engels St., Volzhsky 404120, Russia
E-mail: kondrutsky@gmail.com

M. Konstantinova
Federal State Budgetary Institution of Science, N. M. Emanuel Institute of Biochemical Physics, Russian Academy of Sciences, 4 Kosygina St., 119334 Moscow, Russia

Anna Korneva
Institute of Metallurgy and Materials Science, Polish Academy of Sciences, Krakow, Poland
Department of Anatomy and Physiology, The Alfred Meissner Graduate School of Dental Engineering and Humanities, Ustron, Poland

E. Kostanova
Federal State Budgetary Institution of Science, N. M. Emanuel Institute of Biochemical Physics, Russian Academy of Sciences, 4 Kosygina St., 119334 Moscow, Russia

A. L. Kovarski
Federal State Budgetary Institution of Science, N. M. Emanuel Institute of Biochemical Physics, Russian Academy of Sciences, 4 Kosygina St., Moscow 119334, Russia

G. V. Kozlov
Kabardino-Balkarian State University, KBR, Nal'chik, Chernyshevsky St., 173, 360004, Russia
E-mail: I_dolbin@mail.ru

Mikhail V. Kozlov
N. M. Emanuel Institute of Biochemical Physics, Russian Academy of Sciences, 4 Kosygin St., Moscow, Russia

Z. G. Kozlova
N. M. Emanuel Institute of Biochemical Physics, Russian Academy of Sciences, 4 Kosygin St., Moscow 119334, Russia
E-mail: yevgeniya-s@inbox.ru, Fax: (495) 137-41-01

Wojciech Król
Chair and Department of Microbiology and Immunology in Zabrze, Medical University of Silesia in Katowice, Poland

Tomasz Kupka
Unit of Dental Materials Science of Department of Prosthodontics and Dental Materials Science, Medical University of Silesia, Katowice, Poland

Nodar Lekishvili
Ivane Javakhishvili Tbilisi State University I, Ilia Chavchavadze Ave., 0128 Tbilisi, Georgia
Email: lekino@gmail.com

V. B. Leonova
Federal State Budgetary Institution of Science, N. M. Emanuel Institute of Biochemical Physics, Russian Academy of Sciences, 4 Kosygina St., Moscow 119334, Russia

A. M. Likhter
Astrakhan State University Bld. 20a, Tatischeva St., Astrakhan, 414056 Russian Federation, Russia
E-mail: pjulia@pisem.net, Tel.: (8512) 25-17-18

V. S. Litvishko
Plekhanov Russian University of Economics, Stremyanny per. 36, Moscow 117997, Russia
E-mail: LVS-1@mail.ru

A. S. M. Lomakin
N. M. Emanuel Institute of Biochemical Physics, Russian Academy of Sciences, 4 Kosygin St., 119334 Moscow, Russia

Piotr Malara
Institute of Engineering Materials and Biomaterials, Silesian University of Technology, 18A Konarskiego St., 44-100 Gliwice, Poland
Clinic for Oral and Maxillofacial Surgery, DENTARIS Medical Centre, 12B Lowiecka St., 41-707 Ruda Slaska, Poland

M. M. Malikov
Joint Institute for High Temperatures Russian Academy of Science, Moscow, Russia

E. L. Maltseva
N. M. Emanuel Institute of Biochemical Physics, Russian Academy of Sciences, 4 Kosygin St., Moscow 119334, Russia
E-mail: emal@sky.chph.ras.ru

E. I. Martirosova
N. M. Emanuel Institute of Biochemical Physics, Russian Academy of Sciences, 4 Kosygin St., Moscow 119334, Russia
E-mail: ms_martins@mail.ru

A. Matskiv
Sochi Institute of Russian People`s Friendship University, 32 Kuibyshev St., Russia

Ludmila I. Matienko
N. M. Emanuel Institute of Biochemical Physics, Russian Academy of Sciences, 4 Kosygin St., Moscow 119334 Russia
E-mail: matienko@sky.chph.ras.ru

Anna Mertas
Chair and Department of Microbiology and Immunology in Zabrze, Medical University of Silesia in Katowice, Poland

Viacheslav M. Misin
N. M. Emanuel Institute of Biochemical Physics, Russian Academy of Sciences, Moscow, Russia

Larisa A. Mosolova
N. M. Emanuel Institute of Biochemical Physics, Russian Academy of Sciences, 4 Kosygin St., Moscow 119334 Russia

V. Mottaghitalab
University of Guilan, Rasht P.O. BOX 3756, Guilan, Iran

Izabela Orlicka
Department of Anatomy and Physiology, The Alfred Meissner Graduate School of Dental Engineering and Humanities, Ustron, Poland

Rajmund Orlicki
Department of Anatomy and Physiology, The Alfred Meissner Graduate School of Dental Engineering and the Humanities in Ustron, Poland

N. P. Palmina
N. M. Emanuel Institute of Biochemical Physics, Russian Academy of Sciences, 4 Kosygin St., Moscow 119334, Russia
E-mail: emal@sky.chph.ras.ru

P.V. Pantyukhov
Plekhanov Russian University of Economics, Stremyanny per. 36, Moscow 117997, Russia

G.B. Pariiskii
N. M. Emanuel Institute of Biochemical Physics, Russian Academy of Sciences, 4 Kosygin St., 119334 Moscow, Russia

Grazyna Pawlowska
Department of Chemistry, Czestochowa University of Technology, Poland

E. M. Pearce
Brooklyn Polytechnic University, 333 Jay St., Six Metrotech Center, Brooklyn, NYC, NY, USA
E-mail: EPearceg@gmail.edu

I. G. Plashchina
N. M. Emanuel Institute of Biochemical Physics, Russian Academy of Sciences, 4 Kosygin St., Moscow 119334, Russia

U. A. Pleshkova
Astrakhan State University Bld. 20a, Tatischeva St., Astrakhan, 414056 Russian Federation, Russia
E-mail: pjulia@pisem.net, Tel.: (8512) 25-17-18

T. V. Pokholok
N. M. Emanuel Institute of Biochemical Physics, Russian Academy of Sciences, 4 Kosygin St., 119334 Moscow, Russia

S. Razumovskii
Federal State Budgetary Institution of Science, N. M. Emanuel Institute of Biochemical Physics, Russian Academy of Sciences, 4 Kosygina St., 119334 Moscow, Russia

M. Rosenfeld
Federal State Budgetary Institution of Science, N. M. Emanuel Institute of Biochemical Physics, Russian Academy of Sciences, 4 Kosygina St., 119334 Moscow, Russia
E-mail: markrosenfeld@rambler.ru, Fax: +7 (495) 137-41-01

A. A. Rybalko
Sochi Institute of Russian People`s Friendship University, 32 Kuibyshev St., Russia

A. E. Rybalko
Sochi Institute of Russian People`s Friendship University, 32 Kuibyshev St., Russia

A. Shegolihin
Federal State Budgetary Institution of Science, N. M. Emanuel Institute of Biochemical Physics, Russian Academy of Sciences, 4 Kosygina St., 119334 Moscow, Russia

Lyudmila N. Shishkina
N. M. Emanuel Institute of Biochemical Physics, Russian Academy of Sciences, 4 Kosygin St., Moscow, Russia
E-mail: shishkina@sky.chph.ras.ru

G. B. Shustov
Kabardino-Balkarian State University, KBR, Nal'chik, Chernyshevsky St., 173, 360004, Russia

Vladimir V. Tsetlin
Institute for Biomedical Problems, Russian Academy of Sciences, Moscow, Russia

O. N. Sorokina
Federal State Budgetary Institution of Science, N. M. Emanuel Institute of Biochemical Physics, Russian Academy of Sciences, 4 Kosygina St., Moscow 119334, Russia

D. A. Strikanov
Joint Institute for High Temperatures Russian Academy of Science, Moscow, Russia

Krzysztof Sztwiertnia
Institute of Metallurgy and Materials Science, Polish Academy of Sciences, Krakow, Poland
Department of Anatomy and Physiology, The Alfred Meissner Graduate School of Dental Engineering and Humanities, Ustron, Poland

R. R. Usmanova
Ufa State Technical University of Aviation, 12 Karl Marks St., Ufa 450000, Bashkortostan, Russia
E-mail: Usmanovarr@mail.ru

G. E. Valyano
Joint Institute for High Temperatures Russian Academy of Science, Moscow, Russia

Włodzimierz Więckiewicz
Department of Prosthetic Dentistry, Faculty of Dentistry, Wroclaw Medical University, Wroclaw, Poland

Lidiya A. Zimina
N. M. Emanuel Institute of Biochemical Physics, Russian Academy of Sciences, 4 Kosygin St., Moscow 119334, Russia

Olga Zineeva
Volzhsky Polytechnic Institute (branch of) Volgograd State Technical University, 42-A, Engels St., Volzhsky 404120, Russia

Magdalena Ziqbka
Department of Ceramics and Refractories, AGH University of Science and Technology, Krakow, Poland

LIST OF ABBREVIATIONS

AO	antioxidants
AOA	antioxidant activity
AP	acetophenone
APA	antiperoxide activity
BAL	benzaldehyde
BCS	biocybernetical system
BH	benzyl hydroperoxide
BSA	bovine serum albumin
BSE	backscattered electrons
BZA	benzyl alcohol
CdTe	cadmium telluride
CE	cytotoxic effect
CIGS	copper indium gallium (di)selenide
CTMP	chemithermomechanical pulp
DBM	dibytyl maleate
DC	diene conjugates
DEF	diethylferrocene
DLA	diffusion-limited aggregation
DMPC	dimethyl phenylcarbinol
DPhO	2,5-diphenyl-1,5-oxazole
DSA	doxyl-stearic acids
DSSC	dye senetesized solar cell
DTA	differential thermal analysis
DTA	differential-thermal analysis
EMI	electromagnetic interference
EPR	electron paramagnetic resonance
EPR	electron paramagnetic resonance
ESR	electron spin resonance
ESR	electron spin resonance
EWG	electron-withdrawing groups
FA	fractal aggregates
FF	fill factor
GPi	graphitized polyimide
ICP-MS	inductively coupled plasma mass spectrometry
IM	iod-methylate

IR	Infrared
KD	ketodienes
L	lecithin
LPO	lipid peroxidation
MM	molecular mass
MNPs	magnetic nanoparticles
MNSs	magnetically targeted nanosystems
MOF	metal-organic frameworks
MPC	methylphenylcarbinol
MPP	maximum power point
nAg	nanoparticles
OSCs	organic solar cells
P	product
P3HT	poly(3-hexyl thiophene)
PAni	polyaniline
PBA	perbenzoic acid
PEH	phenylethylhydroperoxide
PET	poly thyleneterphthalate
pFXIII	Plasma fibrin-stabilizing factor
PhOH	phenol
PL	phospholipids
PSM	post-synthetic modification
PSU	polysulfone
PV	photovoltaic
PVP	poly(vinyl pyrrolidone)
ROS	reactive oxygen species
ROS	reactive oxygen species
SBUs	secondary building units
SCE	saturated calomel electrode
SE	secondary electron
SE	secondary electrons
SEM	scanning electron microscope
TGA	thermo gravimetric analysis
TSC	textile solar cell

LIST OF SYMBOLS

d = dimension of Euclidean space
E = total energy
exp = experimental value
h = dimensionalities of energy
h = hour
h = Planck's constant
H_0 = Gamete acidity
K_m = copolymer constant
l = liter
min = minute
n = full earthday number
pK_a = basicity parameter
Q = constant insolation
R = atomic radius
r = ionic radius
R_m = radius of the cylindrical chamber
s = second
t = time
T_g = Gordon-Tailor-Wood equation
W = comonomer molar fraction
wk = week
yr = year
c = velocity

Greek Symbols

Y = dimensionalities of energy
ν = electromagnetic wave frequency
λ = wavelength
η = intrinsic viscosity
τ = estimation of rotational correlation time of labels
ξ = coefficient of hydraulic resistance

PREFACE

Bioscience Methodologies in Physical Chemistry is an essential tool for understanding the complexities and goes beyond the introductory aspects. This volume shows how basic physicochemical principles are important to an understanding of all scientific aspects. Chapters are organized logically, with visual representations to aid in the understanding of difficult concepts. The book is well-established and written by experienced academics.

Bioscience Methodologies in Physical Chemistry covers the current state of theoretical and experimental studies from the physico-chemical standpoint and provides a technologist with the means for a quantitative approach to preparing compositions and choosing proper operating conditions for their processing into materials and articles. Recommendations are given on the practical use of found physic patterns as well as the physico-chemical analysis of the effect of the composition ingredients on the final properties.

In the first chapter of this book, the selective alkylarens oxidations with dioxygen in the presence of catalytical systems is studied. Kinetics and mechanism toluene and cumene oxidation by dioxygen with nickel (II) complexes are presented in the second chapter.

The importance of polymer-based solar cells in today's modern life is reviewed in Chapter 3. The formation of metal organic frameworks are described in Chapter 4. Quantum and wave characteristics of spatial-energy interactions are explained in Chapter 5. The fractal analysis in polycondensation is reviewed in Chapter 6. This chapter explains irreversible aggregation model used for the description of some features of different polymers polycondensation: aromatic copolyethersulfoneformals (APESF), diblockcopolyether of oligoformal 2,2-di-(4-oxiphenyl)-propane and oligosulfone of phenolphthalein (CP-OF-10/OS-10), and also polyarylate on the base of dichloroanhydride of terephthalic acid and phenolphthalein (F-2).

In Chapter 7 the selection of medical preparations for treating lower parts of the urinary system is studied in detail. Improvement of the functional properties of lysozyme by interaction with 5-methylresorcinol is investigated in Chapter 8. The aim of this chapter is to study the effect of MR on the surface activity of lysozyme and rheological properties of its adsorption layers at the air/solution (0.05 M phosphate buffer, pH 6.0) interface.

An introduction to the culture of *in vitro* rare bulbous plants of the Sochi Black Sea coast is presented in Chapter 9.

Chapter 10 studies the effect of cobalt content on the corrosion resistance in acid, alkaline, and Ringer's solution of Fe-Co-Zr-Mo-W-B metallic glasses. The aim of the research is to investigate the corrosion resistance of $Fe_{68-}xCo_xZr_{10}Mo_5W_2B_{15}$ materials in different environments, especially in Ringer's solution simulating electrochemical conditions of the human body. The results presented in this chapter are highly important to the potential application of these materials in medicine.

In Chapter 11 the methods used to study the processes of gaining information on optical biological objects are presented. The effect of free radical oxidation on structure and function of plasma fibrin-stabilizing factor is reviewed in Chapter 12.

Functioning similarity of physicochemical regulatory system of the lipid peroxidation on the membrane and organ levels is investigated in Chapter 13. The aim of this work is to study interrelationships between the physicochemical properties and the composition of lipids of liposomes formed from the different natural lipids and of the organ lipids of mice.

The tensile strength of three new soft silicone elastomers and comparison with results of previously tested material is investigated in Chapter 14. The aim of this chapter is to investigate the tensile strength of three silicone elastomers: A-Soft line 30, Elite Super Soft, and Bosworth Dentusil and compare the findings with previously evaluated material, Ufi Gel SC.

The influence of the earth's electromagnetic radiation on water by electrochemical method is analyzed in Chapter 15. The aim and scope of this analysis are to search for a new concept of mechanism of mild exposure of external factors on physical-chemical properties of water.

The formation of glycoside bonds mechanism of reactions is presented in Chapter 16. In Chapter 17, new additives for burning of the fuels is introduced. The changes of dynamic lipid structure of membranes are studied in Chapter 18.

Magnetic nanoparticles are introduced in Chapter 19. The goals of this chapter is to create a stable protein coating on the surface of individual MNPs using a fundamentally novel approach based on the ability of proteins to form interchain covalent bonds under the action of free radicals and estimate the activity of proteins in the coating.

In Chapter 20 heat resistance of copolymers of vinyl acetate as a component of biodegradable materials is investigated in detail. Chapter 21 presents the degradation and stabilization of polyvinylchloride. The influence of UV and visible laser radiation on layered organic-inorganic nanocomposites zinc and copper is described in Chapter 22. In this chapter, the authors present the results of studying the structural and morphological changes that occur after UV and visible laser irradiation of layered organic–inorganic composites zinc and copper synthesized by laser ablation in liquid.

Modeling and optimization of the design parameters scrubber is presented in Chapter 23. In Chapter 24 biocompatibility and antibacterial properties of polysulphone/nanosilver composites are evaluated. This chapter presents the obtaining and testing of bactericidal composite materials confronted with gram-positive and gram-negative bacteria and, then, the assessment of their biocompatibility under *in vitro* conditions.

The technique of harvesting cancellous bone from the proximal tibia for different applications in maxillofacial surgery is studied in Chapter 25. The purpose of the chapter is to describe the operative technique of cancellous bone harvesting from proximal tibia for filling the bone cavities in the orofacial region on the basis of literature reviews and the author's own experience. The chapter includes a step-by-step description of the operative technique, postoperative care, and early rehabilitation. Possible complications at the donor site, such as fractures due to the weakening of the tibia, donor site morbidity, the necessary period of postoperative hospital stay, and mean volume of cancellous bone obtained, are also described.

Experimental adhesive biomaterial in the development of restorative concept toward the biomimetic dentistry is presented in Chapter 26. Chapter 27 is an investigation on the icrostructure of dental casting alloy Ni-Cr-Mo (rodent). The last chapters cover some new aspects for biochemists.

This new book provides an application of physical principles in explaining and rationalizing chemical and biological phenomena. Chapters of this book do not stick to the classical topics that are conventionally considered as part of physical chemistry; the principles presented in the book are deciphered from a modern point of view, which is the strength of this book. The level of mathematics used in each chapter to formulate and prove the physicochemical principles is remarkably consistent throughout the whole book.

— Alberto D'Amore, A. K. Haghi, and Gennady E. Zaikov

CHAPTER 1

THE SELECTIVE ALKYLARENS OXIDATIONS WITH DIOXYGEN IN THE PRESENCE OF CATALYTICAL SYSTEMS

L. I. MATIENKO, L. A. MOSOLOVA, A. D'AMORE, and G. E. ZAIKOV

CONTENTS

1.1 INTRODUCTION

The major developments in hydrocarbon oxidations have often been motivated by the need for the ever-growing polymer industry. The functionalization of naturally occurring petroleum components through reaction with air or molecular oxygen was naturally seen as the simplest way to derive useful chemicals [1]. The research of N.N. Semenov (gas-phase oxidation reactions) [2] and later N.M. Emanuel (liquid-phase hydrocarbon oxidation with molecular oxygen) [3] and others [4] clarified the concepts of chain reactions and put the theory of free-radical autoxidation on a firm basis. Industrial practice developed alongside. The development of the industrial processes depends mainly on the investigators ability to control these processes. The one of the methods of control of the rate and mechanism of the free-radical autoxidation processes is the change of medium, in which the autoxidation occurs (the pioneer works of Professor G.E. Zaikov) [5], followed by [1, 6]. The homogeneous catalysis of liquid-phase hydrocarbon oxidation has played no fewer roles in the improvement of oxidation processes. The selective oxidation of hydrocarbons with molecular oxygen as an oxidant to desired products is now a foreground line of catalysis and suggests the use of metal-complex catalysts. In the last years the development of investigations in the sphere of homogeneous catalysis with metal compounds occurs in two ways – the chain free-radical catalytic oxidation and catalysis with metal-complexes, modeling the action of ferments. But the most of the reactions performed at the industrial scale are on autoxidation reactions mainly because of low substrate conversions at catalysis by biological systems models [1, 7].

In works of N.M. Emanuel and his school it was established for the first time that transition metals compounds participated in all elementary stages of chain oxidation process with molecular oxygen [8–13]. Later these discoveries were confirmed and described in reviews and monographs [14–20]. However, there is no complete understanding of mechanism yet. Special attention was attended to investigation of role of metals compounds at stages of free radicals generation, in chain initiation reactions (O_2 activation) and hydroperoxides dissociation. Reaction of chain propagation under interaction of catalyst with peroxide radicals (Cat + $RO_2^{\bullet}\rightarrow$) is studied insufficiently. Catalysis by nickel compounds (NiSt,

$Ni(acac)_2$) was studied in details only in works L.I. Matienko together with Z.K. Maizus, L.A. Mosolova, E.F. Brin [12, 21–23].

Solution of the problem of the selective oxidation of hydrocarbons into hydroperoxides, primary products of oxidation is the most difficult one. High catalytic activity of the majority of used catalysts in ROOH decomposition doesn't allow suggesting of selective catalysts of oxidation into ROOH to present day. Application of transition metals salts rarely leads to significant increase in process selectivity, since transformations of all intermediate substances are accelerated not selectively [20]. For alkylarens, hydrocarbons with activated C–H bonds (cumene, ethylbenzene) the problem of oxidation into ROOH at conditions of radical-chain oxidation process with degenerate branching of chain is solvable, since selectivity of oxidation into ROOH at not deep stages (~1–2%) is high enough (S~80–95%). In this case the problem is in increase of reaction rate and conversion of hydrocarbon transformation into ROOH at maintaining of maximum reachable selectivity. Obviously, effective catalysts of oxidation into ROOH should possess activity in relation to chain initiation reactions (activation by O_2) accelerating formation of ROOH and also should be low effective in reactions of radical decomposition of formed during oxidation process active intermediates [20]. It should be noted that except the catalytic systems developed by the authors nobody had proposed effective catalysts for selective oxidation of ethylbenzene into α-phenylethylhydroperoxide (PEH) up to now in spite of the fact that ethylbenzene oxidation process was well studied and a large number of publications and books in the sphere of homogeneous and heterogeneous catalysis were devoted to it [20, 24–27].

At recent decades the interest to fermentative catalysis and investigation of possibility of modeling of biological systems able to carry out selective introduction of oxygen atoms by C–H bond of organic molecules (mono- and dioxygenase) is grown [28–30]. Both the alkylarens oxidations at catalysis by biological systems models, and the traditional transition metal catalyzed liquid-phase radical-chain oxidation of alkylarens with dioxygen occurs mainly into the alcohols and carbonyl compounds. The recently discovered molybdoenzyme ethylbenzene dehydrogenase (EBDH) catalyzes the oxygen-independent oxidation of ethylbenzene to (S)-1-phenylethanol [31]. Unfortunately, dioxygenases able to realize chemical reactions of alkane's dioxygenation are unknown [29].

In addition to the theoretical interest, the problem of selective oxidation of alkylarens (ethylbenzene and cumene) with dioxygen in ROOH is of current importance from practical point of view in connection with ROOH use in large-tonnage productions such as production of propylene and styrene (α-phenylethylhydroperoxide), or phenol and acetone (cumyl hydroperoxide) [1, 32]. The method of transition metal catalysts modification by additives of electron-donor mono- or multidentate ligands for increase in selectivity of liquid-phase alkylarens oxidations into corresponding hydroperoxides was proposed by authors [33] for the first time. The mechanism of ligand modifiers control of catalytic alkylarens (ethylbenzene and cumene) oxidation with molecular oxygen into ROOH was established, and new effective catalysts for ethylbenzene and cumene oxidation in ROOH were modeled [33].

1.2 HOMOGENEOUS CATALYTIC OXIDATIONS OF ALKYLARENS WITH MOLECULAR OXYGEN

The various catalytic systems on the base of transition metal compounds have been used for the alkylarens oxidation with molecular oxygen. And all of them catalyzed alkylarens oxidations mainly to the products of deep oxidation [6, 34]. One of the most striking examples is the oxidation of alkylarens into carbonyl compounds and carbonic acids by dioxygen in the presence of so-called MC-catalysts (Co(II) and Mn (II) acetates, HBr, HOAc) [6].

Cobalt complexes with pyridine ligands, for example, catalyzed the oxidation of neat ethylbenzene to acetophenone in 70% conversion and 90% selectivity [35]. Mn porphyrin complex catalyzes the ethylbenzene oxidation with dioxygen to 3:14 mixture of methylphenylcarbinole and acetophenone in the presence of acetaldehyde [36]. The system $CuCl_2$–crown ether in the presence of acetaldehyde is efficient as catalyst of oxidation of ethylbenzene, indane, and tetralin by dioxygen (70°C) into the corresponding alcohols and ketones with high TON [37]. The oxidations were established to occur via a radical pathway and not by a metal–oxo intermediate. In the absence and in the presence of crown ether the hydroperoxide was established as the main product of the indane oxidation at room temperature [38].

The oxidation of ethylbenzene using iron-haloporphyrins in a solvent-free system under molecular oxygen at 70–110°C gives mixture of α-phenylethylhydroperoxide, methylphenylcarbinole, and acetophenone (1:1:1). The catalyst is (TPFPP=5,10,15,20-tetrakis (pentafluorophenyl) porphyrin). Ethylbenzene conversion does not more than 5%. The oxidation occurs via radical pathway [39].The products of ethylbenzene oxidation with air under mild condition ($T > 60°C$, atmospheric pressure), catalyzed by $[TPPFe]_2O$ or $[TPPMn]_2O$ (μ-oxo dimeric metalloporphyrins, μ-oxo-bis(tetraphenylporphyrinato)iron (manganese)) without any additive are acetophenone and methylphenylcarbinole. The ethylbenzene oxidation is radical chain oxidation in this case also. The ketone/alcohol (mol/mol) rations are 3.76 ($[TPPMn]_2O$, ethylbenzene conversion – 8.08%), 2.74 ($[TPPFe]_2O$, ethylbenzene conversion – 3.73%) [40].

1.3 THE APPLICATION OF THE DIFFERENT METHODS FOR INCREASE IN ACTIVITY AND SELECTIVITY OF HOMOGENEOUS CATALYSTS IN THE OXIDATION PROCESSES

The application of metal-complex catalysis opens possibility of regulation of relative rates of elementary stages Cat–O_2, Cat–ROOH, Cat–RO_2 and in that way to control rates and selectivity of processes of radical-chain oxidation [20]. Varying ligands at the metal center or additives, one can improve yields of the aim oxidation products, and control the selectivity of the reaction. Besides, initial catalyst form is often only the precursor of true catalytic particles and functioning of catalyst is always accompanied by processes of its deactivation. Introduction into reaction of various ligands-modifiers may accelerate formation of catalyst active forms and prevent or trig processes leading to its deactivation. The understanding mechanisms of the additive's action at the formation of catalyst active forms and mechanisms of regulation of the elementary stage of the radical-chain oxidation may be resulted in new efficient catalytic systems and selective catalytic processes.

The methods of heterogeneous catalysts modification with additives of different compounds, which increased catalytic activity and protected catalysts from deactivation, are known for a long time. But researches of action of various ligands-modifiers in homogenous catalysis are often rare and relate mainly to investigating of ligands-modifiers influence on catalyst activity in radical initial stages (O_2 activation, ROOH homolytic decomposition) [20, 24]. Besides this, the reaction of O_2 activation by transition metal complexes in schemes catalytic radical chains oxidation is not taken into consideration in most cases. The additives, often being axial ligands for metal complexes, are considered in models, which mimic enzyme reaction center (mono- and dioxygenase). At now the numerous examples of various catalytic reactions are known when addition of certain compounds in small amounts dramatically enhances the reaction rate and rarely the product yield. As a rule mechanisms of the additives' action are not proved although the authors tentatively propose mechanistic explanations [41]. The works of Ellis and Lyons and more recently that of Gray and Labinger have identified the halogenated metal porphyrins – catalyzed oxidation of alkanes into alcohols by dioxygen at the mild conditions (100°C) [42–45]. However, substituted alkanes such as 2-methylbutane, 3-methylpentane, 2,3-dimethylbutane, and 1,2,3-trimethylbutane are oxidized into a mixture of products due to oxidative cleavage of carbon- carbon bond [43]. The including of halogen, electron withdrawing substituents, into porphyrin ligand increases the stability and as result the activity of halogenated iron porphyrins [42, 43]. Nevertheless the relative low conversion is due to the catalyst decomposition. It is now generally agreed that one-electron redox reactions and oxygen-centered free radical chemistry being about the oxidations in these systems are most probably than mechanisms similar to those proposed for biological oxidations by Cytochrome P-450 and methanemonooxygenase (through two-electron oxygen-transfer processes at participation of high valent metal-oxo oxidant) [44–46]. Perhalogenated iron porphyrins are known to be effective at decomposing alkyl hydroperoxides via free radicals formation [45, 46].

1.3.1 IMMOBILIZATION OF HOMOGENEOUS CATALYST ON HETEROGENEOUS SUPPORT FOR INCREASE IN ACTIVITY AND SELECTIVITY OF CATALYST IN THE ALKYLARENS OXIDATIONS

Several studies are focused on the silica-, zeolite- and polymer – supported metal – catalyzed oxidation [47–54].

The potential advantages of using a solid catalyst include the case of its removal from the oxidation mixture and subsequent reuse, and control of its reactivity through the microenvironment created by the support. The metal complexes, heterogenised in the zeolite pores, are prevented from deactivation; the oxidation of the ligand by another complex cannot be realized. The increase in stability encapsulated salen complex arises from the protection of the inert zeolite framework, making complex degradation more difficult by impeding sterically the attack to the more reactive parts of the ligand, and the life of salen catalyst is prolonging [48]. At the same time the zeolite influences on the formation of products by steric and electronic influences on transition state of the reaction, they also control the entry and departure of reagents and products. The one of the limitations of zeolites are that their tunnel and pore sizes are no large than about 10 Å [50]. The occluded catalytic complexes require a zeolite with caves or intersections, which are large enough to embed them. For these purposes faujasites, containing super cages, are most frequently used [48]. The creation of mesopores in zeolite particles to increase accessibility to internal surface has been the subject of many studies (mesopore – modified zeolites). It is known that postsynthesis hydrothermal dealumination and other chemical treatments form defect domains of 5 – 50 nm (which are attributed to mesopores) in faujasites, mainly zeolite Y [48].

The low activity of these zeolite catalysts is connected with their highly hydrophility as result of low silicon to aluminum ration. The deactivation by sorption of polar products and solvent on pores of zeolite still remained a serious issue for oxidation of alkanes (with low polarity). Even dealumination of the structure up to silicon to aluminium ratio above 100, increased the activity only twice [48]. The creation of a hydrophobic environment around the active site was required to circumvent the activity and sorption problems. In the case of the reaction of cyclohexane oxidation to adipic acid with air in the presence of Fe – aluminophosphate-31 (ALPO-31) (with narrow pore, 0.54-nm diameter) cyclohexane is easily adsorbed in the micro pores [51]. But desorption of initial products such as cyclohexylperoxide or cyclohexanone is slow. Consequently, subsequent radical reactions occur until the cyclohexyl ring is broken to form linear products that are sufficiently mobile to diffuse out of the molecular sieve. In contrast, with a large pore Fe – ALPO-5, cyclohexanol and cyclohexanone account for ~60% of the oxidation products. Thus, localization of a free radical reaction inside micro pores seems to give rise to particular selectivity.

Often the catalytic activity is unchanged practically if supported metal complex is used. So the silica – and polymer – supported iron(III) tetrakis(pentafluorophenyl)porphyrins, $FeTF_8PP$, [49] catalyzed the ethylbenzene oxidation reactions by dioxygen into the same three products, α-phenylethylhydroperoxide, methylphenylcarbinole, and acetophenone (1:1:1), as analogous homogeneous catalyst, suggesting that these catalytic oxidations proceed by the same mechanism. However, in general, the heterogeneous catalytic ethylbenzene oxidation is even slower. The products yields are limited by the stability/activity of iron porphyrin and these in turn are dependent mainly on catalyst loading and microenvironment provided by support.

The "neat" and zeolite-Y-encapsulated copper tri- and tetraaza macrocyclic complexes exhibit efficient catalytic activity in the regioselective oxidation of ethylbenzene using *tert*-butyl hydroperoxide [52]. Acetophenone was the major product; the small amounts of o- and p-hydroxyacetophenones were also formed, revealing that C–H activation occurs at both the benzylic and aromatic ring carbon atoms. The latter is significant over the "neat" complexes in the homogeneous phase, while it is suppressed significantly in the case of the encapsulated complexes. Molecular isolation and the absence of intermolecular interactions (as revealed by EPR spectroscopy), synergism due to interaction with the zeolite framework and restricted access of the active site to ethylbenzene are the probable reasons for the differences in activity/selectivity of the encapsulated catalysts. The differences in selectivity are attributed to the formation of different types of "active" copper–oxygen intermediates, such as side-on peroxide, bis-μ-oxo complexes and Cu-hydroperoxo species, in different proportions over the "neat" and encapsulated complexes.

Water-soluble catalysts combining the properties of metal complexes and surfactants on the basis of terminally functionalized polyethylene glycols and block-copolymers of ethylene oxide and propylene oxide with various combinations of ethylene and propylene oxide fragments were investigated [53]. Polymers, functionalized by dipyridyl and acetyl acetone were used as ligands for preparation Co(II) complexes. Macro complexes PEG-acac-Co turned out to be more active than their non-polymeric analogues in oxidation of ethylbenzene by dioxygen under the same temperature (120°C). The only product was acetophenone. Cobalt remains fixed at the end of the polymer chain with acac-ligand and is surrounded by oxygen atoms of the PEG chain. Such surrounding is labile and does not preclude from activation of dioxygen.

The activity of the liquid phase polyhalogenated metalloporphyrins (Co, Mn, Fe) and supported catalysts (silica, polystyrene) and the cationic metalloporphyrins encapsulated in NaX zeolite are founded to be active for cyclooctane oxidation with molecular O_2 into ketone and alcohol with primary ketone formation. At the last case the ration c-one/c-ol is higher than at the use supported on silica and polystyrene catalysts and in fact coincide with results, which are received with the cationic metalloporphyrins in solution. [54].

1.3.2 MODIFICATION OF METAL COMPLEX CATALYSTS WITH ADDITIVES OF MONODENTATE AXIAL LIGANDS

For the first time the phenomenon of significant rise of not only initial rate (w_0), but also the selectivity (S = [PEH] / Δ[RH]·100%) and conversion degree (C = Δ[RH]/$[RH]_0$·100%) of oxidation of alkylarens (ethylbenzene, cumene,) into ROOH by molecular O_2 under catalysis by transition metals complexes $M(L^1)_2$ (M = Ni(II), Co(II), L^1=acac$^-$) in the presence of additives of electron-donor monodentate ligands (L^2 = HMPA (hexamethylphosphorus triamide), dimethyl formamide (DMF), N-methyl pyrrolidone-2 (MP)), MSt (M = Li, Na, K) was found by authors of the articles [55–57].

On the example of ethylbenzene oxidation (120°C) the mechanism of control of $M(L^1)_2$ complexes catalytic activity by additives of electron-donor monodentate ligands L^2 (L^2 = HMPA, DMF, MP, MSt) was established [58–61].

The coordination of exo ligand L^2 to $M(L^1)_2$ changes symmetry of complex and its oxidative-reductive activity. At that the catalytic activity of formed in situ primary complexes $M(L^1)_2 \cdot L^2$ is increased that is expressed in the rise in the rate of free radical formation in chain initiation (activation by O_2) and PEH homolytic decomposition, and increase in initial oxidation rate (I macro stage) [58, 59]. In this connection at the first macro stage the selectivity of ethylbenzene oxidation into PEH is not high. With process development the increase in S_{PEH} ($S_{PEH,max} \approx 90\%$) in comparison with I macro stage ($S_{PEH,max} = 80\%$), and decrease in reaction w are observed (II macro stage). Ligands L^2 control transformation of $M(L^1)_2$ complexes into more active selective particles. At that the rise in S_{PEH} is reached at the expense of catalyst participation in activation reaction of O_2, and inhibition of chain and heterolytic decomposition of PEH. Beside this the direction of formation of side products, acetophenone (AP) and methylphenylcarbinol, (MPC), is changed from consequent (under hydroperoxide decomposition) to parallel at the expense of modified catalyst in the chain propagation ($Cat + RO_2^{\bullet} \rightarrow$). At the III macro stage the sharp fall of the S_{PEH} is accompanied by the increase in the rate of PhOH formation at the PEH heterolysis, catalyzed by the completely transformed catalyst [59–61].

We have established that in the case of use of nickel complexes $Ni(L^1)_2$ (L^1=acac$^-$) selective catalyst is formed as result of controlled by L^2 ligand regio-selective connection of O_2 to nucleophilic carbon γ-atom of one of the ligands L^1. Coordination of electron-donor exo ligand L^2 by $Ni(L^1)_2$ promoting stabilization of intermediate zwitter-ion $L^2(L^1M(L^1)^+O_2^-)$ leads to increase in possibility of regio-selective connection of O_2 to acetylacetonate ligand activated in complex with nickel(II) ion. Further introduction of O_2 into chelate cycle accompanying by proton transfer and bonds redistribution in formed transition complex leads to break of cycle configuration with formation of (OAc^-) ion, acetaldehyde, elimination of CO and is completed by formation of homo- and hetero poly nuclear heteroligand complexes of general formula $Ni_x(acac)_y(L^1_{ox})_z(L^2)_n$ (L^1_{ox} = $MeCOO^-$)

("A") (Scheme 1-3) [59-61]. Transformation of complexes $Ni(acac)_2 \cdot L^2$ (L^2 = HMPA, DMF, MP, MSt)) leads to formation of homo bi-(L^2 = HMPA, DMF, MP) or hetero three nuclear (L^2 = MSt, M=Na, Li, K) heteroligand complexes "A": $Ni_2(OAc)_3(acac)L^2$ (Scheme 1) [10]. The structure of the complex "A" with L^2 =MP is proved kinetically and by various physical-chemical methods of analysis (mass-spectrometry, electron and IR-spectroscopy, element analysis).

Scheme 1.

$$L^2 \cdot L^1Ni(COMeCHMeCO)_2 + O_2 \rightarrow L^2 \cdot L^1Ni(COMeCHMeCO)^+ ... O_2^-$$
$$L^2 \cdot L^1Ni(COMeCHMeCO)^+ ... O_2^- \rightarrow L^2 \cdot L^1Ni(MeCOO) + MeCHO + CO$$
$$L^1 = (COMeCHMeCO)^-$$
$$2Ni(COMeCHMeCO)_2L^2 \xrightarrow{O_2} \frac{Ni_2(MeCOO)_3(COMeCHMeCO)L^2 + 3MeCHO + 3CO + L^2}{L^2 = N\text{-метилпирролидон-2}}$$

Scheme 2.

$$Ni(COMeCHMeCNH)2 + O2 \longrightarrow .L1 \times Ni(COMeCHMeCNH)+ ... O2-$$
$$L1 \times Ni(COMeCHMeCNH)+ ... O2- \longrightarrow L1 \times Ni(NHCOMe) + MeCHO + CO$$
$$(Q)$$
$$\downarrow H_2O$$
$$L^1 \times Ni(MeCOO) + NH_3$$
$$(P)$$

Transformation of $Ni(L^1)_2$ (L^1=enamac$^-$, chelate group (O/NH)) is realized in the absence of activating ligands (L^2) [60] (L^1_{ox}=NHCOMe$^-$ or MeCOO$^-$) (Scheme 2) by analogy withreactions of oxygenation imitating the action of L-tryptophan-2,3-dioxygenase [62, 63].

The principle scheme of oxygenation of ligand $(acac)^-$ in complex with Ni(II), initiated with exo ligand L^2.

Similar change in 'complexes' ligand environment in consequence of acetylacetonate ligand oxidative cleavage under the action of O_2 was observed in reactions catalyzed of the only known to date a Ni(II)-containing dioxygenase – acireductone dioxygenase, ARD [64], and in reactions of oxygenation imitating the action of quercetin 2,3-dioxygenase (Cu, Fe) [65, 66].

The similarity of kinetic dependences in the parent processes of ethylbenzene oxidation in the presence of {$Fe(III)(acac)_3 + L^2$} and {Ni(II) $(acac)_2 + L^2$} (L^2=DMF) (120°C) is in agreement with assumption that transformation of $Fe(II)(acac)_2 \cdot DMF$ complexes, formed at initial stages

Scheme 3.

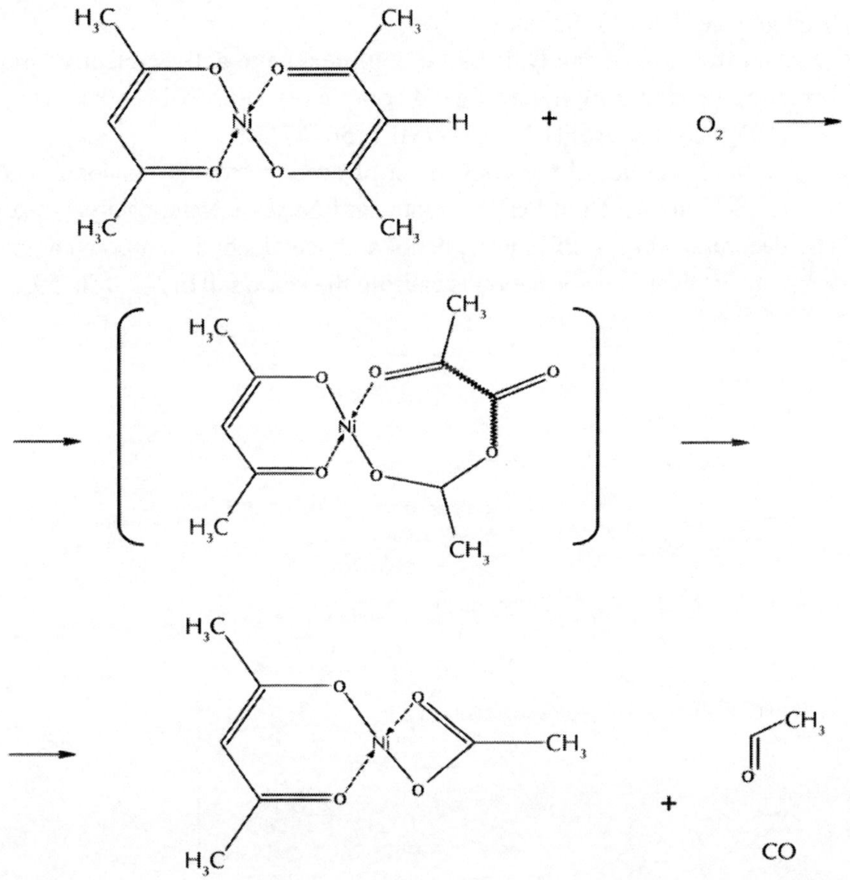

of ethylbenzene oxidation at catalysis by {Fe(III)(acac)$_3$+ DMF}, into more active selective catalytic species can be also the result of the regioselective addition of O_2 to the γ-C atom of acetylacetonate ligand (controlled by L^2 ligand) [67]. However due to the favorable combination of the electronic and steric factors appeared at inner and outer sphere coordination (hydrogen bonding) of ligand DMF with Fe(II)(acac)$_2$ the oxygenation of the acetylacetonate ligand may follow another mechanism. Insertion of O_2 into C–C bond (not the C=C bond as takes place for nickel(II) complexes with consequent break-down of cycle configuration through Criegee mechanism) can lead to the formation of methylglyoxal as the second destruction product in addition to the $(OAc)^-$ ion via 1,2-dioxetane intermediate (by analogy with the action of Fe(II) containing acetylacetone dioxygenase (Dke 1) (Scheme 4) [68].

As in the case of catalysis by Ni complexes the active selective transformation products are hetero ligand complexes of probable structure: $Fe(II)_x(acac)_y(OAc)_z(L^2)_n$ (L^2=DMF) [56, 67].

The final product of the conversion of acetylacetonate ligands is Fe(II) acetate (Scheme 4). Both Fe(II) acetate, and Ni(II) acetate, catalyze heterolytic decomposition of PEH into phenol and acetaldehyde. At both cases the complete catalyst transformation is causing the sharp fall of S_{PEH} [56, 59, 67].

Scheme 4.

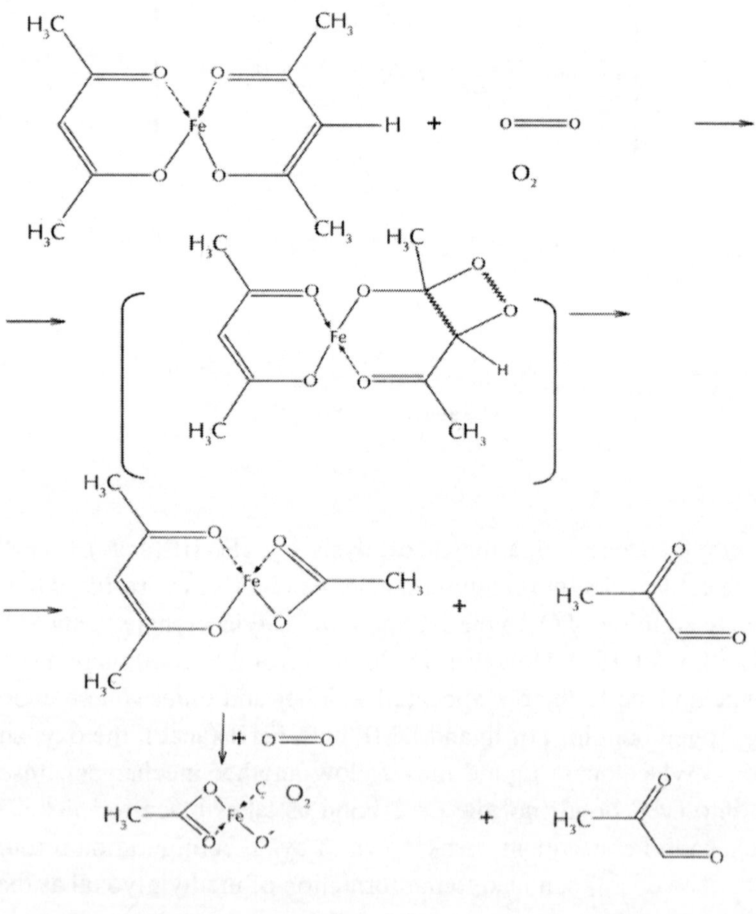

The principle scheme of dioxygen-dependent conversion of 2,4-pentandione catalyzed by acetyl acetone dioxygenase Fe(II).

The enzymatic cleavage of C–C bonds in β-diketones has growing significance for various aspects of bioremediation, biocatalysis, and mammalian physiology, and the mechanisms by which this particular cleavage is achieved are surprisingly diverse [69], ranging from metal-assisted hydrolytic processes [69] to those catalyzed by dioxygenases [68]. Carbon monoxide, one of the products of $(acac)^-$ – ion oxygenate breakdown path catalyzed with the only known to date a Ni(II)-containing dioxygenase – acireductone dioxygenase, ARD, and releasing at the oxygenation of $Ni(L^1)_2 \cdot L^2$ (Scheme 1–3), previously considered biologically relevant only as a toxic waste product, is now considered a candidate for a new class of neural messengers [68].

1.3.3 MODELING OF TRANSITION METAL COMPLEX CATALYSTS UPON ADDITION OF AMMONIUM QUATERNARY SALTS AND MACRO-CYCLE POLYETHERS AS LIGANDS-MODIFIERS. THE ROLE OF HYDROGEN – BONDING INTERACTIONS

Ammonium salts are well-known cationic surfactants. These amphiphilic molecules aggregate in aqueous solution to micelles and at higher concentrations to lyotropic (typical member is CTAB, cetyltimethylammonium bromide) (or thermotropic) mesophases. Beside this ammonium salts are used as phase transfer catalysts and as ionic liquids (ILs) in synthesis of nanopartickle catalysts [70–74].

It was established earlier that quaternary ammonium salts R_4NX can play two different roles in various catalytic reaction in water – organic systems. These salts can act as catalysts of phase transfer but also R_4NX salts are often directly involved in catalytic reaction itself. Thus, for example, in reactions of the oxobromination of aromatic compounds a lipophylic ammonium salt transfers H_2O_2 into the organic phase. At the same time, since it is a Lewis acid it forms $R_4NBr•(Br_2)_n$ or $R_4NBr•(HBr)_n$ adducts thus activating Br_2 or salts of HBr for electrophilic attack on the aromatic ring [75]. In the catalytic oxidation of styrene to benzaldehyde by H_2O_2

in water – organic solvent systems ammonium salts completely transfer H_2O_2 and catalyst (Ru, Pd) into the organic phase by forming hydrogen bonds. Moreover, the complex formation affects the properties of the catalyst by the changing its activity (rate and selectivity of the reaction) [75]. In the oxidation of *p*-xylene in water – organic system in the presence of $CoBr_2$ and R_4NBr the catalytically active species are complexes $CoBr_2$ with R_4NBr [76]. It is known also that catalytic activity of CTAB in the ROOH decomposition in the presence of metals compounds is dependent on structural changes in the formed inverse micelles [77, 78].

The ability of ammonium quaternary salts to complex formation with transition metals compounds was established. It was proved for example that $M(acac)_2$ (M=Ni, Cu) form with R_4NX ($X=(acac)^-$, R=Me) complexes of $[R_4N][M(acac)_3]$ structure. Spectral proofs of octahedral geometry for these complexes were got [79]. Complexes Me_4NiBr_3 were synthesized and their physical properties were studied [80].

The selective complexation ability of crown ethers is one of their most attractive properties. Crown ethers are of considerable interest in biologically modeling of enzyme catalysis, and as phase transfer catalysts [71, 81]. Intermolecular and intramolecular hydrogen bonds and other non-covalent interactions are specific in molecular recognition [81]. Interest in studying of structure and catalytic activity of nickel complexes (especially nickel complexes with macrocycle ligands) is increased recently in connection with discovering of nickel-containing ferments [82–86]. So, they established that active sites of ferment urease are binuclear nickel complexes containing N/O-donor ligands [82]. Cofactor of oxidation-reduction ferment methyl-S-coenzyme-M-reductase in structure of methanogene bacteria is tetra-aza-macrocycle nickel complex with hydrocorfine $Ni(I)F_{430}$ axially coordinated inside of ferment cavity [84].

Inclusion of transition metals cations into cavity of macrocycle polyether is proved by now by various physical-chemical methods. At that the concrete structure of complex is determined not only by geometric accordance of metal ion and crown-ether cavity but by the whole totality of electron and spatial factors created by metal, polyether and other ligand atoms and also by solvent [87].

1.3.4 THE ABILITY OF THE AMMONIUM QUATERNARY SALTS AS WELL AS MACROCYCLE POLYETHER TO FORM COMPLEXES WITH TRANSITION METALS COMPOUNDS WAS USED BY US TO DESIGN EFFECTIVE CATALYTIC SYSTEMS.

It was established by us earlier, that at the relatively low nickel catalyst concentration the selectivity of the ethylbenzene oxidation into PEH, catalyzed by $Ni(L^1)_2$ ($1,5 \cdot 10^{-4}$ mol/l), was sufficiently high: $S_{PEH,max} = 90\%$. This fact may be expected from the analysis of the scheme of catalyzed oxidation including participation of catalyst in chain initiation under catalyst interaction with ROOH, in chain propagation ($Cat + RO_2^{\bullet} \rightarrow$) and assuming the chain decomposition of ROOH. In this case the rate of reaction should be decreased, and $[ROOH]_{max}$ should be increased with decrease of $[Cat]_0$ [8]. But the growth in $S_{PEH,max}$ is not accompanied the growth in the conversion C. The value C into PEH in the ethylbenzene oxidation, catalyzed by $Ni(L^1)_2$ ($1,5 \cdot 10^{-4}$ mol/l), was not exceeded $C = 2–4\%$ [59, 60]. The change in the direction of by products formation is observed. Products AP and MPC are formed in this case not from PEH but parallel with PEH, i.e. $w_P / w_{PEH} \neq 0$ at $t \rightarrow 0$, and furthermore $w_{AP} / w_{MPC} \neq 0$ at $t \rightarrow 0$ that indicates on parallelism of formation of AP and MPC (P = AP or MPC) [33]. At these conditions addition of electron-donor monodentate ligands turned to be low effective [33, 56, 59] and the change of $S_{PEH,max}$ and $C_{S=90\%}$ under introduction of additives L^2 (L^2 = HMPA, MP) into system practically was not observed.

Coordination of 18K6 or R_4NX with $Ni(II)(acac)_2$ was seemed to promote oxidative transformation of nickel (II) complexes (Schemes 1–3) into catalytically active particles and result in increase in C at conservation of $S_{PEH,max}$ not less than 90%. This supposition was based on the next literature data.

For example, the ability of crown-ethers to catalyze electrophilic reactions of connection to γ-C-atom of $acac^-$ ligand is known [71, 88].

It is known that R_4NX in hydrocarbon mediums forms with acetylacetone complexes with strong hydrogen bond $R_4N^+(X...HOCMe{=}CHCOMe)^-$ in which acetylacetone is totally enolyzed [89]. The controlled by R_4NX regio-selective connection of O_2 by γ-C-atom of $(acac)^-$ ligand in complex $M(acac)_n \cdot R_4NX$ is probable enough. Various electrophilic reactions in complexes $R_4N^+(X...HOCMe{=}CHCOMe)^-$ proceed by γ-C-atom of acetylacetone [71, 89].

Obviously, the favorable combination of H-bonding and steric factors, appearing under coordination of 18K6 or R_4NX may not only accelerate the active multi-ligand complex formation (Schemes 1-4) but also hinder the transformation of active catalyst into inactive particles.

At the introduction of 18K6 or Me_4NBr additives into ethylbenzene oxidation reaction catalyzed by complexes $Ni(L^1)_2$ the extraordinary results were received. Really significant increase in conversion degree of oxidation into PEH at maintenance of selectivity on level S_{PEH} ~ 90% occurs. The degree of conversion into PEH is increased from 4–6 up to 12% for complexes of $Ni(II)(acac)_2$ ($Ni(O/O)_2$) with 18C6 (1:1 and 1:2) and from 12 up to 16% for complex of $Ni(II)(enamac)_2$ ($Ni(O/NH)_2$) (1:1). Besides this the increase in the initial rate of reaction w_0 (Fig. 1), and $S_{PEH,max}$ from 90% to 98–99% (18K6) are observed [33, 90, 91].

In the case of additives of Me_4NBr into reaction of ethylbenzene oxidation catalyzed by $Ni(II)(acac)_2$ the value of $S_{PEH,max}$=95% is higher than under catalysis by $Ni(II)(acac)_2$ without addition of L^2. The $S_{PEH,max}$ is reached not at the beginning of reaction of ethylbenzene oxidation, as it occurs at the case of complexes with 18C6, but at C=2–3%. Selectivity remains in the limits $90\% < S \leq 95\%$ to deeper transformation degrees of ethylbenzene $C \approx 19\%$ than in the presence of additives 18K6 ($C \approx 12\%$) [33, 92, 93].

Additives of 18K6 or Me_4NBr to ethylbenzene oxidation reaction catalyzed by $Ni(L^1)_2$ lead to significant hindering of heterolysis of PEH with formation of phenol (PhOH) responsible for selectivity decrease. At that induction period of PhOH formation in the presence of Me_4NBr is significantly higher than in the case of 18K6 [90–93].

Influence of quaternary ammonium salt on catalytic activity of Ni(II)$(acac)_2$ as selective catalyst of ethylbenzene oxidation into PEH extremely depends on radical R structure of ammonium cation. If cetyltimethylammonium bromide (CTAB) is added, $S_{PEH,max}$ is reduced down to 80–82% [92, 93]. The rate w_0 is significantly increased, in ~ 4 times in comparison with ethylbenzene catalysis by $Ni(II)(acac)_2$ complex. The initial rate of PEH accumulation $w_{PEH,0}$ is higher than in the case of ethylbenzene oxidation catalyzed by the system $\{Ni(II)(acac)_2 + Me_4NBr\}$. However initial rates of accumulation of side products of reaction of AP with MPC are also significantly increased (Fig. 1b). The decrease in PEH selectivity connected with heterolysis of PEH. The phenol formation is observed at lower conversions of RH transformation.

Analysis of consequence of ethylbenzene oxidation products formation catalyzed by systems {$Ni(L^1)_2$+18K6} and {$Ni(II)(acac)_2$+R_4NBr} showed that the mechanism of products formation is unchanged as compared with oxidations catalyzed by $Ni(L^1)_2$ or {$Ni(L^1)_2$+HMPA}. The products PEH, AP and MPC were established to be formed parallel ($w_P/w_{PEH} \neq 0$ at $t \to 0$) in the course of the process, AP and MPC were formed also parallel ($w_{AP}/w_{MPC} \neq 0$ at $t \to 0$).

Catalysis of ethylbenzene oxidation initiated by {$Ni(II)(acac)_2$ + CTAB} system is not connected with formation of micro-phase by the type of inverse micelles since the micellar effect of CTAB revealing at $t^\circ < 100^\circ$ [77] is as a rule not important at $t^\circ \geq 120^\circ$. Furthermore, as we saw the system {$Ni(II)(acac)_2$ + CTAB} was not active in decomposition of ROOH.

For estimation of catalytic activity of nickel complexes as selective catalysts of ethylbenzene oxidation into α-phenylethylhydroperoxide we proposed to use parameter $\tilde{S}\cdot C$. The $\tilde{S}$ is mean value selectivity of oxidation into PEH, which evaluated a change of S in the course of oxidation from S_0 at the beginning of reaction to some S_{lim}, conditional value chosen as a standard. For comparable by value systems selectivity as S_{lim} was selected the value S_{lim} = 80% approximately equal to selectivity of non-catalyzed ethylbenzene oxidation into PEH at initial stages of reaction. The value selectivity was carried out in the limits $S_0 \leq S \geq S_{lim}$ (S_{max} > 80%), C – conversion at $S = S_{lim}$ [33, 60]. The system {$Ni(II)(acac)_2$+Me_4NBr} by the value of parameter $\tilde{S}\cdot C$ ($\tilde{S}\cdot C \sim 24\cdot 10^2$ (%,%)) is the most active catalyst of ethylbenzene oxidation into PEH as compared with catalysis by systems {$Ni(II)(L^1)_2$+18C6(HMPA)} [91].

We established that in the presence of 18K6 or R_4NBr only without nickel complex auto-catalytic developing of process with initial rates by order lower was observed. The $S_{PEH,max}$, equal to 85% (18K6) or 95% (Me_4NBr) at the beginning of ethylbenzene oxidation was sharply reduced with the increase of ethylbenzene conversion degree. The PhOH formation is observed from the very beginning of reaction. Synergetic effects of increase in parameter w_0 and $\tilde{S}\cdot C$ and indicated the formation of active complexes $Ni(L^1)_2$ with (L^2) [94] structure 1:1 (L^1= $(acac)^-$, L^2=18C6, R_4NBr), 1:2 (L^1= $(acac)^-$, $(enamac)^-$, L^2=18C6) [90-93] and also the products of their transformation (Figs. 1(a, b), and 2). The stability of homo polynuclear heteroligand complexes $Ni_x(L^1)_y(L^1_{ox})_z(L^2)_n$ (L^1_{ox}= $MeCOO^-$, L^2=18C6, R_4NBr) ("A") formed in the course of oxidation seems to be due to the intermolecular and intramolecular hydrogen bonds. The possibility of supra molecular structures formation in this case is high [95–97].

The formation of complexes $Ni(L^1)_2$ with L^2=18C6 or R_4NBr was also proved by spectrophotometry under analysis of UV spectra of absorption of $Ni(L^1)_2$ and R_4NBr (18C6) mixtures solutions. At that L^2 coordinate with metal ion with preservation of ligand L^1 in internal coordination sphere of complex [90, 92]. Under formation of complexes of $Ni(L^1)_2$ with L^2 in spite of axial coordination by the fifth coordination place of nickel (II) ion the outer sphere coordination of L^2 (H–bonding) with acetylacetonate-ion is also possible.

The possibility of outer sphere coordination of R_4NX with *β*-diketonates (Ni(II), Fe(III)) was demonstrated by us using complexes $Fe(III)(acac)_3$ in the presence of various R_4NBX. In the UV spectrum $Fe(III)(acac)_3$ exhibit an intense absorption band at $\nu = 37\cdot10^{-3}$ cm^{-1} ($CHCl_3$) of the $\pi - \pi^*$ transition of the conjugated cycle of the acetylacetonate ion [92, 93]. In the presence of salts R_4NX (Me_4NBr, CTAB, $(C_2H_5)_4NBr$, $(C_2H_5)_3C_6H_5NCl$ and the other) a decrease in the intensity and a bathochromic shift of the absorption maximum to $\nu = 36\cdot10^{-3}$ cm^{-1} ($\Delta\lambda \approx 10$ nm) are observed. Such a change in spectrum indicates the influence of R_4NX coordinated in the outer sphere on conjugation in the ligand. The change in the conjugation in the chelate ring of acetylacetonate complex, when R_4NX is coordinated in the outer coordination sphere of the metal can be caused by participation of the oxygen atoms of the acetylacetonate ion in the formation of coordination bonds with the ammonium ion or hydrogen bonds with alkyl substituents of the ammonium ion [92, 93].

As in the case of ethylbenzene oxidation catalyzed by nickel complexes ($Ni(L^1)_2$), at the catalysis by $Fe(III)(acac)_3$ the $S_{PEH,max}$ increases as [Cat] is reduced. However, this increase is less significant, from S_{PEH} = 42-46% to S_{PEH} = 65%. We also observed a reduction in the rate of ethylbenzene oxidation as $[Fe(III)(acac)_3]$ decreased. However, the dependence of $[PEH]_{max}$ on $[Fe(III)(acac)_3]$ shows an extremum, suggesting that the mechanism of catalysis is more complicated in this case [67,93,98].

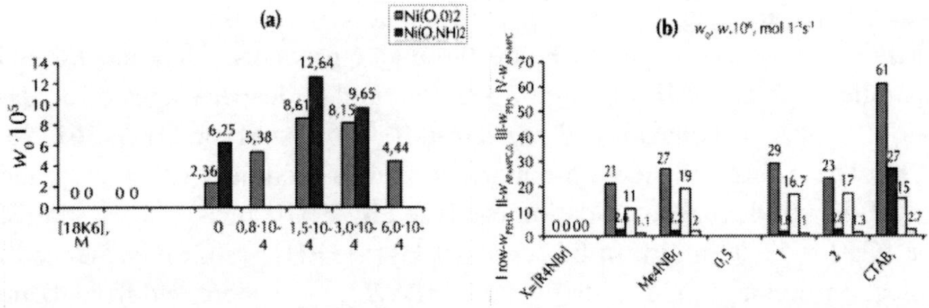

FIGURE 1 **(a)** Dependences of initial rates w_0 (mol l^{-1} s^{-1}) on [18K6] concentration in ethylbenzene oxidation reactions catalyzed by $\{M(L^1)_2$+18K6$\}$ (M=Ni(II), L^1=acac^{-1}, enamac^{-1}). $[M(L^1)_2]$=1,5·10^{-4} mol/l, 120°C.
(b) Dependences of initial rates w_0 (mol l^{-1} s^{-1}) and the rates w (mol l^{-1} s^{-1}) in the process of ethylbenzene oxidations catalyzed by $\{M(L^1)_2$+18K6$\}$ (M=Ni(II), L^1=acac^{-1}, enamac^{-1}) on $[R_4NBr]$ (mol/l). $[M(L^1)_2]$=1,5·10^{-4} mol/l, 120°C.

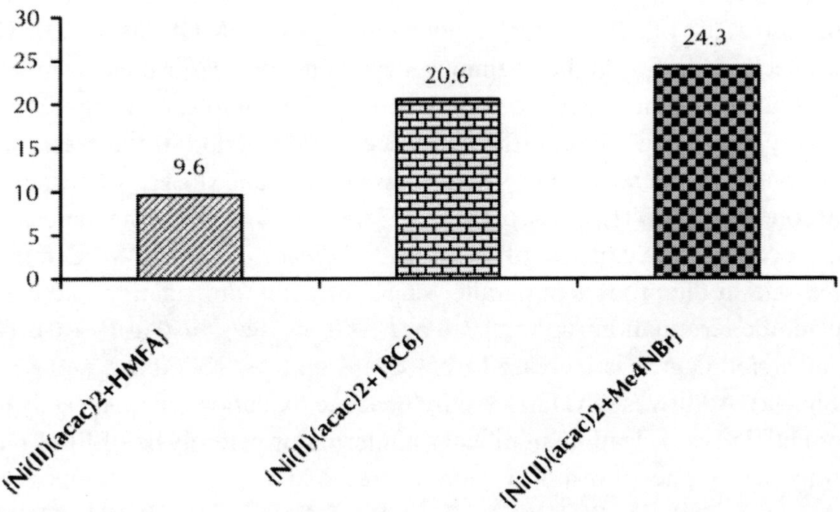

FIGURE 2 Parameter $\tilde{S}\cdot C\cdot 10^{-2}$ (%,%) in the ethylbenzene oxidation upon catalysis by catalytic systems $\{Ni(II)(acac)_2+L^2\}$ with L^2=Me_4NBr, 18C6, HMPA. [Ni(II)(acac)$_2$]=1.5·10^{-4} mol/l, 120°C.

Fe(III)(acac)$_3$, and formed in the course of ethylbenzene oxidation Fe(II)(acac)$_2$, are inactive in PEH decomposition [67,93,98]. In our articles it was established, that in ethylbenzene oxidation catalyzed by Fe(III)

$(acac)_3$ ([Cat]=$(0.5\div5)\cdot10^3$ mol/l, (80, 120°C)) the oxidation products MPC and AP as well as PEH are the major products. They are formed parallel both at the beginning of reaction, and at deeper stages of oxidation: w_P/w_{PEH} is constant and nonzero at t → 0 (P=AP or MPC) [67, 93, 98]. The effects of electron-donor exoligands-modifiers on w, S_{PEH} and C of the ethylbenzene oxidation catalyzed by $Fe(III)(acac)_3$ were studied at [Cat] = $5\cdot10^{-3}$ mol/l. In this case [PEH] = $[PEH]_{max}$. In the presence of electron-donor monodentate ligand HMPA $S_{PEH,max}$ is increased from 42 up to ~57% (80°C), conversion degree C from 5 up to 15% ([Cat] = $5\cdot10^{-3}$ mol/l). These were the maximum effects of electron-donor monodentate ligands on the S_{PEH} and C of ethylbenzene oxidation at catalysis by Fe(III) $(acac)_3$ [67, 93].

In ethylbenzene oxidation in the presence of {$Fe(III)(acac)_3$($5\cdot10^{-3}$ mol/l)+ R_4NBr($0.5\cdot10^{-3}$ mol/l)} (R_4NBr=CTAB) (80°C) the $S_{PEH,max}$=65%, which is reached in the developed process (Fig. 4a(1, 2)), is higher than in the case of use of additives of monodentate ligands HMFA, DMF [93]. The fast decrease in S_{PEH} at the beginning steps of the process is connected with the transformation Fe(III) complexes in Fe(II) complexes in the course of ethylbenzene oxidation (the auto acceleration period of the reaction is observed), the increase in $w_{PEH,0}$, decrease in $w_{P,0}$, and $[PEH]_{max}$ at catalysis by complexes $(Fe(II)(acac)_2)_x\cdot(R_4NBr)_y$ are observed. Then the increase in S_{PEH} occurs at the expense of significant decrease in AP and MPC formation rate in the process at parallel stages of chain propagation and chain quadratic termination ($w_P/w_{PEH} \neq 0$ at $t \to 0$, $w_{AP}/w_{MPC} \neq 0$ at $t \to 0$). The conversion degree is increased from C = 4 up to ~ 8% (at S_{PEH}=40-65%) (Fig. 4a). Additives of CTAB to ethylbenzene oxidation reaction catalyzed by $Fe(III)(acac)_3$ lead to significant hindering of heterolysis of PEH with formation of phenol responsible for decrease in S_{PEH}.

The growth in $\tilde{S}\cdot C$ is ~ 2.6, 2.36, 1.4 times for R_4NBr=CTAB, $(C_2H_5)_4NBr$), Me_4NBr) respectively in comparison with catalysis by $Fe(III)(acac)_3$ ($\tilde{S}\cdot C$=$2.1\cdot10^2$ (%,%)) (Fig. 4b). In given case for value S_{lim} as standard we accept S_{lim} =40%, a value that approximately corresponds to the selectivity of ethylbenzene oxidation in the presence of ligand-free $Fe(III)(acac)_3$ ($5\cdot10^{-3}$ mol/l) (80°C) under the steady-state reaction conditions, C is the conversion for which $S_{PEH} \leq S_{lim}$. In the absence of a catalyst, the addition of L^2 has practically no effect on selectivity of ethylbenzene oxidation reaction and the reaction proceeds in the autocatalytic mode at w_0 below that in the catalysis by Fe(III)(acac)$_3$.

The no additive (synergetic) effects of growth in $\tilde{S}\cdot C$ parameter and w_0 observed in the reactions catalyzed by Fe(III)(acac)$_3$ in the presence of R$_4$NBr, and also obtained kinetic regularities of ethylbenzene oxidation indicate the formation catalytic active complexes [94] presumably of (Fe(II)(acac)$_2$)$_x$·(R$_4$NBr)$_y$ as well as the complexes, produced as a result of the transformation of (Fe(II)(acac)$_2$)$_x$·(R$_4$NBr)$_y$ during oxidation.

The most effect of increase in $\tilde{S}\cdot C$ was got in the case of CTAB additives. As we saw at the catalysis by nickel complexes in the presence of the CTAB additives $S_{PEH,max}$ is reduced down from 90 to 80-82% as compared with increase in $S_{PEH,max}$ (94%) in the case of Me$_4$NBr additives [93].

Due to the favorable combination of the electronic and steric factors appeared at inner and outer sphere coordination (hydrogen bonding) of CTAB with Fe(II)(acac)$_2$ the oxidative degradation of the acetylacetonate ligand may follow "dioxygenase-like" mechanism, described by Scheme 4. There is a high probability of formation of stable complexes of structure Fe(II)$_x$(acac)$_y$(OAc)$_z$(CTAB)$_n$ ("B"). Out-spherical coordination of CTAB evidently creates sterical hindrances for regio-selective oxidation of the (acac)$^-$– ligand, and the transformation of the intermediate complex ("B") into the final product of dioxygenation.

In the case of catalysis by the {Fe(III)(acac)$_3$ + DMF} system the complexes Fe(II)$_x$(acac)$_y$(OAc)$_z$(DMF)$_n$, formed in the process, are not stable, though DMF like CTAB forms H-bonds with acetylacetonate ion [67]. The rapid decrease in S_{PEH} was observed. $S_{PEH,max}$ at the catalysis by the {Fe(III)(acac)$_3$ + DMF} system was not higher in fact than $S_{PEH,max}$ at the catalysis by the {Fe(III)(acac)$_3$ in the absence of the additives. With the use of HMPA as exo ligand, that did not form H – bonds with chelate ring of Fe(II)(acac)$_2$, the transformation of Fe(II)(acac)$_2$)·HMPA was not observed, although HMPA as electron-donor ligand was characterized with a higher DN value (after V. Gutmann) as compared with DMFA [67].

Catalysis of ethylbenzene oxidation initiated by {Fe(III)(acac)$_3$ + CTAB} system (80°C) in the case of application of the small concentrations R$_4$NBr ($0.5\cdot10^{-3}$ mol/l) is not connected with formation of microphase by the type of inverse or sphere micelles. As we saw above the system {Fe(III)(acac)$_3$ + CTAB} was not active in decomposition of PEH. $w_P/w_{PEH} \neq 0$ at $t \to 0$, $w_{AP}/w_{MPC} \neq 0$ at $t \to 0$. Analogous mechanism of formation PEH, AP and MPC is observed at the use of Me_4NBr and $(C_2H_5)_4NBr$ additives which do not form micelles. At the concentration of [CTAB] = $5 \cdot 10^{-3}$ mol/l the rate of the PEH accumulation and $[PEH]_{max}$ decreases significantly, since the probability of micelles formation obviously increases.

Thus, we established the interesting fact – the catalytic effect of small concentrations of quaternary ammonium salts, $[R_4NBr] = 0.5 \cdot 10^{-3}$ mol/l, which in 10 times less than [Fe(III)(acac)$_3$]. It is known that salts QX can form complexes with metal compounds of variable composition which depends on the nature of solvent [93]. The formation of poly nuclear heteroligand complexes $(Fe(II)(acac)_2)_x \cdot (R_4NBr)_y$ (and $Fe(II)_x(acac)_y(OAc)_z(R_4NBr)_n$ also) seems to be probable.

1.3.5 TRIPLE CATALYTIC SYSTEMS INCLUDING BIS (ACETYLACETONATE) NI(II) AND ADDITIVES OF ELECTRON-DONOR COMPOUND L^2 AND PHENOL AS EXO LIGANDS

One of the most effective methods of control of selective ethylbenzene oxidation into α-phenylethylhydroperoxide with dioxygen may be the application of the third component of catalytic system – phenol (PhOH) along with Ni(II)(acac)$_2$ and the additives of electron-donor ligands L^2 (L^2= MSt (M=Na, Li), MP, HMPA) [99].

We discovered phenomenon of the considerable increase in the efficiency of selective ethylbenzene oxidation reaction into α-phenylethylhydroperoxide with dioxygen in the presence of triple systems {Ni(II)(acac)$_2$+L^2+PhOH}, parameters $\tilde{S} \cdot C$, the conversion degree C (at S_{PEH} ~85-90%), and the hydroperoxide contents ($[PEH]_{max}$), in comparison with catalysis by binary systems {Ni(II)(acac)$_2$+L^2}. The obtained synergetic effects of increase in $\tilde{S} \cdot C$ under catalysis by {Ni(II)(acac)$_2$ + L^2} on [MP] ($(\tilde{S} \cdot C)^{max} \sim 17.5 \cdot 10^2$ (%,%) at [MP]=const=$7 \cdot 10^{-2}$ mol/l), in the presence of inhibitor phenol ([PhOH]=const=$3 \cdot 10^{-3}$ mol/l) seems to be due to unusual catalytic activity of formed at mentioned conditions triple complexes $[M(L^1)_2 \cdot (L^2)_n \cdot (PhOH)_m]$. This presumption is confirmed by dependences of $\tilde{S}\cdot C$ on [Ni(II)(acac)$_2$] at [PhOH]=const=$3\cdot10^{-3}$ mol/l and [MP]=const=$7\cdot10^{-2}$ mol/l (($\tilde{S}\cdot C)^{max}=17.47\cdot10^2$ (%,%), [Ni(II)(acac)$_2$]=$3\cdot10^{-3}$ mol/l) and also of $\tilde{S}\cdot C$ on [PhOH] at [Ni(II)(acac)$_2$]=const=$3\cdot10^{-3}$ M and [MP]=const=$7\cdot10^{-2}$ mol/l. In last case $\tilde{S}\cdot C$ reaches extremum ($\tilde{S}\cdot C)^{max}$=17.5 and $18.12\cdot10^2$ (%,%) at two values of [PhOH] = $3\cdot10^{-3}$ и $4.6\cdot10^{-4}$ mol/l accordingly (Fig. 5a).

Comparison of kinetic regularities of ethylbenzene oxidation catalyzed by triple systems {Ni(II)(acac)$_2$($3\cdot10^{-3}$ mol/l) + MP($7\cdot10^{-2}$ mol/l) + PhOH} ([PhOH]= $3\cdot10^{-3}$ or $4.6\cdot10^{-4}$ mol/l) was carried out. Obtained data testify on the fact that in both cases selective catalysis of ethylbenzene oxidation into PEH is connected with formation in the course of oxidation of catalytically active complexes with structure 1:1:1 [94, 99]. The some differences observed at the initial stages of two reactions are caused obviously by the different initial conditions of triple complexes Ni(II)(acac)$_2$·(L^2)·(PhOH) formation in the course of catalytic ethylbenzene oxidation in these cases. At catalysis by triple system {Ni(II)(acac)$_2$+MP+PhOH} with small [PhOH] =$4.6\cdot10^{-4}$ mol/l the fast increase in the concentration of PhOH right up to [PhOH] = (3–5)$\cdot10^{-3}$ mol/l (at t=0–5 h) is observed. [PhOH] = (3–5)$\cdot10^{-3}$ mol/l ~ corresponds to [PhOH] for the first combination {Ni(II)(acac)$_2$ ($3.0\cdot10^{-3}$ mol/l) + MP ($7.0\cdot10^{-2}$ mol/l) + PhOH ($3.0\cdot10^{-3}$ mol/l)} and to the formation of complexes of structure [M(L^1)$_2$·(L^2)·(PhOH)].

The increase in the rate of PhOH accumulation at the beginning of the process may be due to the function of PhOH as acid that becomes stronger because of outer sphere coordination of PhOH with nickel complex Ni(II)(acac)$_2$·MP, and this effect favors to heterolysis of PEH with the formation of phenol [100]. This supposition is confirmed by the following facts. So the accumulation of PhOH, but not the consumption, at the maximum initial rate $w_{PhOH,0}=w_{PhOH,max}$ is observed upon addition of PhOH ($3.0\cdot10^{-3}$ mol/l) into the reaction of ethylbenzene oxidation catalyzed by coordinated saturated complexes Ni(II)(acac)$_2$·2MP ([Ni(II)(acac)$_2$] = $3.0\cdot10^{-3}$ mol/l, [MP] = $2.1\cdot10^{-1}$ mol/l), and also in the case of the ethylbenzene oxidation catalyzed by binary system {Ni(II)(acac)$_2$($3.0\cdot10^{-3}$ mol/l) + PhOH($4.6\cdot10^{-4}$ mol/l)} at [MP] = 0 [99].

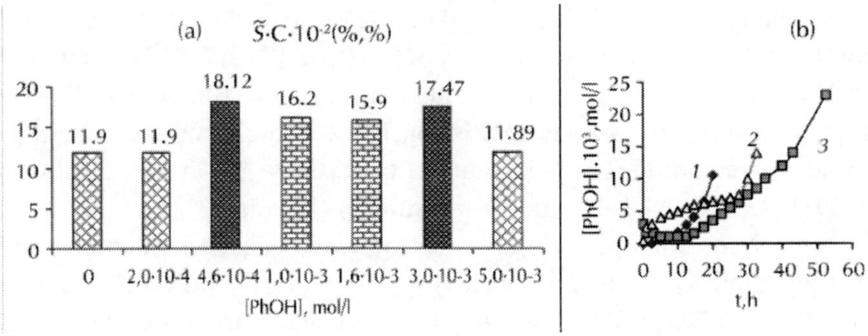

FIGURE 3 **(a)** Dependence of parameter $\tilde{S}C\cdot10^{-2}$(%,%) on [PhOH] in reaction of ethylbenzene oxidation catalyzed by {Ni(II)(acac)$_2$+MP+PhOH}. [Ni(II)(acac)$_2$] = const = $3\cdot10^{-3}$ mol/l, [MP] = const = $7\cdot10^{-2}$ mol/l. 120°C.
(b) Kinetics of accumulation of PhOH in reaction of ethylbenzene oxidation catalyzed by binary system {Ni(II)(acac)$_2$+MP} (*1*) and two triple systems {Ni(II)(acac)$_2$ + MP + PhOH} with variable values of [PhOH] = $4.6\cdot10^{-4}$ mol/l (*2*) or $3\cdot10^{-3}$ mol/l (*3*) and [Ni(II)(acac)$_2$] = const = $3\cdot10^{-3}$ mol/l, and [MP] = const = $7\cdot10^{-2}$ mol/l. 120°C.

Phenomenal results were obtained by us in the case of application of system, including NaSt as L^2 {Ni(II)(acac)$_2$ ($3.0\cdot10^{-3}$ mol/l) + NaSt ($3.0\cdot10^{-3}$ mol/l) + PhOH ($3.0\cdot10^{-3}$ mol/l)}. Parameters C > 35% at the $S_{PEH,max}$ = 85–87%, concentration $[PEH]_{max}$ = 1.6 – 1.8 mol/l (~27 mass), $\tilde{S}\cdot C \sim 30.1\cdot10^2$ (%,%) are much higher, than in the case of the other triple systems and the most active binary systems [99]. These data and some of other effective triple systems (L^2=LiSt, MP, HMPA) are protected by patent RU (2004); the authors are L.I. Matienko, L.A. Mosolova, patent holder is Emanuel Institute of Biochemical Physics, Russian Academy of Sciences.

Similarity of phenomenology of ethylbenzene oxidation in the presence of {Ni(II)(acac)$_2$ ($3.0\cdot10^{-3}$ mol/l) + MP ($7.0\cdot10^{-2}$ mol/l) + PhOH ($3.0\cdot10^{-3}$ mol/l)} and {Ni(II)(acac)$_2$ ($3.0\cdot10^{-3}$ mol/l) + NaSt (LiSt) ($3.0\cdot10^{-3}$ mol/l) + PhOH ($3.0\cdot10^{-3}$ mol/l)} allows assuming analogous mechanism of selective catalysis realizing by triple complexes formed in the course of oxidations. Also, the parallel formation of PEH and side products AP and MPC is established in these two cases: $w_P/w_{PEH} \neq 0$ at $t\to0$ (P=AP or MPC) and $w_{AP}/w_{MPC} \neq 0$ at $t\to0$ at the beginning of reaction and in developed reaction of ethylbenzene oxidation catalyzed by {Ni(II)(acac)$_2$+L^2+ PhOH} (L^2 = NaSt (LiSt), MP). Increase in S_{PEH} during the catalysis by complexes Ni(II)

$(acac)_2 \cdot L^2 \cdot PhOH$ (L^2 = NaSt, MP) in comparison with non-catalyzed oxidation is connected with the change of direction of the formation of side products AP and MPC (AP and MPC are not formed from PEH, as it takes place in non-catalyzed oxidation) and also with hindering of heterolytic decomposition of PEH [99].

The advantage of the triple systems consists in the fact that the formed *in situ* complexes $Ni(II)(acac)_2 \cdot L^2 \cdot PhOH$ are active for a long time, and are not transformed in the course of the process into inactive particles. Thus, the application of triple systems, including $Ni(II)(acac)_2$, electron donor ligand L^2 and PhOH, as homogeneous catalysts is one of the most effective methods of control of selective ethylbenzene oxidation by dioxygen into PEH.

1.4 THE ROLE OF HYDROGEN–BONDING INTERACTIONS IN MECHANISMS OF HOMOGENOUS CATALYSIS

As a rule reactivity and selectivity of metal complex homogenous catalysts have been controlled by variations in axial ligands used, focusing mainly on steric and electronic properties of the latter. At that the interactions in the secondary coordination sphere and the role of hydrogen bonds are investigated least of all. An increasing number of synthetic catalysts and related systems show the benefits of secondary interactions [101, 102], which are generally difficult to control.

While nature's metalloenzymes use secondary interactions, hydrogen bonding or proton transfers in active site. The importance of H-bonds in dioxygen O_2 binding and O_2 activation is well documented in metalloproteins [103]. For instance, removal of residues within the active sites that H-bond to the Fe–O_2 unit in hemoglobins causes a loss in respiration [104]. In addition, dioxygen affinity in hemoglobins has been correlated with the H-bond network surrounding the iron center. Protein dysfunction is observed in Cytochrome P450 when the active site H-bond network proximal to the Fe–O_2 moiety is disrupted [105].

H–bonding interactions are useful for design of catalytic systems, imitating enzymes activity. The participation of H-bonds in dioxygen binding to cobalt complexes and O_2 activation was studied in [106]. The cobalt complexes $[Co^{II}H_2 2^{iPr}]^-$ (Potassium {Bis[(N'-tert-butylureayl)-N-ethyl]-(N'-isopropylcarbamoylmetyl)-aminato-cobaltate(II)}) and $[Co^{II}H_1{}^{iPr}]^-$ (Potassium {[(N'-tert-butylureayl)-N-ethyl]-bis(N'-isopropylcarbamoylmetyl)-aminato-cobaltate(II)}) with multiple H-bond donors readily bind (Co–O_2) and activate dioxygen (Co(III)–OH complexes). The greater number of intramolecular H-bonds produces the more stable Co(III)–OH complex. The complex $[Co^{II}H0^{iPr}]^-$ (Potassium {bis[((N'-isopropylcarbamoylmetyl)-aminato]cobaltate(II)}) with no intramolecular H-bond donors does not react with dioxygen. Investigated Co(II) complexes with rigid H-bond framework are not able to form intermolecular H-bonds.

Mononuclear non-heme iron proteins are involved in various biological processes. Iron centers with terminal hydroxo ligands (Fe–OH) are proposed to be the active species in many catalytic cycles of enzymes including for example protocatechuate 3,4-dioxygenase [107]. Synthesizing the mononuclear Fe–OH moiety is a challenge because of its strong propensity to form multinuclear hydroxo- and oxo-bridged complexes and the different methods are used for synthesis of stable mononuclear Fe–OH complexes [108,109,110]. So, bulky deprotonated urea-derived ligand (H_3buea) stabilizes the Fe–OH units by steric hindrance and intramolecular hydrogen bonds forming a protective cavity [110]. In [107] it was reported the first mononuclear iron(II) hydroxo (1) and iron(III) dihydroxo (2) complexes (1=$[Fe(II)(L)_2(OH)](BF_4)$, 2=$[Fe(III)(L)_2(OH)_2](BF_4)$, L=bis(N-methylimidazol-2-yl)-3-methylthiopropanol) stabilized by an *intermolecular* hydrogen-bonding assembly.

The principle of acceleration of chemical transformations by preliminary favorable orientation of reagent by virtue of molecular forces of the hydrogen bond type or hydrophobic interactions is widely involved in enzymatic catalysis [111]. An example of catalytic process for which the orientation of reagents in the outer coordination sphere of metal complexes is to be important is the formation of urethanes in the coordination sphere of $Fe(III)(acac)_3$ [112]. The role of the catalyst in this process consists in the creation of favorable conditions for the formation of a co-planar complex between the reagents in which an optimal mutual orientation of isocyanide and alcohol, providing a noticeable decrease in activation energy, is accomplished.

The numerous transition-metal β-diketonates undergo a wide range of substitution reactions common to aromatic systems. The methine protons on the complexes' chelate rings can be displaced by many different unsaturated electrophilic groups "E" [114]. This is metal-controlled process of C–C bond formation [115]. The most effective catalyst of those reactions is $Ni(II)(acac)_2$. The reactions formally are analogy to electrophilic reactions of Michael addition [114]. The limiting stage of those reactions is the formation of resonance stabilized zwitter-ion $\{M(II)(L^1)_n^{+}\bullet E^{-}\}$, in which proton transfer occurs [114, 115]. The appearance of new absorption in electron specters of absorption of mixtures $\{Ni(II)(acac)_2 + L^2 + E\}$, which may be explained as charge transfer from ligand donor systems of complex $Ni(II)(acac)_2\bullet L^2$ to π-acceptors E= tetrtacyanethylene or chloranil, testify in the CTC $L^2Ni(II)(acac)_2\bullet E$ favor. The outer sphere reaction of connection of E to γ-C atom of acetylacetonate ligand is followed by the formation of $L^2Ni(II)(acac)_2\bullet E$ [33, 59].

As above mentioned the axial coordination of electron-donor exo ligands L^2 by $M(II)(L^1)_2$ controls the formation of primary $M(II)(L^1)_2\bullet O_2$ complexes and the following outer sphere coordination interactions. The coordination of L^2 by $Ni(L^1)_2$ (L^1=acac$^-$) promoting stabilization of intermediate zwitter-ion $L^2(L^1M(L^1)^+O_2^-)$ leads to increase in possibility of regio-selective connection of O_2 to C–H methine bond of acetylacetonate ligand activated by the coordination with metal. The outer sphere reaction of insertion of O_2 into chelate cycle is dependent on the metal and ligand-modifier L^2. So the reaction of L^1 oxygenation occurs by analogy with the action of the Ni(II)-containing dioxygenase – acireductone dioxygenase, ARD [64], with reactions of oxygenation imitating the action of quercetin 2,3-dioxygenase (Cu, Fe) [65, 66], with the action of L-tryptophan-2,3-dioxygenase (nickel complexes (Scheme 1–3)) or by analogy with the action of Fe(II)-containing Acetylacetone dioxygenase (Dke 1) (iron complexes, Scheme 4).

The role of H-bonding interaction in mechanism of catalytic active complexes formation was investigated by us by means of the introduction of small amounts of H_2O into catalytic reaction [116, 117].

In the last years the interactions between enzyme molecules and surrounded H_2O molecules, because of its importance for enzyme activity is in the centre of attention of many investigators [118, 119].

Water in active site of protein can play more than a purely structural role: as a nucleophile and donor proton, it can be a reagent in biochemical processes [118].

So proton transfer facilitated by a bridging water molecule also seems to occur in horseradish Peroxidase, where it enables the transfer of a proton from iron-coordinated H_2O_2 to a His residue in the active site [120] – the first step in cleavage of the O–O bond. Ab Initio simulations without this bridging water arrive at an energy barrier considerably greater than that found experimentally, because of the large separation of the proton source and sink.

In Ref. [121] the role of H-bonds of water in mechanism of action of Heme oxygenases (HO) is investigated. HO uses Heme (iron-protoporphyrin IX) as substrate and cofactor to cleave it at one meso position into biliverdin, CO, and free iron [122].

The additives of water can serve as mechanistic probes and aid in obtaining true mechanistic understanding in some organocatalytic reactions [123]. The water is nucleophile in palladium-catalyzed oxidative carbohydroxylation of allene-substituted conjugated dienes [124]. This is an example of Pd- catalyzed oxidation leading to C –C bond formation in water with subsequent water attack on a (π-allyl) palladium intermediate. The different effect of the water concentration on the intra- and extra-diol oxygenations of 3,5-di-tert-butylcatechol with O_2, catalyzed by $FeCl_2$ in tetrahydrofuran-water indicates that the intermediates for two reactions are different (model for Catechol-2,3-dioxygenases) [125].

We considered the possibility of the positive effect of small amounts of water on the rate of the transformation of iron complexes with L^2 (L^2 = R_4NBr, 18C6) and, probably, on the parameters S_{PEH} and C in the ethylbenzene oxidation, catalyzed $\{Fe(III)(acac)_3 + L^2\}$. Outer sphere coordination of H_2O molecules may promote the stabilization of intermediate zwitter-ion $L^2(L^1M(L^1)^+O_2^-)$ and as a consequence the increase in the probability of the regioselective addition of O_2 to nucleophilic γ-C atom of $(acac)^-$ ligand one expected. It is well-known that the stability of zwitter-ions increases in the presence of the polar solvents [114]. The H – bond formation between H_2O molecule and zwitter-ion may also promote the proton transfer inside of the zwitter-ion followed by the zwitter-ion conversion into the products via Scheme 4 [114, 115]. It is known the cases of the increase in the ratio of alkylation's products on γ-C atom of the R_4N(acac) in the presence of insignificant additives of water (~10^{-3} моль/л) as compared to the alkylation's reaction in the non proton solvents [126]. It is mentioned analogy facts of increase in catalytic activity of 18C6 reference to electrophilic reactions on γ-C-atom of $(acac)^-$ ligand in THF in the presence of mill moles of water [126].

Monodiamine complex Ni(II)-[(R,R)-N,N-dibenzylcyclohexane-1,2-diamine]Br_2 catalyzes the Michael addition reactions in the presence of water [127].

Few examples are known to date about the influence of small additives of H_2O (~10^{-3} mol/l) on the homogeneous catalysis by transition metal complexes in the hydrocarbon oxidation with molecular O_2. The role of H_2O as a ligand in metal complex-catalyzed oxidation has not been practically investigated [111, 121, 128]. And it is unknown examples of catalytic reactions, when addition of water in small amounts enhances the reaction rate and the product yield. Some of known facts are concerned of the use of onium salts QX together with metal catalyst. So the decrease in the rate of the tetralin oxidation, catalyzed with onium-decavanadate (V) ion-pair complexes in the presense of ~10^{-3} mol/l H_2O was observed [129]. The oxidations are dependent on structural changes in the inverse micelles, in response to concentration changes of ion-pair complexes existing only in the presence of small amounts of H_2O [130]. The most known facts are connected with the influence of small concentrations of H_2O on the catalysis of the ROOH homolysis by onium salts (including quaternary ammonium salts). The acceleration of ROOH homolysis may be the consequence of H – bond formation between ROOH, H_2O and QX [131]. Also it is important to understand the role of small amounts of water because some water is always formed during the catalytic oxidation of the hydrocarbon. The homogeneity of the hydrocarbon solution remains upon addition of small amounts H_2O ($[H_2O]$ ~ 10^{-3} mol/l) [130].

1.4.1 THE EFFECT OF SMALL AMOUNTS OF H2O IN THE ETHYLBENZENE OXIDATION, CATALYZED WITH {FE(III)(ACAC)3+R4NBR} SYSTEM

We first established the increase in catalytic activity of system on basis of transition metal complex and donor ligand-modifier, namely, system

{Fe(III)(acac)$_3$ + CTAB}, as catalyst of the ethylbenzene oxidation to PEH at the addition of small amounts of H_2O (~10^{-3} mol/l) [116]. It was found that the admixtures of H_2O caused no additive (synergistic) effects of growth in selectivity $S_{PEH,max}$, conversion degree C ($S_{PEH,max} \approx 78.2\%$, $C \approx 12\%$) (parameter $\tilde{S}{\cdot}C$). $\Delta S_{PEH,max} \approx 14\%$ and $\Delta C \approx 4\%$, as compared with catalysis by {Fe(III)(acac)$_3$ + CTAB} and $\Delta S_{PEH,max} \approx 40\%$ and $\Delta C \approx 8\%$ as compared with catalysis by Fe(III)(acac)$_3$ (Fig. 4a,b).

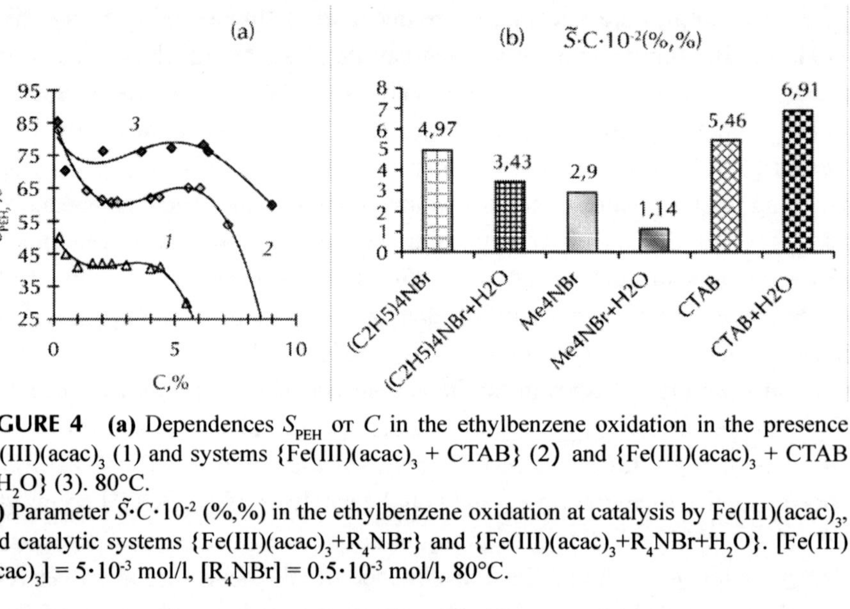

FIGURE 4 **(a)** Dependences S_{PEH} от C in the ethylbenzene oxidation in the presence Fe(III)(acac)$_3$ (1) and systems {Fe(III)(acac)$_3$ + CTAB} (2) and {Fe(III)(acac)$_3$ + CTAB + H_2O} (3). 80°C.
(b) Parameter $\tilde{S}{\cdot}C{\cdot}10^{-2}$ (%,%) in the ethylbenzene oxidation at catalysis by Fe(III)(acac)$_3$, and catalytic systems {Fe(III)(acac)$_3$+R_4NBr} and {Fe(III)(acac)$_3$+R_4NBr+H_2O}. [Fe(III)(acac)$_3$] = $5{\cdot}10^{-3}$ mol/l, [R_4NBr] = $0.5{\cdot}10^{-3}$ mol/l, 80°C.

The effect of small additives H_2O depends significantly on radical R structure of ammonium cation. The decrease in the selectivity of systems {Fe(III)(acac)$_3$ + R_4NBr} (R = Me or C_2H_5) as catalysts of the ethylbenzene oxidation to PEH was observed. The dependence of S_{PEH} on the C in the discussed ethylbenzene oxidation, catalyzed with iron complexes in the presence of small amounts of H_2O has extremum as well in the absence of H_2O additives. The decrease in values of $S_{PEH,max}$ was observed. Thus, $S_{PEH,max} \approx 43\%$ ({Fe(III)(acac)$_3$ + $(C_2H_5)_4$NBr + H_2O}) < $S_{PEH,max}$ = 48% (at the catalysis with {Fe(III)(acac)$_3$ + $(C_2H_5)_4$NBr ($5{\cdot}10^{-4}$ M)} in the absence of H_2O) and $S_{PEH,max} \approx 43\%$ ({Fe(III)(acac)$_3$ + Me_4NBr + H_2O}) < $S_{PEH,max}$ = 64% ({Fe(III)(acac)$_3$ + Me_4NBr}). The fall in the parameters $S_{PEH,max}$ and

$\tilde{S}\cdot C$ one can explain by high rates of oxygenation of intermediate complexes $Fe(II)_x(acac)_y(OAc)_z(R_4NBr)_p(H_2O)_q$ into final product (Fig. 4b).

The discovered fact of increase in parameters $S_{PEH,max}$, C (and $\tilde{S}\cdot C$) at catalysts of the ethylbenzene oxidation by {$Fe(III)(acac)_3$ + CTAB + H_2O($\sim 10^{-3}$ mol/l)} seems to be due to the increase in stationary concentration of active selective heteroligand intermediate $Fe(II)_x(acac)_y(OAc)_z(CTAB)_n(H_2O)_m$, formed in the course of the ethylbenzene oxidation (Scheme 4). The out-spherical coordination of CTAB, may create sterical hindrances from H_2O coordination and regio-selective oxidation of $(acac)^-$ – ligand by the described above mechanism, and the rate of oxygenation of intermediate "B" to inactive final product $Fe(OAc)_2$ reduced. Besides that the part of H_2O molecules may be absorbed by hydrophilic cation n-$C_{16}H_{33}Me_3N^+$, and as a result the lowering of the rate of the intermediate heteroligand complex "B" conversion to the end products was realized.

In the oxidation in the presence of {$Fe(III)(acac)_3$ + CTAB($5\cdot 10^{-4}$ mol/l) + H_2O ($3.7\cdot 10^{-3}$ mol/l)} system the PhOH as oxidation product was not found right up to 50 h of the ethylbenzene oxidation. This fact is connected probably with the significant decrease in the activity of formed catalyst in the heterolytic decomposition of PEH with formation of phenol (PhOH) and also the inhibition of the rate of particles formation ($Fe(OAc)_2$), responsible for PEH heterolysis [116].

The all reactions to investigate proceed in autocatalytic mode due to the transition Fe(III) to Fe(II). The products were formed with auto acceleration period longer than in the case of the H_2O additives – free process. The reaction rates (as well as in the absence of the H_2O additives [22]) rapidly becomes equal to $w = w_{lim} = w_{max}$ (w_0). Under these steady – state reaction conditions the changes in oxidation rates in the both cases were due to the changes in PEH or P (AC+MPC) accumulations.

The increase in w_0 at catalysis by complexes $(Fe(II)(acac)_2)_n(Me_4NBr)_m$ in the presence of the H_2O is observed (as compared with catalysis by $Fe(II)(acac)_2$, and catalysis by complexes $(Fe(II)(acac)_2)_n(Me_4NBr)_m$ without H_2O [93]).

The rate w_0 decreases insignificantly in the case of catalysis by $(Fe(II)(acac)_2)_n((C_2H_5)_4NBr)_m$ and H_2O additives (as compared with catalysis by complexes $(Fe(II)(acac)_2)_n((C_2H_5)_4NBr)_m$ without H_2O additives [93].

These are unusual results as compared with known facts of significant decrease in the hydrocarbon oxidation rate due the solvatation of radicals $RO_2^{\bullet}$ by molecules of water, forming in the course of the oxidation in the absence of Cat [24], or the deactivation of Cat with water in the processes of chain-radical catalytic hydrocarbon oxidation by O_2 in no polar medium [25].

Unlike the catalysis by {Fe(III)(acac)$_3$+CTAB+H_2O} fall in initial rate w_0, (in ~ 2 times) is observed (Table).

It was established that at the addition of $3.7\cdot10^{-3}$ mol/l H_2O into the ethylbenzene oxidation, catalyzed with {Fe(III)(acac)$_3$+ R_4NBr} (R_4NBr= CTAB, Me_4NBr, $C_2H_5)_4$NBr} the mechanism of products formation is obviously unchanged. As in the absence of H_2O the products AP and MPC formed parallel to PEH formation, at parallel stages of chain propagation and chain quadratic termination, AP and MPC formed in parallel stages also ($w_P/w_{PEH} \neq 0$ at $t \rightarrow 0$, $w_{AP}/w_{MPC} \neq 0$ at $t \rightarrow 0$ (here P= AP or MPC)) [116]. These data differed from known facts of catalysis with CTAB and systems, including CTAB, and transition metal complexes, consisting in the acceleration of PEH decomposition in the micelles of CTAB [78].

1.4.2 THE EFFECT OF SMALL AMOUNTS OF H2O IN THE ETHYLBENZENE OXIDATION, CATALYZED WITH {FE(III) (ACAC)3+18C6} SYSTEM

In reaction of the oxidation of ethylbenzene with dioxygen in the presence of catalytic system {Fe(III)(acac)$_3$($5\cdot10^{-3}$ mol/l)+18K6} (80°C) in the absence of water the dependence of S_{PEH} от C has extremum, as in the case of use of additives of ligands DMF or R_4NBr. $S_{PEH,max}$ = 70% ($[18K6]^0$ = $0.5\cdot10^{-3}$ mol/l) and $S_{PEH,max}$ = 75.7% ($[18K6]_0$ = $5\cdot10^{-3}$ mol/l) in the process are higher than $S_{PEH,max}$ = 65% in the case of use of CTAB as exo ligand-modifier [91, 93, 117] (Fig. 5a,b).

The addition of 18C6 in the ethylbenzene oxidation with dioxygen catalyzed by Fe(III)(acac)$_3$ result in the redistribution of the major oxidation products. The significant increase in $[PEH]^{max}$ is observed ~ 1.6 or 1.7 times at [18C6] = $5\cdot10^{-4}$ mol/l, $5\cdot10^{-3}$ mol/l, at that decrease in [AP] and

[MPC] ~ 4, 5 times accordingly occurs (Fig. 6a,b). Additives of 18C6 lead to significant hindering of heterolysis of PEH with of the formation of phenol responsible for decrease in selectivity.

In the presence of catalytic system {Fe(III)$(acac)_3$+18K6} synergetic effect of increase in $\tilde{S}\cdot C$ parameter ~ 2,5 and 2.8 times at $[18K6]_0=0.5\cdot10^{-3}$ mol/l and $[18K6]_0=5\cdot10^{-3}$ mol/l correspondingly is observed in comparison with catalysis by Fe(III)$(acac)_3$ ($\tilde{S}\cdot C=2{,}1\cdot10^2$ (%,%)) [91,117].

Obtained kinetic regularities of the oxidation of ethylbenzene testify to formation presumably of $(Fe(II)(acac)_2)_p\cdot(18K6)_q$ complexes and products of their transformation in the course of oxidation. It is known that Fe(II) and Fe(III) halogens form complexes with crown-ethers of variable composition (1:1, 1:2, 2:1) and structure dependent on type of crown-ether and solvent [87].

Supposedly, due to the favorable combination of the electronic and steric factors appeared at inner and outer sphere coordination (hydrogen bonding) of 18C6 with Fe(II)$(acac)_2$ (as also in the case of catalysis with complexes with CTAB [13]) there is a high probability of formation of sufficiently stable hetero ligand complexes of the common structure $Fe(II)_x(acac)_y(OAc)_z(18C6)_n$, the intermediate products of the oxygenation of $(acac)^-$ ligands in the $(Fe(II)(acac)_2)_p\cdot(18C6)_q$ complexes by analogy to the catalysis by acetyl acetone dioxygenase (M = Fe(II)) (Dke 1) (Scheme 4) that results in the $S_{PEH,max}$ and C increase:

$$\{Fe(III)(acac)_3+18K6\}\rightarrow Fe(II)(acac)_p\cdot(18K6)_q+O_2\rightarrow Fe(II)_x(acac)_y(OAc)_z(18K6)_n.\ (I)$$

At the addition of $3.7\cdot10^{-3}$ mol/l H_2O into the ethylbenzene oxidation, catalyzed with {Fe(III)$(acac)_3$($5\cdot10^{-3}$ mol/l)+18C6} system (80°C) the efficiency of system as selective catalyst, evaluated with parameters $S_{PEH,max}$ and $\tilde{S}\cdot C$, decreases (Fig. 5a,b). But the increase in C (at the $S_{PEH,max}\geq40\%$) from 4% into ~ 6,5% ({Fe(III)$(acac)_3$+18K6($5\cdot10^{-4}$ моль/л)+H_2O}) and 9% ({Fe(III)$(acac)_3$+18C6($5\cdot10^{-3}$ моль/л)+H_2O}) is observed.

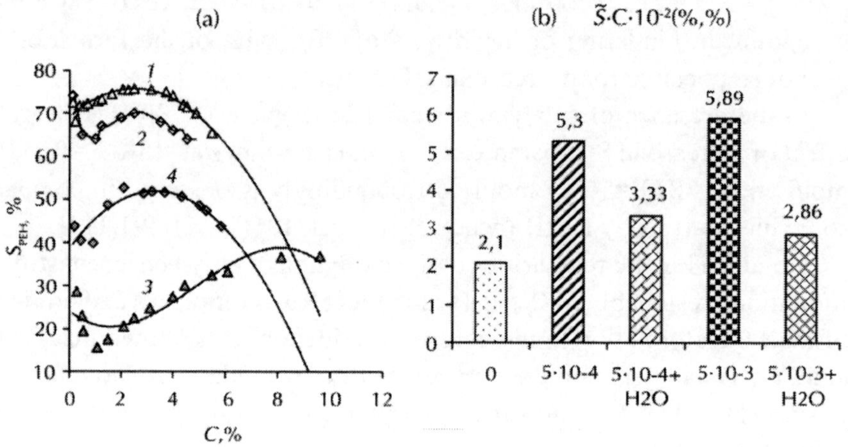

FIGURE 5 **(a)** Dependence of S_{PEH} от C in the reactions of the oxidation of ethylbenzene catalyzed catalytic systems {Fe(III)(acac)$_3$+18C6(5·10^{-3} mol/l)} and {Fe(III)(acac)$_3$+18C6(0,5·10^{-3} mol/l)} without additives of water (1,2), and in the presence of 3.7·10^{-3} mol/l H_2O. [Fe(III)(acac)$_3$]=5·10^{-3} mol/l. 80°C.
(b) The values of parameter $\tilde{S}\cdot C\cdot 10^{-2}$ (%,%) in the reactions of the oxidation of ethylbenzene catalyzed by Fe(III)(acac)$_3$ or catalytic systems {Fe(III)(acac)$_3$+18C6} without additives of water (1,2), and in the presence of 3.7·10^{-3} mol/l H_2O. [Fe(III)(acac)$_3$]=5·10^{-3} mol/l, 80°C.

Analogy facts of decrease in parameters $S_{PEH,max}$ and $\tilde{S}\cdot C$ in the presence small amounts of water< as mentioned above were got by us at the use of systems ({Fe(III)(acac)$_3$+ R$_4$NBr(Me$_4$NBr, (C$_2$H$_5$)$_4$NBr)+H$_2$O}as catalysts [116].

The observed kinetic regularities (Figs. 5, and 6; Table 1) seems to be due to (Fe(III,II)(acac)$_n$)$_m$·(18K6)$_n$ complexes and dioxygenation products formations.

Obviously at the coordination of H_2O molecules the shift of 18K6 into outer coordination sphere of iron complex does not take place [132], as the fall of activity of catalytic system does not occur. It is known also that H_2O molecules may form with crown-ethers the inclusion complexes through H-bonding [133], but enthalpy of its formation is small: ~ 2–3 kcal/mol [134].

The outer coordination of H_2O molecules with iron complex with 18C6 [117] seems to promote the transformation of Fe(II)(acac)$_2$)$_p$·(18K6)$_q$ into particles of "B" type (Scheme 4), catalyzed the ethylbenzene oxidation to PEH, as the growth in S_{PEH} occurs in the ethylbenzene oxidation.

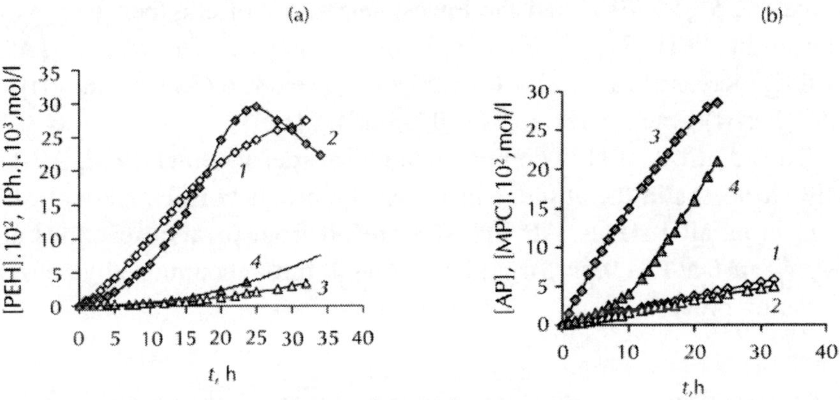

FIGURE 6 **(a)** Kinetics of accumulation of PEH (1,2) and Ph (3,4) in the reactions of the oxidation of ethylbenzene catalyzed {Fe(acac)$_3$+18C6} (1.3) and {Fe(acac)$_3$+18C6+H_2O} (2.4) [Fe(acac)$_3$]=[18K6]=5·10^{-3} mol/l. [H_2O]=3,7·10^{-3} mol/l. 80°C.
(b) Kinetics of accumulation of AC (1, 3) и MPC (2, 4) in the reactions of the oxidation of ethylbenzene catalyzed {Fe(acac)$_3$+18C6} (1.2) and {Fe(acac)$_3$+18C6+H_2O} (3.4). [Fe(acac)$_3$]=[18K6]=5·10^{-3} mol/l. [H_2O]=3,7·10^{-3} mol/l. 80°C.

The fall in the $S_{PEH,max}$ at the use of water as ligand-modifier in this case seems to be due to the growth in the transformation rate of the active intermediate, polynuclear heteroligand complexes Fe(II)$_x$(acac)$_y$(OAc)$_z$(18K6)$_m$(H_2O)$_n$, into Fe(II) acetate [68], and the decrease in steady-state concentration of Fe(II)$_x$(acac)$_y$(OAc)$_z$(18K6)$_m$(H_2O)$_n$.

The facts, which testify favor the increase in catalytic activity of {Fe(III)(acac)$_3$ + 18C6} system as catalyst of the ethylbenzene oxidation into PEH in the presence of 3.7·10^{-3} mol/l H_2O, were obtained.

The increase in the rate of the ethylbenzene oxidation catalyzed with ({Fe(III)(acac)$_3$+18C6+H_2O}) both at the initial stages of the process and in the course of the oxidation was observed as compared with the catalysis in the absence of water. The growth in the initial rate is connected mainly with increase in the [AP] accumulation.

The redistribution of the oxidation products takes place in the presence of water. At the early stages of the ethylbenzene oxidation in the presence of catalytic system {$Fe(III)(acac)_3$($5.0\cdot10^{-3}$ mol/l)+18C6($5.0\cdot10^{-3}$ mol/l)+H_2O($3.7\cdot10^{-3}$ mol/l)} AP but not PEH becomes the major product of the reaction. The highest value of ethylbenzene oxidation selectivity to the AP, $S_{AP,0}$ = 70%, and the lowest selectivity of ethylbenzene oxidation to the PEH, $S_{PEH,0}$ ≈ 25%, are got in this case. The relation [AΦ]/[MΦK] increased from ~ 1.2 ({$Fe(III)(acac)_3$ + 18C6 ($5.0\cdot10^{-3}$ mol/l)}) to 6.5 ({$Fe(III)(acac)_3$ + 18C6 ($5.0\cdot10^{-3}$ mol/l) + H_2O}).

The additives of H_2O do not change the order in which PEH, AP, and MPC form. As in the absence of water, the products PEH, AP and MPC form in parallel stages, AP and MPC result from parallel reactions also ($w_P/w_{PEH} \neq 0$ at $t \to 0$, $w_{AP}/w_{MPC} \neq 0$ at $t \to 0$) throughout the ethylbenzene oxidation process.

1.4.3 PARTICIPATION OF ACTIVE FORMS OF IRON AND NICKEL CATALYSTS IN ELEMENTARY STAGES OF RADICAL-CHAIN ETHYLBENZENE OXIDATION

We suggest the original method for evaluation of catalytic activity of complexes formed *in situ* at the beginning of reaction and in developed process, at elementary stages of oxidation process [33, 90–93] by simplified scheme assuming quadratic termination of chain and equality to zero of rate of homolytic decomposition of ROOH. In the framework of radical-chain mechanism the chain termination rate in this case will be Eq. (1):

$$w_{term} = k_6[RO_2^{\times}]^2 = k_6\left\{\frac{w_{PEH}}{k_2[RH]}\right\}^2 \tag{1}$$

where w_{PEH} – rate of PEH accumulation, k_6 – constant of reaction rate of quadratic chain termination; k_2 – constant of rate of chain propagation reaction $RO_2^{\cdot}$ + RH→.

We established that complexes $M(L^1)_n$ (M=Ni(II) ([Cat]=$(0.5–1.5)\cdot10^{-4}$ mol/l), Fe(III) ([Cat]=$(0.5–5)\cdot10^{-3}$ mol/l)) were inactive in PEH homolysis, products MPC and AP were formed at stages of chain propagation Cat + $RO_2^{\cdot}$→ and quadratic termination of chain. Actually, w_0~$[Cat]^{1/2}$ and $w_{i,0}$~[Cat] and linear radicals termination on catalyst may be not taken into account. In the case of quasi-stationarity by radicals $RO_2^{\bullet\times}$ the values $w_{term.} = w_i$ are the measures of nickel(II) and iron(II) complexes activity in relation to molecular O_2. Discrepancy between w_{AP+MPC} and w_{term} in the case of absence of linear termination of chain is connected with additional formation of alcohol and ketone at the stage of chain propagation Cat + $RO_2^{\bullet}\rightarrow$ (2):

$$w_{pr.} = w_{AP+MPC} - w_{term} \quad (2)$$

The direct proportional dependence of $w_{pr,0}$ on [Cat] testifies in favor of nickel(II) and iron(II) complexes participation at stage of chain propagation Cat + $RO_2^{\bullet}\rightarrow$.

We suggest that these conditions $w_0 \sim [Cat]^{1/2}$ and $w_{i,0} \sim [Cat]$ will be fulfilled also in the presence of additives of R_4NBr, 18C6 and small amounts of water in the case of catalysis with iron complexes, as mechanism of catalytic reaction is not changed in the all examined cases $w_p/w_{PEH} \neq 0$ at $t \rightarrow 0$, $w_{AP}/w_{MPC} \neq 0$ at $t \rightarrow 0$.

Except theoretical consideration in Ref. [24], activity of transition metals complexes $M(L^1)_n$ (M=Ni, Co, Fe, L^1 = acac$^-$, enamac$^-$) at stage of chain propagation (Cat + $RO_2^{\bullet}\rightarrow$) of ethylbenzene oxidation is estimated only in our works [33, 90–93, 116, 117].

Investigation of reaction ability of peroxide complexes [LM−OOR] (M=Co, Fe) preliminary synthesized by reactions of compounds of Co and Fe with ROOH or $RO_2^{\bullet}$ -radicals [135–137] confirms their participation as intermediates in reactions of hydrocarbons oxidation. Obviously, the schemes of radical-chain oxidation including intermediate formation [LM−OOR] [135–138] with further homolytic decomposition of peroxocomplexes ([LM−OOR]→R'C=O (ROH) + $R^{\bullet}$) may explain parallel formation of alcohol and ketone under ethylbenzene oxidation in the presence of $M(L^1)_2$ (L^1 = acac$^-$, enamac$^-$) and their complexes with 18C6 (R_4NBr).

We established that mechanism of selective catalysis of complexes $M(L^1)_n$ ($M(L^1)_2{\bullet}(L^2)_n$) (M=Ni, Fe) and products of their transformation depended on both ratio of rates of chain initiation w_i (activation by O_2) and propagation (w_{pr}) and on activity of Cat in PEH decomposition (homolysis, heterolysis of PEH, chain decomposition of PEH).

In non-catalytic ethylbenzene oxidation at high temperatures the formation of active free radicals occurs in reaction of chain initiation ($RH+O_2\rightarrow$) and under chain decomposition of PEH, the value S_{PEH} to a significant extent should be determined by factor of instability of PEH $\beta = w_{PEH}^{-} / w_{PEH}^{+}$ (w_{PEH}^{-} – sum rate of PEH decomposition (thermal (molecular) and chain), w_{PEH}^{+} – rate of chain PEH formation). Actually, it turned out that value β in the course of non-catalyzed process of ethylbenzene oxidation is increased at the expense of rise of PEH chain decomposition rate that leads to reduction of S_{PEH} [20, 24].

At conditions of ethylbenzene oxidation catalyzed by $M(L^1)_2$ (M=Ni(II), Fe(II)) and $M(L^1)_2$ complexes with R_4NBr, 18C6 the value is $\beta = w_{PEH}^{-}/w_{PEH}^{+} \rightarrow 0$ as at the beginning, so in developed process, the direction of AP and MPC formation is changed (consequent (from PEH decomposition) → parallel), S_{PEH} depends on catalyst activity at stages of chain initiation (activation by O_2) and chain propagation ($Cat + RO_2^{\cdot}\rightarrow$).

1.4.3.1 CATALYSIS WITH NICKEL COMPLEXES

Calculation by Eqs. (1) and (2) show that high activity of "primary" complexes $Ni(II)(L^1)_2\cdot 18C6_n$ as selective catalysts of ethylbenzene into PEH oxidation is connected with five-(chelate group (O/NH)) and twenty-(chelate group (O/O))-fold growth of rate $w_{i,0}$ in comparison with catalysis by $Ni(L^1)_2$, hindering of rate of chain propagation ($w_{pr,0}$) ($Cat + RO_2^{\cdot}\rightarrow$). Under catalysis by complexes $Ni(acac)_2\cdot 18C6_n$ (n=1,2) and $Ni(enamac)_2 18C6_n$ (n=1) experimentally determined $w_{AP+MPC,0}$ completely coincide with calculated ones by Eq. (1) values of $w_{term,0}$.

Complexes $Ni(O,NH)_2\cdot 18C6_n$ are twice as more active than $Ni(O,O)_2\cdot 18C6_n$ at stage of free radicals origin ($w_{i,0}$), although the "crown-effect" (increase of w_0 and $w_{i,0}$ under the effect of 18K6 additives) observed in the case of catalysis by $Ni(II)(O,NH)_2\cdot 18K6_n$ is lower. It may be explained by reduction of acceptor properties of complex $Ni(II)(O,NH)_2$ in comparison with $Ni(II)(O,O)_2$ in relation to coordination 18K6 that is caused by covalent character of bonds Ni-NH and reduction of effective charge of metal ion.

Conditions allowing estimation of w_{pr} and w_i (Eqs. (1) and (2)) in developed process under catalysis $Ni_x(L^I)_y(L^I_{ox})_z$ и $Ni_x(L^I)_y(L^I_{ox})_z$ $18K6_n$ (L^I= enamac^{-1}) are fulfilled. It turned out that the role of reaction of chain propagation in developed oxidation reaction of ethylbenzene is increased.

In contrast to catalysis by complexes $Ni(II)(L^1)_2$ (L^1=acac^{-1}, enamac^{-1}) with 18K6 in reaction of ethylbenzene oxidation catalyzed by $Ni(II)(L^1)_2$ in the absence of crown-ethers additives increase of initial rate of oxidation is connected mainly with participation of catalyst at stage of chain propagation. At that under catalysis by $Ni(O,NH)_2$ complex the value $w_{pr,0}$ is twice as much than under catalysis by $Ni(O,O)_2$. at the same time the rate of chain initiation almost in order exceeds $w_{i,0}$ in oxidation reaction catalyzed by $Ni(II)(acac)_2$. As it obvious, presence of donor NH-groups in chelate group of nickel complex promotes significant increase of role of activation reaction of molecular oxygen in catalysis mechanism [33].

High activity of "primary" complexes $Ni(II)(acac)_2{\cdot}R_4NBr$ as selective catalysts of ethylbenzene into PEH oxidation as well as the activity of $Ni(L^I)_2{\cdot}18C6_n$ is connected with growth of rate $w_{i,0}$ in comparison with catalysis by $Ni(L^I)_2$, hindering of rate of chain propagation ($w_{pr,0}$) Cat + $RO_2^{\bullet}\rightarrow$ (Fig. 7). As is seen from Fig. 7 that the minimum value of $w_{pr,0}$ is at $[Me_4NBr] = 1{\cdot}10^{-3}$ mol/l, which corresponds to the formation of the complex in 1:1 ratio. As $[Me_4NBr]$ increases further $w_{pr,0}$ increases too. At $[Me_4NBr] = 2{\cdot}10^{-3}$ mol/l the latter reaches the value of $w_{pr,0}$ observed in the presence of $Ni(II)(acac)_2$ only. It is evident that at the Me_4NBr coordination (in the inner and in outer spheres) steric hindrances to coordination of $RO_2^{\bullet}$ with the metal ion can appear ($w_{pr,0}$ drops). If $[Me_4NBr]$ is rather large, the probability of opening of the chelate ring of the $(acac)^-$ ion increases and coordination of the radical $RO_2^{\bullet}$ with the metal center becomes possible ($w_{pr,0}$ increases).

Under substitution of radical CH_3 (Me_4NBr) in cation R_4N^+ by radical n-$C_{16}H_{33}$ (CTAB) the activity of formed complexes $Ni(II)(acac)_2{\cdot}$CTAB at stages of chain initiation and propagation is increased in 4.6 and in 20.5 times correspondingly. At that the rate of PEH accumulation ($w_{PEH,0}$) is increased only in 2 times, and $w_{AP+MPC,0}$ in 15.4 times in comparison with catalysis by $Ni(II)(acac)_2{\cdot}Me_4NBr$ (Fig. 1b).

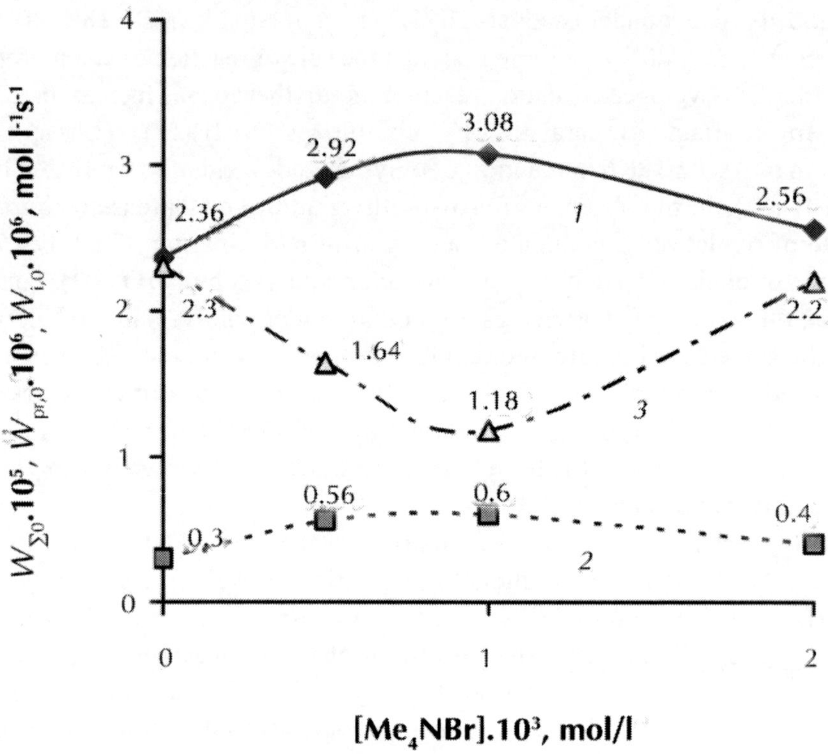

FIGURE 7 The ethylbenzene oxidation rates at the beginning of reaction $w_{\Sigma,0}$ (*1*), and calculated rates of chain initiation $w_{i,0}$ (*2*), and propagation $w_{пp,0}$ (*3*) as a function of [Me$_4$NBr] in the ethylbenzene oxidation upon catalysis by {Ni(II)(acac)$_2$+Me$_4$NBr}. [Ni(II)(acac)$_2$]=1.5·10^{-4} mol/l, 120°C.

1.4.3.2 CATALYSIS WITH IRON COMPLEXES

The chain initiation in ethylbenzene oxidation by dioxygen in the presence of Fe(III)(acac)$_3$ and {Fe(III)(acac)$_3$+ L^2} (L^2 = R$_4$NBr, HMPA, DMF) can be represented by the following reaction:

Fe(III)(acac)$_3$ ((Fe(III)(acac)$_2$)$_m$·(L^2)$_n$)+ RH →
→ Fe(II)(acac)$_2$ ((Fe(II)(acac)$_2$)$_x$·(L^2)$_y$) … Hacac + R$^\cdot$ (I)

The reaction of Fe(III) with RH and interaction of the resulting Fe(II) complex with dioxygen appear to be responsible for chain initiation in the reaction catalyzed by $Fe(III)(acac)_3$ and $\{Fe(III)(acac)_3 + R_4NBr\}$.

As is seen from the Table, the rate of chain initiation in the presence of $\{Fe(III)(acac)_3 + R_4NBr(CTAB)\}$ is higher than in the reaction catalyzed by $Fe(III)(acac)_3$ and much higher than in the no catalytic reaction ($w_{i,0} \approx 10^{-9}$ mol l^{-1} s^{-1}).

$S_{PEH,0}$ depends on the catalyst activity at stages of chain initiation (activation by O_2) and propagation Cat + $RO_2^{\bullet} \rightarrow$ at catalysis by $Fe(III)(acac)_3$ in the presence of R_4NBr. Thus, the rise of $S_{PEH,0}$ from 45 up to ~ 65% in ethylbenzene oxidation catalyzed by $\{Fe(III)(acac)_3 + R_4NBr\}$ (R_4NBr=CTAB) system (Table) is connected with an increase in $w_{i,0}$ (in 2 times) and a small decrease in $w_{pr,0}$. The rate of PEH formation increases in ~1.5 times and the rate AP and MPC formation at stages of chain propagation Cat + $RO_2^{\bullet} \rightarrow$ increases insignificant. The analogous effects are observed with R_4NBr=Me_4NBr. In the case of $(C_2H_5)_4NBr$ the decrease in the $S_{PEH,0}$ is due to the growth of the rates $w_{AP+MPC,0}$ ($w_{pr,0}$) as compared with an increase in the rates $w_{PEH,0}$. The complexes $Fe(II)_x(acac)_y(OAc)_z(R_4NBr)_n$ formed in the process of ethylbenzene oxidation catalyzed by system $\{Fe(III)(acac)_3 + R_4NBr\}$ are likely to be inactive in the heterolytic decomposition of PEH and in the reaction with the $RO_2^{\bullet}$ radicals (Figs. 3, and 4).

The reaction (I) appears to be responsible for chain initiation in the reaction catalyzed by $(Fe(II,III)(acac)_2)_p \cdot (18C6)_q$ complexes.

As one can see from the Table the increase in $S_{PEH,0}$ from ~ 40–50 to ~ 65–70% at the initial stages of the oxidation of ethylbenzene catalyzed by system $\{Fe(III)(acac)_3 + 18K6\}$ ($[18K6]_0 = 0.5 \cdot 10^{-3}$ mol/l, $5.0 \cdot 10^{-3}$ mol/l), at catalysis by $(Fe(II)(acac)_2)_p \cdot (18C6)_q$ complexes, is connected mainly with decrease in the rates of the formation of AP and MPC at micro steps of chain propagation (Cat + $RO_2^{\bullet} \rightarrow$). The drop of $w_{pr,0}$ and $w_{i,0}$ occurs evidently due to electron and steric factors, appeared at the inner and outer sphere coordination (H-bonding) of 18C6 with $Fe(II)(acac)_2$, that may decrease the probability of the coordination of $RO_2^{\bullet}$ and O_2 with the metal ion and formation primary complexes with O_2 and / or $RO_2^{\bullet}$ following the formation of the active particles of superoxide and peroxide types.

The ratio $w_{i,0} / w_{pr,0} \approx 2$–5% ($\{Fe(III)(acac)_3 + L^2\}$ $L^2 = R_4NBr$, HMPA, DMF, and also 18C6 [15]) means that the iron complexes are more active in chain propagation (Cat + $RO_2^{\bullet}$→) than in chain initiation. Furthermore, this value indicates that the reaction (Cat + $RO_2^{\bullet}$→) plays a greater role in ethylbenzene oxidation catalyzed by the iron complexes than in the same reaction catalyzed by the Ni(II) complexes. In the latter case, the ratio $w_{i,0}/ w_{pr,0} \approx 11–50\%$, and depends on the nature of the ligand environment of metal ion [33, 91, 93]. The estimated ratio $w_{i,0}/ w_{pr,0} \approx 2-5\%$ suggests that AP and MPC form mainly in the chain propagation step (conceivably, through the homolytic decomposition of the intermediate complex [$L^2Fe(L^1)_2$—OOR] [135–138]).

In the case of catalysis by iron complexes with HMPA, which does not transformed in the course of oxidation, it is possible to estimate the apparent activation energies for micro steps of ethylbenzene oxidation − chain initiation (activation by O_2) and propagation (Cat + $RO_2^{\bullet}$→) at two temperatures, 80 and 120°C. These are $E_a(w_i)$ =24.53 and 13.03 kcal/mol and $E_a(w_{pr})$ =21.46 and 17.63 kcal/mol in the absence and presence of HMPA, respectively. The gain in activation energy of the initiation reaction 11.5 kcal/mol via the coordination of HMPA is approximately equal to the energy $E_a \sim 10$ kcal/mol of ligand addition to metal acetylacetonates [139]. The difference in E_a between the initiation and propagation reactions in the presence of HMPA is presumably responsible for tendency of oxidation selectivity S_{PEH} to increase with decrease in temperature.

The higher activity of Fe(II)(acac)$_2$)·DMF complexes as compared with Fe(II)(acac)$_2$)·HMPA complexes at the initiation (w_i) and propagation (w_{pr}) steps seems to be to π-donor properties of DMF and its ability to form H-bonds [140]. The coordination of DMF may increase the probability of formation of the primary O_2 complexes O_2·Fe(II)(acac)$_2$)·DMF [141, 142] and, presumably, enhance the activity of the nascent superoxide complexes [DMF·Fe(II)(acac)$_2$) $O_2^{-\bullet}$] at the radical generation step (w_i). As above the the schemes of radical-chain oxidation including intermediate formation [LM-OOR] peroxo complexes at the propagation step followed by the homolytic dissociation of the peroxo complexes ([LM-OOR]→R'C=O (ROH) + R$^{\bullet}$) are a likely explanation of the observed increase in the rate and MPC selectivity of the Fe(II)(acac)$_2$)·DMF catalyzed ethylbenzene oxidation in the initial stages of the reaction $S_{MPC,0} \approx 58\%$. It is quite likely that the coordination of π-donor DMF will facilitate the stabilization of DMF·Fe(II)(acac)$_2$)–O$^{\bullet}$, an oxo species that is produced upon the degradation of the intermediate [ROO-Fe–DMF] peroxo complexes via the homolytic O–O bond dissociation in propagation step ($[L^2Fe\text{-}O^{\bullet}\text{–}^{\bullet}OR] \rightarrow R'C{=}O$ (ROH) + $R^{\bullet}$), and a growth in the probability that $RO^{\bullet}$ radicals escape from the solvent cage (cage "latent radical" mechanism).

As mentioned above the mechanism of the ethylbenzene oxidation catalyzed with {$Fe(III)(acac)_3$+ R_4NBr (18C6)} is obviously unchanged at the addition of $3.7 \cdot 10^{-3}$ mol/l H_2O. So we proposed that catalysis by complexes $(Fe(II)(acac)_2)_x \cdot (R_4NBr)_y \cdot (H_2O)_n$, $(Fe(II)(acac)_2)_p \cdot (18C6)_q \cdot (H_2O)_n$ satisfied the conditions $w_0 \sim [Cat]^{1/2}$ and $w_{i,0} \sim [Cat]$ that allowed $w_{i,0}$ and $w_{pr,0}$ to be calculated by Eqs. (1) and (2) and the catalytic activity of complexes $(Fe(II)(acac)_2)_x \cdot (R_4NBr)_y \cdot (H_2O)_n$ at the micro stages of chain initiation (activation of O_2, $w_{i,0}$) and chain propagation (Cat + $RO_2^{\bullet} \rightarrow$, $w_{pr,0}$) can be evaluated.

As follows from the data in Table the growth in $S_{PEH,0}$ at the catalysis by complexes $Fe(II)(acac)_2)_x \cdot (CTAB)_y \cdot (H_2O)_n$ is connected mainly with the considerable fall in the value of $w_{pr,0} \sim 3.2$ times. The value of $w_{i,0}$ decreases by a factor of ~ 1.3. At that the rate of {AP+MPC} accumulation $w_{p,0}$ decreases ~ 3 times, and $w_{PEH,0}$ decreases only by a factor of ~ 1.26.

The decrease in the rate of chain propagation $w_{pr,0}$ at the catalysis by complexes $(Fe(II)(acac)_2)_x \cdot (CTAB)_y \cdot (H_2O)_n$ seems to be caused by unfavorable steric factors for the $RO_2^{\bullet}$ coordination with metal centre appeared in this case. Beside this the part of H_2O molecules may be absorbed with hydrophilic cation $n\text{-}C_{16}H_{33}Me_3N^+$, creating unfavorable conditions for $RO_2^{\bullet}$ coordination with metal ion.

TABLE 1 The initial rates w_0 and calculated rates of micro stages of chain initiation ($w_{i,0}$), chain propagation ($w_{pr,0}$) and $(w_{i,0}/w_{pr,0}) \cdot 100\%$ in the ethylbenzene oxidations catalyzed with $Fe(III)(acac)_3$ and {$Fe(III)(acac)_3 + L^2$} or {$Fe(III)(acac)_3 + L^2 + L^3$} systems. $[Fe(III)(acac)_3] = 5 \cdot 10^{-3}$ mol/l. L^2 = [CTAB] = $0.5 \cdot 10^{-3}$ mol/l. L^2 = [18C6] = $5 \cdot 10^{-3}$ mol/l. L^3 = $[H_2O]$ = $3.7 \cdot 10^{-3}$ mol/l. 80° C.

L^2, L^3	$w_0 \cdot 10^6$	$w_{i,0} \cdot 10^7$	$w_{pr,0} \cdot 10^6$	$(w_{i,0}/w_{pr,0})$ 100%
—	6.30	0.79	3.32	2.38
CTAB	7.65	1.63	3.14	5.19
CTAB+H_2O	4.85	1.21	0.98	12.24
18C6	2.63	0.24	0.93	2.58
18C6 + H_2O	6.94	0.15	5.58	0.27

At the catalysis with complexes $(Fe(II)(acac)_2)_x{\cdot}(CTAB)_y{\cdot}(H_2O)_n$ the growth in ratio $w_{i,0}/ w_{pr,0}$ to a grate extent (by a factor of ~ 2.35 as compared with $(Fe(II)(acac)_2)_x{\cdot}(CTAB)_y$ was received. In the case of the use of the other R_4NBr as ligand-modifier L^2 the decrease in parameter $w_{i,0}/ w_{pr,0}$ was observed at the H_2O addition :~ 1.4 times ($L^2 = (C_2H_5)_4NBr$ (mainly in consequence of the decrease in $w_{i,0}$ ~ 1.5 times)); ~ 1.22 times (L^2 = Me_4NBr (mainly in consequence of the increase in $w_{pr,0}$~ 1.7 times ($w_{i,0}$ increases ~ 1.4 times))) as compared with catalysis by systems without admixed H_2O.

As seen from the data presented in Table, the reaction of the chain propagation ($Cat + RO_2^{\cdot} \rightarrow$) is evidently the principal reaction of the AP and MPC formation in the ethylbenzene oxidation in the presence of systems $\{Fe(III)(acac)_3 + L^2 + H_2O\}$ (L^2=CTAB, 18C6). It took place also in the cases of use of composition of $\{Fe(III)(acac)_3 + L^2\}$ or only $Fe(III)(acac)_3$. The contribution of the reaction of chain quadratic termination in the mechanism of AP and MPC formation is inessential.

In the ethylbenzene oxidation, catalyzed with $(Fe(II)(acac)_2)_x{\cdot}(18C6)_y{\cdot}(H_2O)_n$ complexes, the significant decrease in parameter $w_{i,0}/ w_{pr,0}$ is observed ~ 10 times as compared with catalysis with $(Fe(II)(acac)_2)_p{\cdot}(18C6)_q$. This fall of $w_{i,0}/ w_{pr,0}$ is evidently caused mainly with growth in the $w_{pr,0}$ value. In this case the lowest $S_{PEH,0} \approx 25\%$ and the highest selectivity of the ethylbenzene oxidation into AP $S_{AP}{}^0 \approx 70\%$ values are got. The [AP]/[MPC] ration increases from 1.2 ($(Fe(II)(acac)_2)_p{\cdot}(18C6)_q$) to 6.5 ($(Fe(II)(acac)_2)_x{\cdot}(18C6)_y{\cdot}(H_2O)_n$).

It is known for example that the ethylbenzene oxidation (70° C, CH_2Cl_2, CH_3CHO) upon catalysis by complexes of Cu(II) with 18C6 $[(CuCl_2)_4(18C6)_2(H_2O)]$ occurs mainly with the formation of ketone (AP). The [alcohol]/ [ketone] ration is 6.5 [37]. As shown above, at catalysis by the system $\{Fe(III)(acac)_3 + 18K6\ (5{,}0{\cdot}10^{-3}\ M) + H_2O\}$ (80°C) the oxidation selectivity $S_{AP,0}$ =70 is sufficiently high at C=1%. But $S_{AP,0}$ is lower than that at the catalytic oxidations, which imitate the action of monooxygenases, for example in the case of oxidations in the presence of Sawyer's system, one of the MMO models [143,144].

At the ethylbenzene oxidation with O_2 (24°C), catalyzed with Sawyer's system $\{[(Fe(II)(Mn(III))L_x(L_x$=bpy, py))+HOOH(R)], polar solvents

MeCN, py+CH_3COOH} S_{AP} ~ 100%, but the conversion only C=0,4%. The catalytic particles were presented by the next hypothetic structure [ ]:

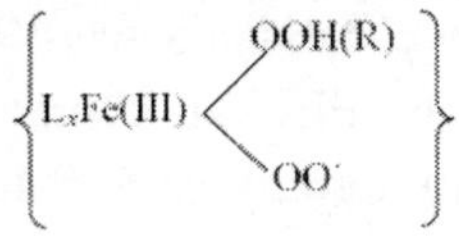

(*).

which oxygenate the ethylbenzene oxidation with the mainly formation of acetophenone. The particles (*) are formed at the reaction of O_2 with peroxo complexes {L_xFe(II)OOH(R) + pyH^+}, which are formed at the initial stage of reaction as result of nucleophilic addition of HOOH(R) to L_xFe(II).

1.5 CONCLUSION

The problem of lowering in homogenous catalyst activity in the oxidation process is the major one, because the functioning of catalyst is always accompanied by processes of its deactivation. The homogenous catalysts heterogenised on silica-, zeolite-, polymer supports are prevented from deactivation and lifetime of catalyst increases. At that the increase in activity, mainly in the oxidation rate, and also in selectivity is observed. But these methods have the rage of limitations, connected with structure peculiarity of supports. As a rule mechanisms of the ligand – modifiers' action in homogenous catalytic oxidation are not proved although the authors tentatively propose mechanistic explanations.

The various catalytic systems on the base of transition metal compounds have been used for the alkylarens oxidation with molecular oxygen. And all of them catalyzed alkylarens oxidations mainly to the products of deep oxidation.

The method of transition metal catalysts modification by additives of electron-donor mono- or multidentate ligands for increase in selectivity of liquid-phase alkylarens oxidations into corresponding hydroperoxides was proposed by us for the first time. On the basis of established (Ni) and assumed (Fe) mechanisms of formation of catalytic active particles and mechanisms of catalyst actions more active catalytic systems $\{M(L^1)_2 + L^2\}$ (L^2 = crown-ethers or ammonium quaternary salts) for the ethylbenzene oxidation into α-phenylethylhydroperoxide were modeled by us and so the mechanisms of selective catalysis were confirmed. Values of selectivity, S_{PEH}, conversion C, and PEH yield reached at application of L^2= crown-ethers or ammonium quaternary salts (Me_4NBr) exceed analogous parameters in the presence of the other $\{Ni(II)(L^1)_2+L^2\}$ systems [33] and known catalysts of ethylbenzene oxidation into PEH [24–27].

The high activity of $\{M(L^1)_2 + L^2\}$ (L^2 = crown-ethers or ammonium quaternary salts) as catalysts of the ethylbenzene oxidation into α-phenylethylhydroperoxide is connected with the formation active primary complexes $(M(II)(L^1)_2)_x\cdot(L^2)_y$, and homo poly nuclear hetero ligand complexes $M(II)_x(L^1)_y(L^1_{ox})_z\cdot L^2_n$, ("A" (Ni), "B" (Fe)) (formed through "dioxygenase-like" mechanisms). The stability of "A" ("B") to the L^1 dioxygenation seems to be due to the intramolecular and intermolecular H-bonding interactions.

The additions of small amounts of water into catalytic systems (M=Fe) are used as mechanistic probe. Results exceed all expectations. Not only is the role of H-bonding interactions in the mechanism of "B" formation confirmed. The increase in the catalytic activity of Fe systems at the addition of small additives of H_2O are established also. The growth in the selectivity of the ethylbenzene oxidation into α-phenylethylhydroperoxide is observed at catalysis by $\{Fe(III)(acac)_3+CTAB+H_2O\}$ system.

The significant increase in the oxidation rate and the selectivity of the ethylbenzene oxidation into acetophenone at catalysis by $\{Fe(III)(acac)_3+18C6+H_2O\}$ system are received.

It is discovered unusual activity of mono- or hetero bi nuclear heteroligand $Ni(II)(acac)_2\cdot L^2\cdot PhOH$ (L^2=MSt (M=Na, Li), MP, HMPA) complexes, including phenol, as the very active catalysts of the ethylbenzene oxidation into α-phenylethylhydroperoxide. The H-bonding interactions are assumed in mechanism of formation of these catalytic complexes. The formation of stable supra molecular structures on the base of $\{Ni(II)(acac)_2\cdot NaSt\cdot PhOH\}$ as a result of intramolecular and intermolecular H-bonds is very probably [95–97].

The catalytic activity of nickel and iron complexes with 18K6 or R_4N-Br at micro stages of chain initiation (w_i, activation O_2) and propagation at the participation of Cat (w_{pr}, Cat+$RO_2^{\cdot}\rightarrow$) in the ethylbenzene oxidation process is evaluated in the framework of radical-chain mechanism with original method, proposed by authors.

KEYWORDS

- **Ammonium quaternary salts**
- **Ethylbenzene oxidation process**
- **Gas-phase oxidation**
- **Homogeneous catalytic oxidations**
- **Liquid-phase hydrocarbon oxidation**
- **Monodentate axial ligands**
- ***p*-xylene**

REFERENCES

1. Suresh, A. K.; Sharma, M. M.; Sridhar, T. *Ind. Eng. Chem. Res.* **2000**, *39*, 3958.
2. Semenov, N. N. *On Some Problems of Chemical Kinetics and Reaction Ability*, Moscow: Iz-vo AN SSSR, 1958 (*in Russian*).
3. Emanuel, N. M.; Denisov, E. T.; Maizus, Z. K. *Liquid-Pase Oxidation of Hydrocarbons*, New York: Plenum Press, 1967; translated by Hazzard, B. J.
4. Sheldon, R. A.; Kochi, J. K. *Metal-Catalyzed Oxidations of Organic Compounds*, New York: Acad. Press, 1981.
5. Emanuel, N. M.; Zaikov, G. E.; Maizus, Z. K. *The Role of Medium in Radical-Chain Reactions of Organic Compounds Oxidation*, Moscow: Nauka, 1973 (*in Russian*).
6. Partenheimer, W. *Catal. Today* **1991**, *23*, 69.
7. Meunier, B. *Chem. Rev.* **1992**, *92*, 1411.
8. Denisov, E. T.; Emanuel, N. M. *Usp. Khim+* **1960**, *29*, 1409 (*in Russian*).
9. Berezin, I. V.; Denisov, E. T.; Emanuel, N. M. *Cyclohexane Oxidation*, Moscow: MSU 1962, (*in Russian*).
10. Matienko, L. I.; Goldina, L. A.; Skibida, I. P.; Maizus, Z. K. *Izv. AN SSSR, Ser. Khim.* **1975**, 287 (*in Russian*).
11. Matienko, L. I.; Skibida, I. P.; Maizus, Z. K. *Izv. AN SSSR, Ser. Khim.* **1975**, 1322 (*in Russian*).
12. Matienko, L. I. *Dissertation of candidate of science*, Moscow: Institute of chemical physics AN SSSR, 1976, (*in Russian*).
13. Skibida, I. P. *Usp. Khim+* **1975**, *44*, 1729 (*in Russian*).

14. Sheldon, R. A.; Kochi, J. K. In *Advances in Catalysis*; Eley, D. D., Pines, H., Weiz P. B., Eds.; New York, San Francisco, London: Acad. Press, 1976.
15. Sheldon, R. A.; Kochi, J. K. In *Metal-Catalyzed Oxidation of Organic Compounds*, New York, London: Acad. Press, 1981.
16. Sheldon, R. A. *J. Mol. Catal.* **1983**, *20*, 1.
17. Mimoun, H. *Chem. Phys. Aspects of Catal. Oxid.* Paris: CNRS, 1981; p. 1.
18. Sheldon, R. A. In *The Activation of Dioxygen and Homogeneous Catalytic Oxidation*, Barton, D. H. R.; Martell, A. E.; Sawyer, D. T., Eds.; New York: Plenum Press, 1993.
19. Mlodnicka, T. In *Metalloporphyrins in Catalytic Oxidation*, Sheldon, R. A. Ed.; New York, Basel, Hong Kong: Marcel Dekker, Inc., 1994; p. 261.
20. Emanuel, N. M. *Usp. Khimi+*, **1978**, *47*, 1329 (*in Russian*).
21. Matienko, L. I.; Maizus, Z. K. *Kinetika I kataliz*, 1974; *15*, 317 (*in Russian*).
22. Maizus, Z. K.; Matienko, L. I. In *Cinetique et Mecanisme des Reactions d'Oxydation Degradation et Stabilisation des Polimers*. 1-er Symposium Franco-Sovietique. Moscou. 1977; S. 128.
23. Mosolova, L. A.; Matienko, L. I.; Maizus, Z. K.; Brin, E. F. *Kinetika i kataliz*, **1980**, *21*, 657 (*in Russian*).
24. Emanuel, N. M.; Gal, D. *Ethylbenzene Oxidation. Model Reaction*, Moscow: Nauka, 1984, (*in Russian*).
25. Norikov, Yu-D.; Blyumberg, E. A.; Salukvadze, *Problemy kinetiki i kataliza*, Moscow: Nauka, **1975**, *16*, 150 (*in Russian*).
26. Nesterov, M. V.; Ivanov, V. A.; Potekhin, V. M.; Proskuryakov, V. A.; Lysukhin, M. Yu. *Zh. Prikl. Khimii,* **1979**, *52*, 1585 (*in Russian*).
27. Toribio, P. P.; Campos-Martin, J. M.; Fierro, J. L. G. *J. Mol. Catal. A: Chem.*, **2005**, *227*, 101 .
28. In *The Activation of Dioxygen and Homogeneous Catalytic Oxidation*, Barton D.H.R.; Martell A. E.; Sawyer D. T., Eds.; New York: Plenum Press 1993.
29. Mansuy, D. In *The Activation of Dioxygen and Homogeneous Catalytic Oxidation*, Barton D. H. R.; Martell A. E.; Sawyer D. T., Eds.; New York: Plenum Press 1993; p. 347.
30. Karasevich, E. I. In *Reactions and Properties of Monomers and Polymers,* D'Amore, A.; Zaikov, G.; Eds.; New York: Nova Sience Publ. Inc., 2007; p.43.
31. Szaleinec, M.; Hagel, C.; Menke, M.; Novak, P.; Witko, M.; Heider, J. *Biochemistry*, **2007**, *46*, 25.
32. Weissermel, K.; Arpe, H.-J. In *Industrial Organic Chemistry*, 3rd ed., transl. by Lindley, C. R.; New York: VCH 1997.
33. Matienko, L. I. In *Reactions and Properties of Monomers and Polymers,* D'Amore, A.; Zaikov, G. Eds.; New York: Nova Sience Publ. Inc., 2007; p. 21.
34. Yoshikuni, T. *J. Mol. Catal. A: Chem.*, **2002**, *187*, 143.
35. Qi, J.-Y.; Ma, H.-X.; Li, X.-J.; Zhou, Z.-Y.; Choi, M.C.K.; Chan, A. S. C.; Yang, Q.-Y. *Chem. Commun.*, **2003**, 1294.
36. Murahashi, S.-I.; Naota, T.; Komiya, N. *Tetrahedron Lett.*, **1995**, *36*, 8059.
37. Komiya, N.; Naota, T.; Murahashi, S.-I. *Tetrahedron Lett.*, **1996**, *37*, 1633.
38. Rudler, H.; Denise, B. *J. Mol. Catal. A: Chem.*, **2000**, *154*, 277.
39. Evans, S.; Smith, J. R. L. *J. Chem. Soc., Perkin Trans.* **2000**, *2,* 1541.

40. Guo, C.; Peng, Q.; Lie, Q.; Jiang, G. *J. Mol. Catal. A: Chem.*, **2003**, *192*, 295.
41. Shul'pin, G. B.; *J. Mol. Catal. A: Chem.*, **2002**, *189*, 39.
42. Ellis, Jr., P. E.; Lyons, J. E.; *Coord. Chem. Rev.*, **1990**, *105*, 181.
43. Lyons, J. E.; Ellis, Jr., P. E.; Myers, H. K. *J. Catal.*, **1995**, *155*, 59.
44. Grinstaff, M. W.; Hill, M. G.; Labinger, J. A.; Gray, *Science*, **1994**, *264*, 1311.
45. Grinstaff, M. W.; Hill, M. G.; Birnbaum, E. R.; Schaefer, W. P.; Labinger; Gray, H. B.; *Inorg. Chem.*, **1995**, *34*, 4896
46. Haber, J.; Matachowski, L.; Pamin, K.; J. Poltowicz, J., *J. Mol. Catal. A: Chem.*, **2003**,*198*, 215.
47. Punniyamurthy, T.; Velusamy, S; Iqbal, J. *Chem. Rev.*, **2005**, *105*, 2329.
48. Tao Y.; Kanoh, H.; Abrams, L.; Kaneko, K. *Chem. Rev.*, **2006**, *106*, 896.
49. Evans, S.; Smith, J. R. L. *J. Chem. Soc. Perkin. Trans.*, **2001**, *2*, 174.
50. Wu, J.; Hou, H.; Han, H.; Fan, Y. *Inorg. Chem.*, **2007**, *46*, 7960.
51. Sankar, G.; Raja, R.; Thomas, J. M. *Catal. Lett.*, **1998**, *55*, 15.
52. Bennur, T. H.; Srinivas, D.; Sivasanker, S. *J. Mol. Catal. A: Chem.*, **2004**, *207*, 163.
53. Karakhanov, E. A.; Kardasheva, Yu. S.; A.L. Maksimov, A. L.; Predeina, V. V.; Runova, E. A.; Utukin, A. M. *J. Mol. Catal. A: Chem.*, **1996**, *107*, 235.
54. Poltowicz, J.; Haber, J. *J. Mol. Catal. A: Chem.*, **2004**, *220*, 43.
55. Mosolova, L. A.; Matienko, L. I.; Maizus, Z. K. *Izv. AN SSSR, Ser. Khim.* **1980**, 278 (*in Russian*).
56. Mosolova, L. A.;Matienko, L. I. *Neftekhimiya*, **1985**, 540 (*in Russian*).
57. Mosolova, L. A.;Matienko, L. I.; Skibida, I. P. *Kinetika i kataliz*, **1988**, *29*, 1078 (*in Russian*).
58. Mosolova, L. A.; Matienko, L. I.; Maizus, Z. K. *Izv. AN SSSR, Ser. Khim*, **1981**, *731*, 1977 (*in Russian*).
59. Mosolova, L. A.; Matienko, L. I.; Skibida, I. P. *Kinetika i kataliz*, **1987**, *28*, 479 (*in Russian*).
60. Matienko, L. I.; Mosolova, L. A. *Izv. AN, Ser. Khim.*, **1999**, 55 (*in Russian*).
61. Mosolova, L. A.; Matienko, L. I.; Skibida, I. P. *Kinetika i kataliz*, **1987**, *28*, 484 (*in Russian*).
62. Sagawa, T.; Ohkubo, K. *J. Mol. Catal. A: Chem.*, **1996**, 269.
63. Zhang, Y.; Kang, S. A.; Mukherjee, T.; Bale, S.; Crane, B. R.; Begley, T. P.; Ealick, S. E. *Biochemistry*, **2007**, *46*, 145.
64. Dai, Y.; Pochapsky, Th. C.; Abeles, R. H. *Biochemistry*, **2001**, *40*, 6379.
65. Gopal, B.; Madan, L. L. ; Betz, S. F.; Kossiakoff, A. A. *Biochemistry*, **2005**, *44*, 193.
66. Balogh-Hergovich, É.; Kaizer, J.; Speier, G. *J. Mol. Catal. A: Chem.*, 2000, *159*, 215.
67. Matienko, L. I.; Mosolova, L.A. *Neftekhimiya*, **2007**, *47*, 42 (*in Russian*).
68. Straganz, G.D.; Nidetzky, B. *J. Am. Chem. Soc.*, **2005**, *127*. 12306.
69. Bennett, J. P.; Whittingham, J. L.; Brzozowski, A. M.; Leonard, Ph. M.; Grogan, G. *Biochemistry*, **2007**, *46*, 137.
70. Binnemans, K. *Chem. Rev.*, **2005**, *105*, 4148.
71. Demlov, E. V. *Izv. AN, Ser. Khim.*, **1995**, 2094 (*in Russian*).
72. Pârvulescu, V. I.; Hardacre, Ch. *Chem. Rev.*, **2007**, *107*, 2615.
73. Dyson, P. J. *Trans. Met. Chem.* **2002**, *27*, 353.
74. Dakka, J.; Sasson, Y. *J. Chem. Soc. Chem. Commun.*, **1987**, 1421.
75. Barak, J.; Sasson, Y. *J. Chem. Soc. Chem. Commun.*, **1987**, 1266.

76. Haruštiak, M.; Hrones, M.; Ilavsky, J. *J. Mol. Catal.*, **1989**, *53*, 209.
77. Panicheva, L. P.; Tret'jakov, N. Ju.; Berezina, S. B.; Juffa, A. Ja.; *Neftekhimiya,* **1994**, *34*, 171 (*in Russian*).
78. Maximova, T. V.; Sirota, T. V.; Koverzanova, E. V.; Kasaikina, O. T. *Neftekhimiya,* **2001**, *41*, 289 (*in Russian*).
79. Satpathy, Ms. M.; Pradhan, B. *Asian J. Chem.*, **1997**, *9*, 873.
80. Yamaguchi, H.; Katsumata, K.; Steiner, M.; Miketa, H.J. *J. Magn. Magn. Mater.*, **1998**, *750,* 177–181.
81. Buschmann, H.-J.; Mutihac, L. *J. Incl. Phenom. Macrocycl. Chem.* **2002**, *42*, 193.
82. Cammack, R.; In *Advanced. Inorganic. Chemistry.,* Sykes,A. G., Ed.; New York, London, Tokyo, Toronto: Acad. Press, Inc., 1988; p.297.
83. Halcrow, M.A.; Chistou, G. *Chem. Rev.*, **1994**, *94. (8),* 2421–2481.
84. Kolodziej, A. F.; In *Progress in Inorganic Chemistry*, Karlin, K. D.; Ed.; New York, Chichester, Brisbaue, Toronto, Singapore: Wiley J. and Sons, Inc., 1994; p. 493.
85. Lamble, S. E.; Albracht, S. P. J; Armstrong, F.A. *J. Am. Chem. Soc.*, **2004**, *126*, 14899.
86. Pelmenschikov, V.; Siegbahn Per, E. M. *J. Am. Chem. Soc.*, **2006**, *128*, 7466.
87. Belskii, V. K.; Buleachev, B. M. *Uspekhi khimii*, **1999**, *68*, 136 (*in Russian*).
88. Schneider, H.-J.; Busch, R.; Kramer, R. In *Nucleophility. Advanced. Chemistry. Ser.*, Harris, J. M.; McManus, S. P., Eds.; Washington: Am. Chem. Soc., 1987; p. 482.
89. Clark, J. H.; Miller, J. M.; *J. Chem. Soc. Perkin Trans.*, **1977**, 1743.
90. Mosolova, L. A.; Matienko, L. I.; Skibida, I. P. *Izv. AN, Ser. Khim,* **1994**, 1406 (*in Russian*).
91. Matienko, L. I.; Mosolova, L. A. In *The successes in the sphere of the heterogeneous catalysis and heterocycles*, Rachmankulov, D. L.; Ed.; Moscow: Chemistry, 2006; 235 (*in Russian*).
92. Mosolova, L. A.; Matienko, L. I.; Skibida, I. P. *Izv. AN, Ser. Khim,* **1994**, 1412 (*in Russian*).
93. Matienko, L. I.; Mosolova, L. A. In *New Aspects of Biochemical Physics. Pure and Applied Sciences*, Varfolomeev, S. D.; Burlakova, E. B.; Popov, A. A.; Zaikov, G.E., Eds.; New York: Nova Science Publ. Inc., 2007; p.95.
94. Golodov, V. A.; *Ross. Khim. Zh,* **2000**, *44*, 45 (*in Russian*).
95. Basiuk, E. V.; Basiuk, V. V.; Gomez-Lara, J.; Toscano, R. A. *J. Incl. Phenom. Macrocycl. Chem.* **2000**, *38*, 45.
96. Wang, L.; Cai, J.; Mao, Z. W.; Feng, X. L.; Huang, J. W.; *Trans. Met. Chem.* **2004**, *29*, 411.
97. Li, Q.; Mak, T.C.W.; *J. Incl. Phenom. Macrocycl. Chem.* **1999**, *35*, 621.
98. Matienko, L. I.; Mosolova, L. A. *Kinetika i kataliz*, **2005**, *46*, 354 (*in Russian*).
99. Matienko, L.I.; Mosolova, L.A.; In *Chemical Physics and Physical Chemistry: Step Into The Future*, Zaikov,G. E.;Kirshenbaum, G.; Eds.; New York: Nova Science Publ. Inc., 2007; p.57.
100. Gutmann, V. *Coord. Chem. Revs.*, **1976**, 8, 225.
101. Borovik, A. S. *Acc. Chem. Res.*, **2005**, 38, 54.
102. Holm, R. H.; Solomon, E. I. *Chem. Rev.*, **2004**, *104*, 347.

103. Tomchick, D. R.; Phan, P.; Cymborovski. M.; Minor, W.; Holm, T. R. *Biochemistry*, **2001**, *40*, 7509
104. Perutz, M. F.; Fermi, G.; Luisi, B.; Shaanan, B.; Liddington, R.C. *Acc. Chem. Res*., **1987**, *20*, 309.
105. Schlichting, I.; Berendzen, J.; Chu, K. *Science*, **2000**, *287*, 165.
106. Lucas, R. L.; Zart, M. K.; Murkerjee, J.; Sorrell, T. N.; Powell, D. R.; Borovik, A. S. *J. Am. Chem. Soc*., **2006**, *128*, 15476.
107. Benisvy, L.; Halut, S., Donnadieu, B.; Tuchagues, J. P.; Chottard, J. C.; Li, Y. *Inorg. Chem*., **2006**, *45*, 2403.
108. Hikichi, S.; Ogihara, T.; Fujisawa, K. *Inorg. Chem*., **1997**, *36*, 4539.
109. Ogo, S.; Yamahara, R.; Roach, M. *Inorg. Chem*., **2002**, *41*, 5513.
110. MacBeth, C. E.; Gupta, R.; Mitchell-Koch, K. R. *J. Am. Chem. Soc*., **2004** *126*, 2556.
111. . Nekipelov, V. M; Zamaraev, K. I. *Coord. Chem. Revs*., **1985**, 61, 185.
112. Nesterov, O. V., Nekipelov, V. M.; Chirkov, Yu. N.; Kitaigorodskii, A. N.; Entelis, S. G. *Kinetika i kataliz*, **1980**, *21*, 1238 (*in Russian*).
113. Uehara, K.; Ohashi, Y. M. *Bull. Chem. Soc. Jap*., **1976**, *49*, 1447.
114. Nelson, J. H.; Howels, P. N.; Landen, G. L.; De Lullo, G. S.; Henry, R. A. In *Fundamental research in homogeneous catalysis*. N.Y.-L.: Plenum Press, 1979; *3*, 921.
115. Daolio, S.; Traldi, P.; Pelli, B.; Basato, M.; Corain, B. *Inorg. Chem*., **1984**, *23*, 4750.
116. Matienko L. I; Mosolova, L. A. In *Modern Tendencies in Organic and Bioorganic Chemistry: Today and Tomorrow*,. Mikitaev, A.; Ligidov, M. Kh. Zaikov, G. E. Eds New York: Nova Sience Publ. Inc., 2008; p. 33.
117. Matienko, L. I.; Mosolova, L.A. *Neftekhimiya*, **2008**, *48*, 360 (*in Russian*).
118. Klibanov, A. M. *Acc. Chem. Res.* **1990**, 23, 114.
119. Ball, Ph.; *Chem. Rev*., **2008**, *108*, 74.
120. Derat, E.; Shaik, S.; Rovira, C.; Vidossich, P.; Alfonso-Prieto, M. *J. Am. Chem. Soc*., **2007**, *129*, 6346.
121. Hua, Li; Yangzhong, L.; Xuhong, Z. *J. Am. Chem. Soc*., **2006**, *128*, 6391.
122. Matsui, T.; Omori, K.; Jin, H.; Ikeda-Saito, M. *J. Am. Chem. Soc*., **2008**; *130*, 4220.
123. Zotova, N.; Franzke, A.; Armstrong, A.; Blackmond, D.G. *J. Am. Chem. Soc*., **2007**, *129*, 15100.
124. Piera, J.; Persson, A.; Cadentey, X.; Bäckvall, J. E. *J. Am. Chem. Soc*., **2007**, *129*, 14120.
125. Funabiki, T.; Yoneda, I. , Ishikawa, M. *J. Chem. Soc. Chem. Commun*., **1994** 1453.
126. Demlov, E.; Demlov, Z. Phase-transfer catalysis. M.: Mir Press, 1987.
127. Evans, D. A.; Mito, Sh.; Seidel, D. *J. Am. Chem. Soc*., **2007**, *129*, 11583.
128. Pardo, L.; Osman, R.; Weinstein, H.; Rabinowitz, J. R. *J. Am. Chem. Soc*., **1993**, *115*, 8263.
129. Csanyi, L. J.; Jaky, K.; Galbacs G. *J. Mol. Catal. A: Chem*., **2000**, *164*, 109.
130. Csanyi, L. J.; Jaky, K.; Dombi, G. *J. Mol. Catal. A: Chem*., **2003**, *195*, 109.
131. Csanyi, L. J.; Jaky, K.; Palinko, I. *Phys. Chem. Chem. Phys.* **2000**, *2*, 3801.
132. Kuramshin, E. M.; Kochinashvili, M. V.; Zlotskii, S. S.; Rachmankulov, D. P. *Dokl. AN SSSR*, **1984**, *279*, 1392 (*in Russian*).
133. Goldberg, I. *Acta Cryst*., **1978**, *34*, 3387.
134. Jaguar-Grodzinnski, J. *Isr. J. Chem*., **1985**, *25*, 39.

135. Arasasingham, R. D.; Cornman, Ch. R.; Balch, A.L. *J. Am. Chem. Soc.*, **1989**, *111*, 7800.
136. Chaves, F. A.; Rowland, J. M.; Olmstead, M. M.; Mascharak, P. K. *J. Am. Chem. Soc.*, **1998**, *120*, 9015.
137. Solomon-Rapaport, E.; Masarwa, A.; Cohen, H.; Meyerstein, D. *Inorg. Chim. Acta.*, **2000**, 41-46..
138. Semenchenko, A. E.; Solyanikov, V. M.; Denisov, E. T.; *Zh. Phiz. Khimii*, **1973**, *47*, 1148 (*in Russian*).
139. Morassi, R.; Bertini, I.; Sacconi, L. *Coord. Chem. Rev.*, **1973**, *11*, 343.
140. Alekseevski, V.A. In. *The problems of chemistry and use of metal β-diketonates*,. Spizin, V.I.; "Nauka", M. Eds; 1982; 65.
141. Stynes, D. V.; Stynes, H. C.; James, B. R; Ibers, J. A. *J. Am. Chem. Soc.*, **1973**, *95*, 1796.
142. Dedieu, A.; Rohmer, M.-M.; Benard, M.; Veillard, A. *J. Am. Chem. Soc.*, **1976**, *98*, 3717.
143. Kang, Ch. K.; Redman, Ch.; Cepak, V.; Sawyer, D.T. *Bioorg. and Medic. Chem.*, **1991**, *1*, 125.
144. Matsushita, T.; Sawyer, D.T.; Sobkowiak, A. *J. Mol. Catal. A: Chem.*, **1999**, *137*, 127.

CHAPTER 2

KINETICS AND MECHANISM TOLUENE AND CUMENE OXIDATION BY DIOXYGEN WITH NICKEL (II) COMPLEXES

L. I. MATIENKO, L. A. MOSOLOVA, A. D'AMORE, and G. E. ZAIKOV

CONTENTS

2.1 INTRODUCTION

The effect of modifying ligands, which usually occupy the axial position in a complex, was as a rule studied using model compounds that mimic active sites of enzymes, namely, mono- and dioxygenases. At present, examples of catalytic systems are known that substantially increase the rates of catalyzed reactions and, sometimes, the yields of products upon the addition of modifiers. As a rule, the mechanisms of action of additives were not established, although possible schemes were put forward to explain the action of modifying ligands [1].

Oxidation of alkylarenes with dioxygen catalyzed by model systems mimicking biological systems (mono- and dioxygenases) that favors the selective insertion of oxygen atoms into C–H bonds of organic molecules affords alcohols and carbonyl compounds [2–6]. For example, toluene 2- or 4-monooxygenase performs aromatic hydroxylation of toluene to corresponding cresols [7]. Unfortunately, dioxygenases capable of effecting alkane dioxygenation to hydroperoxides are so far unknown [3].

The method of transition metal catalysts modification by additives of electron-donor monodentate ligands for increase in selectivity of liquid-phase alkylarens oxidations into corresponding hydroperoxides was proposed by us [8] for the first time.

Based on the established (for Ni complexes) and hypothetical (for Fe complexes) mechanisms of formation of catalytically active species and their operation, efficient catalytic systems $\{ML^1_n + L^2\}$ (L^2 are crown ethers or quaternary ammonium salts) of ethylbenzene oxidation to α-phenyl ethyl hydroperoxide were modeled. Selectivity $S_{PEH,max}$, conversion, and yield of PEH in ethylbenzene oxidation catalyzed by these systems were substantially higher than those observed with conventional catalysts of ethylbenzene oxidation to PEH [8]. The high activity of systems $\{ML^1_n + L^2\}$ (L^2 are crown ethers or quaternary ammonium salts) is associated with the fact that during the ethylbenzene oxidation, the active primary $(M^{II}L^1_2)_x(L^2)_y$ complexes (I macro step of oxidation) and heteroligand $M^{II}_xL^1_y(L^1_{ox})_z(L^2)_n$ complexes (II macro step of oxidation) are formed to be involved in the oxidation process [8].

The involvement of transition metal complexes in the chain propagation step of ethylbenzene oxidation was considered exclusively in our studies [8–14] and in theoretical publications [15, 16]. The attention of researchers is concentrated mainly on the role of transition metal compounds in the steps of generation of free radicals (in chain initiation and hydroperoxide decomposition), while the reaction of a catalyst with peroxide radicals (chain propagation) in the catalytic hydrocarbons oxidation with dioxygen still remains insufficiently explored.

In Ref. [17], it has been established, that electron-donating exo ligand-modifiers L^2 can be used for modification of Ni(II)(acac)$_2$ activity as catalyst of oxidation with dioxygen not only of ethylbenzene, but and other alkylarens with primary and tertiary α-CH-bonds, namely, toluene and cumene, aimed the increase in yield of ROOH. However ligand-modifier influence on the mechanism of toluene and cumene oxidation, catalyzed with Ni(II)(acac)$_2$, has not been established. Here the analysis of the mechanism of oxidation of toluene and cumene with dioxygen, catalyzed by catalytic system $\{ML^1_n + L^2\}$ (L^2 = N-methylpirrolidon-2) is resulted.

Earlier by us it has been established that oxidation reactions of considered hydrocarbons (ethylbenzene, toluene and cumene) by molecular oxygen, catalyzed with system $\{Ni(II)(L^1)_2+L^2\}$ (L^2 = N-methylpirrolidon-2 (MP)) are characterized with general kinetic regularities. As well as at oxidation of ethylbenzene, synergic effects of growth of initial rates of oxidation of toluene and cumene, and also S_{ROOH} and degree of conversion of oxidation in ROOH in presence of system $\{Ni(II)(L^1)_2+L^2\}$ are observed, and maximum possible values of these parameters are received in the optimal conditions of researched oxidations. However, there are differences, which are connected with specificity of each hydrocarbon [17].

In this work features of oxidation of toluene and cumene together with features of oxidation of ethylbenzene are considered in more details with addition of new our and literature data. For the first time the effect of coordination of ligand-modifier to metal (Ni(II)) complex catalyst on mechanism of formation of all analyzed products (toluene, ethylbenzene and cumene) is established and the schemes of oxidation resulted in initial research [17] are added.

It is known that aliphatic C–H bond energy is 87 kcal/mol (ethylbenzene, cumene), and 90 kcal/mol (toluene) [18].

It is well-known, what toluene, ethylbenzene and cumene, which are monoalkylbenzenes with primary, secondary and tertiary α-CH-bonds, on "Oxidability" with molecular oxygen, are situated in a number [19] (Table 1):

toluene < ethylbenzene < cumene

TABLE 1 The values of parameters of $k_2/\sqrt{k_5}$ ((l /mol)$^{1/2}$/s$^{1/2}$) and chain propagation ($RO_2^{\cdot}$ +RH→) k_2 and chain termination ($RO_2^{\cdot}$ + $RO_2^{\cdot}$→) k_5 rate constants (l/mol/s) for monoalkylbenzenes oxidations at 30°C.

Hydrocarbon	$k_2/\sqrt{k_5}.10^3$	k_2	$k_5.10^{-6}$
Toluene	0.014	0.24	150
Ethylbenzene	0.21	1.3	20
Cumene	1.5	0.18	0.0075

2.2 EXPERIMENTAL METHODS

Alkylarenes (RH) oxidation was studied at 100°C (toluene, cumene), at 120°C (ethylbenzene), in glass bubbling-type reactor at the absence catalyst and in the presence of Ni(II)(acac)$_2$ and {Ni(II)(acac)$_2$+L^2}(L^2= N-methylpirrolidone-2). The selectivity S_{ROOH} and the RH conversion C of ethylbenzene into PEH in the alkylarenes oxidation were determined using the formulas: S_{ROOH} = [ROOH]/Δ[RH]·100% and C = Δ[RH]/[RH]$_0$·100%.

2.2.1 ANALYSIS OF OXIDATION PRODUCTS

Hydroperoxides of α-phenylethyl and cumyl, and also the sum of benzyl hydroperoxide (BH) and perbenzoic acid (PBA) at toluene oxidation were determined by iodometry titration [9–14, 17].

At separate definition of benzyl hydroperoxide (BH) and perbenzoic acid (PBA) have taken advantage of their various activities in reaction with KJ. Thus sum of BH and PBA was determined using ethanol solution of KJ after keeping in the dark within several hours. Aqueous KJ was used for determination for perbenzoic acid, thus J_2, formed on reaction PBA with KJ, was titrated right after. Accuracy of definition of PBA was ±5%.

The total maintenance of benzoic and perbenzoic acids was determined alkalimetrcally [17].

Benzaldehyde (BAL), benzyl alcohol (BZA), acetophenone (AP), methylphenylcarbinol (MPC), dimethyl phenylcarbinol (DMPC), and phenol (PhOH), as well as the RH content in cumene and ethylbenzene oxidation processes were examined by GLC [9–14, 17]. The overall rate of the process was determined from the rate of accumulation of all oxidation products (toluene, ethylbenzene), or consumption of starting hydrocarbon (cumene). A correlation between RH consumption and product (P) accumulation was established: $\Delta[RH] = \Sigma [P]$.

The catalytic alkylarens oxidation with dioxygen was carried out in the O_2 – solution two phase systems under kinetic control.

The order in which products P formed was determined from the time dependence of product accumulation rate rations at $t \rightarrow 0$. The variation of these rations with time was evaluated by graphic differentiation $\Delta[P_1]_{ij}/\Delta[P_2]_{ij}$ on t_j, where $\Delta[P_1]_{ij}$ and $\Delta[P_2]_{ij}$ – increase in concentration P_1 и P_2 during $\Delta t = t_j - t_i$ [8, 39].

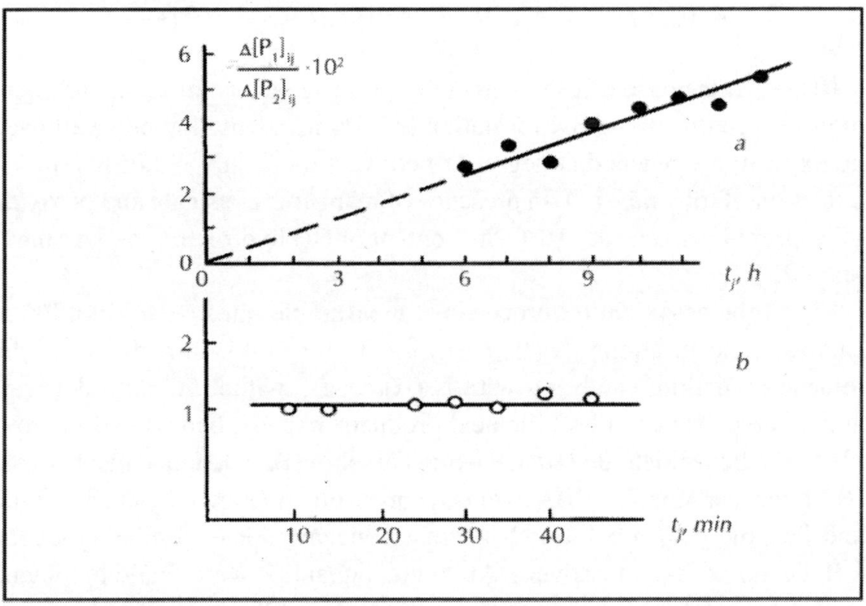

2.3 RESULTS AND DISCUSSION

2.3.1 TOLUENE OXIDATION, CATALYZED WITH SYSTEM {NI(II)(ACAC)$_2$ + MP}

Due to the low values of kinetic parameter $k_2/\sqrt{k_5}$ (Table 1) toluene is virtually not oxidized in the absence of catalyst or initiator. The major products of the initiated and most catalytic oxidations of toluene are benzaldehyde and *benzoic acid* [20–23]. For example, high-selective catalysts of oxidation of methylbenzenes (in that number of toluene) are so called *MC*-systems (Co-Mn-Br-catalytic systems) [20].

The high recombination rates of benzylperoxy radicals (Table of VII. 1), resulting in the formation of benzyl alcohol and benzaldehyde (∗) [24, 25], and low stability of benzyl hydroperoxide, BH, do not permit significant selectivity in BH formation already at low extents of oxidation. Benzaldehyde, formed at decomposition of BH and in chain termination reaction (∗), is then oxidized at high rates to benzoic acid.

$$2PhCH_2O_2^{\cdot} \rightarrow PhCHO + PhCH_2OH + O_2 \qquad (*)$$

Thus, in the case of use of tert.butylperbenzoate as initiator, sufficient high selectivity suff of BH formation (62.5% in calculation on swallowed up oxygen) are reached at degrees of conversion C only ~ 0.05% [26]. At oxidation of toluene by O_2 in presence of manganese resinate and peroxide of isopropyl benzene at 110°C the content of BH hydroperoxide was 0.6% only [23].

The toluene oxidation proceeds at measurable rates (≥10^{-7} mol/l/s) at catalyst concentrations [Ni(II)(acac)$_2$] ≥ $3\cdot10^{-3}$ mol/l. In addition, in the toluene oxidation, catalyzed with Ni(II)(acac)$_2$, without donor additives, and in the presence of MP, the next products, namely, benzyl hydroperoxide (BH), benzaldehyde (BAL), benzyl alcohol (BZA), and acids: benzoic (BA) and perbenzoic (PBA) acids, – are formed (Fig. 1). Benzaldehyde and benzoic acid are formed in great concentrations. Similar products, BH, BAL and BA, moreover BAL in great quantity, were found in toluene oxidation, catalyzed with resinate of manganese [23].

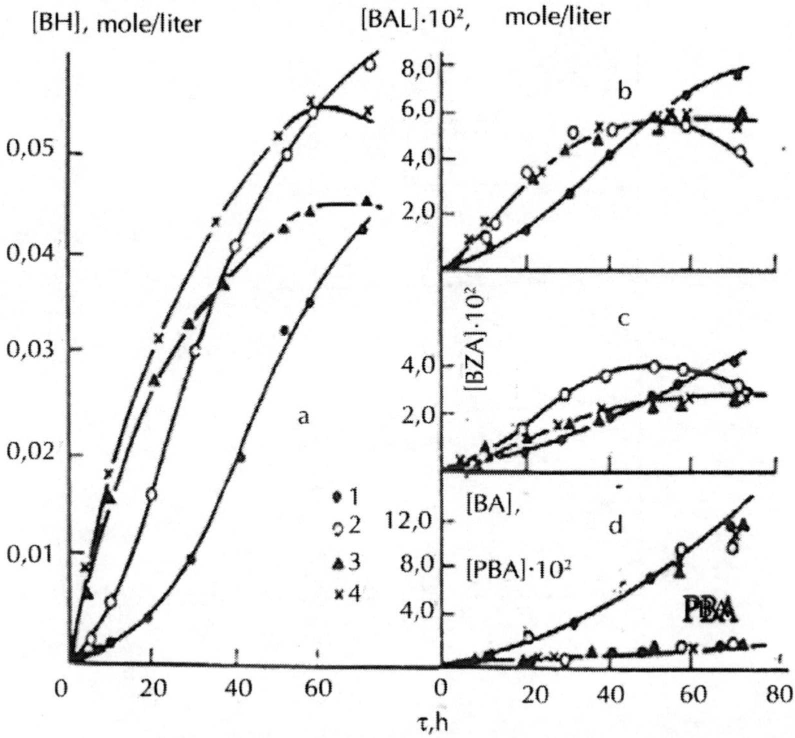

FIGURE 1 Kinetic curves for the accumulation of the oxidation products: BH (mol/l) (a); and BAL (b), BZA (c), BA and PBA (d) (X10², mol/l) in the oxidation of toluene at 100°C in the presence of Ni(II)(acac)$_2$ with 0 (1), 0.03 (2), 0.078 (3), and 0.15 mol/l MP.

In Ref. [21] is established that with an anisole-containing polypyridylamine potential ligand °L, a μ-1,2-peroxodicopper(II) complex $[\{°LCu^{II}\}_2(O_2^{2-})]^{2+}$ forms from the reaction of the mononuclear compound $[CuI(°L)(MeCN)]B(C_6F_5)_4$ ($°LCu^I$) with O_2 in no coordinating solvents at -80°C. Thermal decay of this peroxo complex in the presence of toluene or ethylbenzene leads to rarely seen C-H activation chemistry; benzaldehyde and acetophenone/ 1-phenylethanol mixture (~1:1), respectively, are formed. Very similar toluene oxygenation chemistry occurs with dicopper(III)-bis-μ-oxo species $[\{^{Bz}LCu^{III}\}_2(\mu\text{-}O^{2-})_2]^{2+}$.

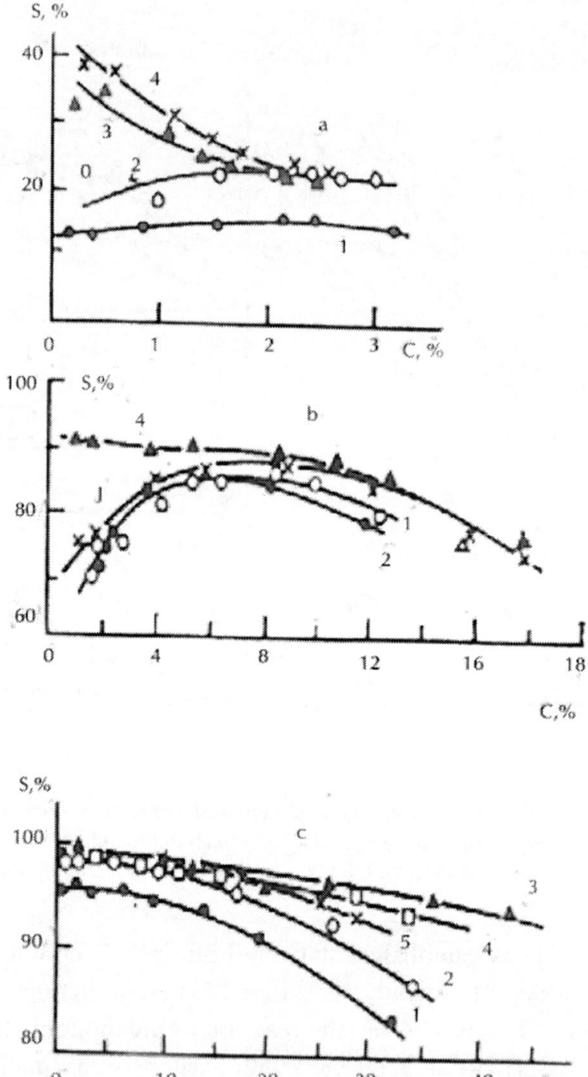

FIGURE 2 Dependence of the selectivity S_{ROOH} of the oxidation of monoalkylbenzenes on hydrocarbon conversion C: a) toluene (designations and conditions in Fig. 1), b) ethylbenzene at 120°C in the presence of Ni(II)(acac)$_2$ ($3 \cdot 10^{-3}$ mol/l), and 0.03 (1), 0.05 (2), 0.07 (3), and 0.15 mol/l MP (4), and c) cumene (1), cumene in the presence Ni(II)(acac)$_2$ ($6 \cdot 10^{-6}$ mol/l) without donor additives (2) and with $6 \cdot 10^{-5}$ (3), $1.5 \cdot 10^{-4}$ (4), $3 \cdot 10^{-4}$ (5) mol/l MP (100°C).

It can how see from Fig. 1(a, в, c), additives MP into oxidation of toluene in presence $Ni(acac)_2$ result in increase in initial rates of accumulation of BH, BZA and BAL. This allows to assume increase in the activity of $Ni(acac)_2$ upon its complexation with L^2 in chain initiation (1.1), BH decomposition (1.3-1.3'), and free radical termination (1.5.) (Scheme 1).

Scheme 1.

$PhCH_3 \xrightarrow{O_2Ni(acac)_2L2} PhCH_2O_2^{\cdot}$ (1.1)

$PhCH_2O_2^{\cdot} + PhCH_3 \rightarrow PhCH_2OOH + PhCH_2^{\cdot}$ (BH) (1.2)

$PhCH_2OOH + Ni(acac)_2\,L^2 \leftrightarrows K \rightarrow PhCH_2O^{\cdot} + Ni(acac)_2OH\cdot L^2$ (1.3)

$PhCH_2OOH + Ni(acac)_2OH\,L^2 \leftrightarrows K \rightarrow PhCH_2O_2^{\cdot} + Ni(acac)_2\,L^2 + H_2O$ (1.3')

$PhCH_2O^{\cdot} + PhCH_3 \rightarrow PhCH_2OH + PhCH_2^{\cdot}$ (BZA) (1.4)

$PhCH_2O_2^{\cdot} + Ni(acac)_2\,L^2 \rightarrow PhCHO + Ni(acac)_2OH\cdot L^2$, (BAL) (1.5)

Scheme 2.

$PhCH_2OH + Ni(acac)_2OH \leftrightarrows K \rightarrow PhC^{\cdot}HOH + H_2O + Ni(acac)_2$ (1.6)

$PhC^{\cdot}HOH + Ni(acac)_2OH \rightarrow PhCHO + Ni(acac)_2 + H_2O$ (BAL) (1.7)

$PhCHO + Ni(acac)_2OH \rightarrow PhCO^{\cdot} + Ni(acac)_2 + H_2O$ (1.8)

$PhCO^{\cdot} + Ni(acac)_2OH \rightarrow PhCO_2H + Ni(acac)_2$ (BA) (1.9)

$PhCO^{\cdot} + O_2 \rightarrow PhCOO_2^{\cdot}$ (1.10)

$PhCOO_2^{\cdot} + PhCH_3 \rightarrow PhCO_3H$ (PBA) (1.11)

$PhCOO_2^{\cdot} + PhCHO \rightarrow PhCO_3H + PhCO^{\cdot}$ (PBA) (1.12)

$PhCO_3H + Ni(acac)_2L^2 \rightarrow PhCO_2^{\cdot} + Ni(acac)_2OH\cdot L^2$ (1.13)

$PhCO_3H + Ni_2(L^1)(L_{ox}^1)_3L^2 \rightarrow PhCO_2^{\cdot} + Ni_2(L^1)(L_{ox}^1)_3OH\cdot L^2$ (1.13')

$PhCO_2^{\cdot} + PhCH_3 \rightarrow PhCO_2H + PhCH_2^{\cdot}$ (BA) (1.14)

$PhCO_3H + PhCHO \rightarrow 2PhCO_2H$ (BA) (1.15)

Reaction steps (1.1), (1.3), (1.5) in mechanism for catalyzed oxidation are given by analogy with results of works [22, 27-29].

Additives of MP in the beginning stages of the oxidation of toluene in presence $Ni(acac)_2$ may accelerate the oxidation of BZA to BAL and BAL in BA, which proceeds on the nickel catalyst by reactions (1.6) – (1.7) and (1.8) – (1.9) [24, 25] (Scheme 2).

The maximum selectivity in the oxidation of toluene to BH in presence of $Ni(II)(acac)_2$, without the addition of donor ligand MP S_{BH} is about ~ 15% at depth of reaction to ~ 2–3% (Fig. 2 (a)).

Additives of MP enhance the selectivity S_{BH} of oxidation of toluene to BH, catalyzed with $Ni(II)(acac)_2$, to ~ 27% at [MP] = $3.0 \cdot 10^{-2}$ mol/l (Fig. 2(a)) with the same extent of oxidation. Thereby on initial reaction steps the yield of BH significantly increases (~ in 4 times). With concentration of [MP] increase, when [MP] ≥ $7 \cdot 10^{-2}$ mol/l, the character of dependence S_{BH} from degree of conversion C is changed: the value of S_{BH} is maximal on initial reaction steps: S_{BH} ≈30-40% at 0.5% of conversion C, and selectivity of BH formation is reduced at $C > 0.5\%$.

Introduction of MP in the reaction of toluene oxidation, catalyzed $Ni(II)(acac)_2$, results to significant synergic effect of increase of initial rate of toluene oxidation w_0, by more than a factor of 3 as the concentration [MP] is increased from 0 to 0.15 mol/l [17]. The dependence of w_0 on [MP] has maximum. It testifies in favor of formation catalytic coordinated unsaturated complexes $Ni(II)(acac)_2 \cdot MP$ and coordinated saturated complexes $Ni(II)(acac)_2 \cdot 2MP$ [30]. The additives of MP in the absence of $Ni(II)(acac)_2$ does not practically influence on initial rate w_0 of reaction of toluene oxidation.

Similar dependences of initial rate on concentration of ligand-modifier at constant concentration of $Ni(II)(acac)_2$ that testified in favor of formation of catalytic active complexes $Ni(II)(acac)_2 \cdot MP$ and $Ni(II)(acac)_2 \cdot 2MP$ [30] have been received by us in case of ethylbenzene oxidation in the presence of system {$Ni(II)(acac)_2$ ($3 \cdot 10^{-3}$ mol/l) + MP}[8, 39].

2.3.2 OXIDATION OF CUMENE, CATALYZED WITH SYSTEM {$NI(II)(ACAC)_2$ + MP}

The selectivity of cumene oxidation to cumyl hydroperoxide (CH) in the absence of catalyst at 90–100^0 is as a rule high, namely, S_{CH} = 96–98% [31–33].

The stability of CH, high values of rate constants for chain propagation and low yield of chain termination products due to low k_5 values facilitate the high concentrations of the hydroperoxide (Table 1).

High selectivity of cumene oxidation to hydroperoxide S_{CH} can be connected with possibility of formation of cumyl hydroperoxide, CH, in chain termination reaction [33].

It is assumed, what recombination of cumyl peroxo radicals ROO• occurs through intermediate formation of tetroxide RO_4R, decomposition of which is realized on Russell mechanism [34], (through cyclic six-member transient state) and results to formation of decay products: CH, α-methyl styrene and O_2,on scheme:

$$2C_6H_5C(CH_3)O_2(CH_3)CC_6H_5 \rightarrow C_6H_5(CH_3)_2OOH+C_6H_5(CH_3)=CH_2+O_2$$

However, the cumene oxidation reaction at 90–100^0 without catalyst, selective on hydroperoxide, is usually proceeds only to a low conversion due to self-inhition by the accumulated products of the extensive oxidation, – phenol and benzoquinone.

Thus, in the oxidation of cumene, we are faced with the problem of increasing the conversion and rate of cumene oxidation while retaining the high selectivity 97–99%.

The use of variable-valence metal compounds as catalysts, as a rule, is accompanied by a decrease in the selectivity of the oxidation of cumene to the hydroperoxide, mainly due to the formation of cumyl alcohol upon the decomposition of CH [35, 36]. An increase in the reaction rate with retention of high selectivity up to cumene conversion of 35–40% is observed at oxidation of cumene when zinc and cadmium complexes of N-heterocyclic bases are used as catalysts [37].

These works appeared actually simultaneously with ours. In presence of heterogenised variable-valence metal compounds the values $S_{CH,max}$ and C_{max} of cumene oxidations to hydroperoxide are lower. For example, at catalysis by silica – and polymer (poly-4-vinylpyridine) – supported $Cu(OAc)_2$, the increase in the rate of cumene oxidation by 68% was observed, and $S_{CH,max}$ was equal 92% at C = 22% [36].

Since the rate of cumene oxidation at 100^0 is rather high, on this system, on-visible, it is necessary to work with small concentrations of nickel catalyst, which can facilitate a high yield of the hydroperoxide [8, 39]. It is evident from analysis of scheme of catalyzed hydrocarbons oxidation, including participation of catalyst in chain initiation reaction under catalyst interaction with ROOH and also in chain propagation ($Ct + RO_2^{\bullet} \rightarrow$), that with decrease in $[Ct]_0$ the rate of reaction should be decreased, and $[ROOH]_{max}$ should be increased [8, 15]. It has appeared that together with increase in concentration $[ROOH]_{max}$ selectivity of ethylbenzene oxidation $S_{ROOH,max}$ increased also at decrease in $[Ni(II)(acac)_2]_0$ [8, 39].

Really, it was established by us earlier that selectivity of cumene oxidation to hydroperoxide CH increased from S_{CH} = 95% (in the uncatalyzed oxidation) up to = 97–98% in the presence of $6\cdot10^{-6}$ mol/l $Ni(II)(acac)_2$, while the conversion with the retention of $S_{CH,max}$ is increased from 10 to 15% [17].

At this $Ni(II)(acac)_2$ catalyst does not affect the composition of the oxidation products, but alters their ratio, increasing the rate of accumulation of CH and its maximum concentration while virtually not affecting the kinetics of the accumulation of the other products of oxidation of cumene, namely, dimethyl phenylcarbinol (DMPC), acetophenone (AP) and phenol (PhOH). To that DMPC is formed in concentration, on order exceeding concentrations [AP] and [PhOH].

The value of $S_{CH,max}$ upon the introduction of MP to the cumene oxidation catalyzed by $Ni(II)(acac)_2$, is virtually unaltered (97–99%) but the cumene conversion is markedly increased to C~ 35% (and to C ≈ 40–45% at selectivity on the level not lower than $S_{CH,max}$ = 95%).

The addition of MP further accelerates the accumulation of CH and has a weak effect on the kinetics of the accumulation of other products [17].

In presence of $Ni(II)(acac)_2$ ($6\cdot10^{-6}$ mol/l) initial rate of oxidation of cumene increases by a factor of ~ 1.7 in comparison with the rate of uncatalyzed oxidation of cumene. Catalytic system $\{Ni(II)(acac)_2+MP\}$ increases initial rate of oxidation in 3 times in comparison with uncatalyzed oxidation, and in 1.6 times in comparison with the oxidation of cumene, catalyzed by $Ni(II)(acac)_2$.

These data indicate, what system $\{Ni(II)(acac)_2+L^2\}$ based on nickel complex catalyst, may serve as catalyst for the alkylarens oxidation: not only ethylbenzene but also toluene and cumene, - to the corresponding hydroperoxides.

Thus the received kinetic laws of catalytic alkylarene oxidations testify in favor of a mechanism generality of catalysis of the investigated hydrocarbons by system $\{Ni(II)(acac)_2+L^2\}$:

Dependence of the oxidation selectivity S_{ROOH} on the conversion *C* for the oxidation ethylbenzene (and of toluene also) at defined values $[L^2]$ (L^2=MP]) (Fig. 2) traverse a maximum. The S_{ROOH} value increases with increasing conversion, reaches a maximum value, and then decreases. Such character of dependence S_{ROOH} on C, obviously, is connected with transformation of complexes Ni(II)$(acac)_2$·MP during reaction by effect of O_2 to a new selective catalyst, heteroligand macro complex $Ni_2(OAc)_3(acac)$·MP ("A") [8, 39].

At MP concentrations, exceeding some defined value, the nature of dependence of S_{ROOH} on C changes: S_{ROOH} is maximal in the beginning of reaction, and is then gradually reduced with increasing C (Fig. 2). Such dependence is characteristic for oxidation of cumene at all the MP concentrations examined.

Complexes Ni(II)$(acac)_2$·2MP may become responsible for high values of selectivity of alkylarenes oxidation to ROOH at sufficiently high MP concentration ([MP]). During alkylarenes oxidation complexes Ni(II)$(acac)_2$·2MP, on-visible, are converted in coordinate unsaturated Ni(II)$(acac)_2$·MP (according to the reaction with O_2 or ROOH) [38], namely, the next reaction becomes possible:

$$Ni(acac)_2 \cdot 2MP + O_2 \rightarrow Ni(acac)_2 \cdot MP \cdot O_2 + MP \quad (**)$$

Complexes $Ni(acac)_2$·MP·O_2 are subsequently transformed in heteroligand macro complexes $Ni_2(OAc)_3(acac)$·MP ("A").

So, the obtained data testifies to that that the next complexes formed *in situ* in the course of alkylarenes oxidations with dioxygen, catalyzed by system {Ni(II)$(L^1)_2$+L^2}, are responsible for observable kinetic laws: primary complexes Ni(II)$(L^1)_2 \cdot L^2_n$ at I macro step of oxidation, and new catalytic heteroligand macro complexes of general formula $M^{II}_x L^1_y (L^1_{ox})_z (L^2)_n$ (L^1_{ox}=$(OAc)^-$) at II macro step of oxidation) [8, 39].

It was shown [39] that the selective catalyst heteroligand complex Ni(acac)(OAc)·L^2 was formed as a result of the ligand L^2-controlled regioselective addition of O_2 to a nucleophilic carbon atom (γ-C) of the $(acac)^-$ ligand. The coordination of the electron-donating axial ligand L^2 with the $Ni(acac)_2$ complex stabilized the intermediate zwitter-ion [L^2(acac)Ni$(acac)^+ \cdots O_2^-$] and increased the probability of regioselective insertion of O_2 to the acetylacetonate ligand activated by coordination with the nickel(II) ion. Further incorporation of O_2 into the chelate acac-ring was accompanied by the proton transfer and the redistribution of bonds in the transition complex "A" leading to the scission of the cyclic system C(O)

MeC(O)OCH(Me)O to form a chelate ligand OAc^-, acetaldehyde and CO (in the Criegee rearrangement) (Scheme 3).

The only known Ni(II)-containing dioxygenase ARD (acireducton dioxygenase), which catalyzes the oxidative decomposition of β-diketones, operates in the analogous way [40]. This applies to the functional enzyme models, namely, Cu(II)- and Fe(II)-containing quercetin 2,3-dioxygenases, which catalyze the decomposition of β-diketones in the enol form to carbonyl compounds with CO evolution [41–42].

Scheme 3.

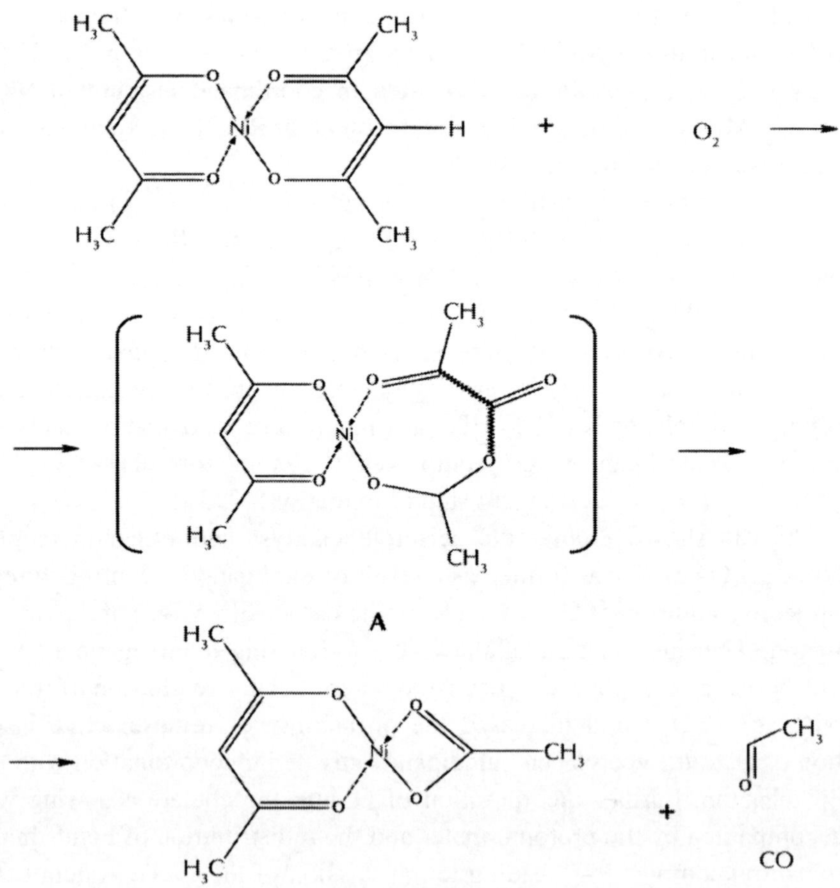

As a result of this process, reactive mono- and poly-nuclear heteroligand complexes with the general composition $Ni_x(acac)_y(AcO)_z(L^2)_n$ L^2 = HMPA, DMF, MP, M'St (M' = Na, Li, K), 18C6, R_4NBr were formed [8].

$$Ni(acac)_2\ L^2 + O_2 \longrightarrow [L^2(acac)Ni(acac)^+ \cdots O_2^-] \longrightarrow L^2(acac)\ Ni(AcO) + MeCHO + CO,$$

$$2Ni(acac)_2\ L^2 + O_2 \longrightarrow Ni_2(AcO)_3(acac)L^2 + 3MeCHO + 3CO + L^2.$$

L^2 = HMPA, DMF, MP, M'St (M' = Na, Li, K).

2.3.2.1 MECHANISM OF FORMATION OF PRODUCTS IN TOLUENE, CUMENE (AND ETHYLBENZENE) OXIDATIONS

For reactions of oxidation of toluene (and ethylbenzene), catalyzed $Ni(acac)_2$ changes in the mechanism of formation of the some products of the reaction it was observed in the presence of additives ligand MP, connected with changes of activity of various forms of the catalytic complexes, formed during oxidation process.

In the case of ethylbenzene oxidation in the beginning of reaction of oxidation in the presence of complexes $Ni(II)(acac)_2$ ($3.0\cdot10^{-3}$ mol/l) with monodentate ligand-modifier MP, by-products AP and MPC (P=AP, or MPC) are formed consistently, at decomposition of PEH, as well as at catalysis $Ni(II)(acac)_2$ without additives MP: $w_P/w_{PEH} \rightarrow 0$ at $t \rightarrow 0$. However unlike uncatalyzed ethylbenzene oxidation and oxidation in the presence of $Ni(II)(acac)_2$, parallel formation AP and MPC is observed at catalysis with $Ni(II)(acac)_2\cdot MP$: $w_{AP}/w_{MPC} \neq 0$ at $t \rightarrow 0$. In the developed reaction at catalysis with transformed form $Ni(II)(acac)_2\cdot MP$ ($Ni_2(OAc)_3(acac)\cdot MP$ ("A")) the direction of formation of products varies: AP and MPC are formed in parallel with PEH, and AP and MPC – also in parallel: $w_P/w_{PEH} \neq 0$ at $t \rightarrow 0$, and, besides, $w_{AP}/w_{MPC} \neq 0$ at $t \rightarrow 0$ in the chain propagation ($Ct + RO_2^{\bullet} \rightarrow$) and quadratic termination ($2RO_2^{\bullet} \rightarrow$) (as well as in the case of catalysis with complexes of $Ni(II)(acac)_2$ with macrocyclic polyethers and quaternary ammonium salts ($[Ni(II)(acac)_2] = 1.5\cdot10^{-4}$ mol/l) [8, 39].

Here it has been established by us for the first time that, as well as in the case of ethylbenzene oxidation, in toluene oxidation in the presence of system $\{Ni(II)(acac)_2 + MP\}$ the route of formation major toluene oxidation products, benzaldehyde (BAL), benzyl alcohol (BZA) and benzoic acid (BA) is changed: from successive (at catalysis by complexes $Ni(II)(acac)_2 \cdot MP$, formed *in situ* at the beginning of toluene oxidation, I macro step) on parallel (at catalysis by complexes $Ni_2(OAc)_3(acac) \cdot MP$ ("A"), formed during oxidation).

At catalysis with complexes $Ni_2(OAc)_3(acac) \cdot MP$ ("A"), benzaldehyde BAL and benzyl alcohol BZA are formed parallel each other ($w_{BAL}/w_{BZA} \neq 0$ at $t \to 0$), apparently, in the steps of chain termination ($2PhCH_2O_2^{\bullet} \to (*)$ and reaction 1.5 (Scheme 1) (BAL)); BAL and BA – also are formed in parallel reactions ($w_{BA}/w_{BAL} \neq 0$ at $t \to 0$) (Fig. 3). At this BA is formed successively from PBA ($w_{BA}/w_{PBA} \to 0$ at $t \to 0$) (Scheme 2, reactions 1.13'–1.14). Unlike this, a successive formation BAL from BZA and BA from BAL (Fig. 3) (and PBA)) occurs at catalysis with primary complexes $Ni(II)(acac)_2 \cdot MP$ (Scheme 2).

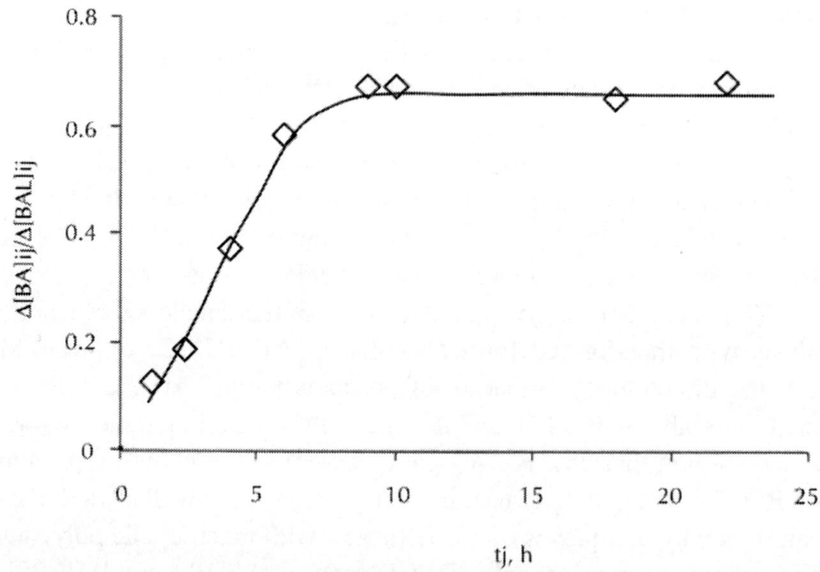

FIGURE 3 Dependence $\Delta[BA]_{ij}/\Delta[BAL]_{ij}$ on time t_j in the course of toluene oxidation, catalyzed by $\{Ni(II)(acac)_2 + MP\}$. $[Ni(II)(acac)_2] = 3 \cdot 10^{-3}$ mol/l, $[MP] = 3 \cdot 10^{-2}$ mol/l, 100°C.

The route of benzaldehyde BAL and benzyl alcohol BZA formation is changed from successive (*via* benzyl hydroperoxide BH decomposition at catalysis with Ni(II)(acac)$_2$·MP (Scheme 1, reactions 1.3–1.5) on parallel with BH – at catalysis with $Ni_2(OAc)_3(acac)$·MP ("A") complexes, formed during oxidation.

The interesting result was received in the case of research of cumene oxidation. At introduction of catalyst (with and without of ligand-modifier) in reaction of oxidation of cumene changes in the mechanism of formation of products it was not observed, as opposed to oxidation of ethylbenzene and toluene. So as well as in uncatalyzed oxidation or at catalysis by Ni(II)(acac)$_2$, and by system {Ni(II)(acac)$_2$ + MP} (complexes Ni(II)(acac)$_2$·2MP and $Ni_2(OAc)_3(acac)$·MP), ketone AP and hydroperoxide CH are formed parallel ($w_{AP}/w_{CH} \neq 0$ at $t \to 0$), but DMPC is product of CH decomposition ($w_{DMPC}/w_{CH} \to 0$) at $t \to 0$) during all reaction of oxidation (Fig. 4). At that AP and DMPC are formed parallel each other ($w_{AP}/w_{DMPC} \neq 0$ at $t \to 0$).

In [43] mechanism of AP formation on reaction of quadratic chain termination of cumyl peroxo radicals ROO• is offered:

$$2PhC(CH_3)_2O_2^{\bullet} \to [PhC(CH_3)_2O_4(CH_3)_2CPh] \to 2PhCOCH_3 + CH_3O_2^{\bullet} + CH_3^{\bullet}$$

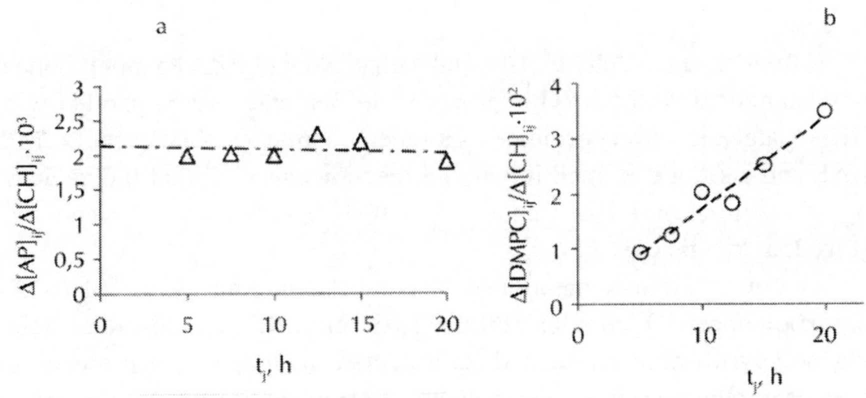

FIGURE 4 Dependences of product accumulation rations on time in the course of cumene oxidation, catalyzed by {Ni(II)(acac)$_2$+MP}:
a. $\Delta[AP]_{ij}/\Delta[CH]_{ij}$ on t_j,
b. $\Delta[DMPC]_{ij}/\Delta[CH]_{ij}$ on t_j. [Ni(II)(acac)$_2$] $=6 \cdot 10^{-6}$ mol/l, [MP] $=1.5 \cdot 10^{-4}$ mol/l, 100°C.

The growth of selectivity S_{CH} from 95% (uncatalyzed reaction) to 99% (catalysis by complexes of nickel with MP) in this case seems to be connected with acceleration of formation ROOH for the account of participation of catalyst in a step of chain initiation (O_2 activation).

Here it has been established by us parallel formation of major products of ethylbenzene and toluene oxidation at catalysis by coordinated saturated complexes $Ni(II)(acac)_2 \cdot 2MP$ as well as at catalysis with $Ni_2(OAc)_3(acac) \cdot MP$ ("A"). So in the beginning of ethylbenzene oxidation, catalyzed with $Ni(II)(acac)_2 \cdot 2MP$ (Fig. 2, b: kinetic curve 4), the oxidation products PEH, AP, and MPC are formed parallel, AP and MPC – also parallel each other. At that the value of S_{PEH} is maximal on initial stages, and $S_{PEH,0} \sim 90\%$ at conversion $C \leq 8\%$. As it has been established by us earlier at catalysis with small concentrations of $Ni(II)(acac)_2$ ($1.5 \cdot 10^{-4}$ mol/l) without additives of ligands-modifiers and in the presence exo ligand L^2=18C6, namely at catalysis with $Ni(II)(acac)_2 \cdot L^2_n$ and heteroligand macro complexes of general formula $M^{II}_x L^1_y (L^1_{ox})_z (L^2)_n$ (L^1_{ox}=$(OAc)^-$), formed in the course of oxidation, products AP and MPC (P) are formed also not from ROOH (PEH) but in parallel with PEH, in steps of chain propagation ($Ct + RO_2^{\cdot} \rightarrow$) and chain termination ($2RO_2^{\cdot} \rightarrow$), $w_P/w_{PEH} \neq 0$ at $t \rightarrow 0$, and, besides, $w_{AP}/w_{MPC} \neq 0$ at $t \rightarrow 0$ [8, 39]. At that high values of selectivity $S_{PEH,max}$ = 90% ($Ni(II)(acac)_2$ ($1.5 \cdot 10^{-4}$ mol/l)), and $S_{PEH'max}$ = 98% at catalysis with $Ni(II)(acac)_2 \cdot L^2_n$ are observed from the very beginning of reaction.

Similarly, benzaldehyde BAL and benzyl alcohol BZA are not products of decomposition of benzyl hydroperoxide BH, and formed parallel with BH at catalysis with coordinated saturated complexes $Ni(II)(acac)_2 \cdot 2MP$. BAL and BZA are formed in parallel reactions also. At that the value of S_{BH} is maximal on initial stages. S_{BH} ≈30–40% at 0.5% of conversion C (Fig. 2, a: kinetic curves 3, 4).

As already it was mentioned above [21], thermal decay of μ-1,2-peroxodicopper(II) complex $[\{^{\circ}LCu^{II}\}_2(O_2^{2-})]^{2+}$ ($^{\circ}L$ = an anisole-containing polypyridylamine potential tetra dentate ligand) in the presence of toluene or ethylbenzene leads to mainly benzaldehyde (40%) (toluene), or acetophenone/ 1-phenylethanol mixture (~1:1) (ethylbenzene), respectively formations. Very similar toluene oxygenation occurs with dicopper(III)-bis-μ-oxo species $[\{^{Bz}LCu^{III}\}_2(\mu\text{-}O^{2-})_2]^{2+}$ (^{Bz}L is tridentate chelate, possessing the same moiety found in °L, but without the anisole O-atom donor). Authors suppose that alkylperoxo radical chemistry (e.g., the Russell mechanism, $RO_2\cdot$ radical disproportionation process) well explains the nearly equal yield of acetophenone and 1-phenylethanol produced from the ethylbenzene oxidation. This reaction mechanism was also reported by Que and co-workers to interpret nonheme diiron catalyzed alkyl benzene oxidation reactions [44].

It was established in Ref. [21] that $[\{°LCu^{II}\}_2(O_2^{2-})]^{2+}$ does not convert benzyl alcohol to benzaldehyde. Instead, the formation of benzaldehyde BAL and benzyl alcohol BZA occur at the same time during the decay process of $[\{°LCu^{II}\}_2(O_2^{2-})]^{2+}$.

$([\{°LCu^{II}\}_2(O_2^{2-})]^{2+} \rightarrow 2[°LCu^{I}]^{+}+O_2)$ and toluene oxidation.

When the formation and decay of $[\{°LCu^{II}\}_2(O_2^{2-})]^{2+}$ was carried out in the presence of excess O_2, the yield of BAL increased to ~ 75%.

2.4 CONCLUSION

One way of retention and increasing the activity of a homogeneous metal complex catalyst is the introduction of modifying ligands. The mechanism of action of the modifying ligands (monodentate, polydentate) in radical-chain ethylbenzene oxidation, catalyzed with transition metal (Ni(II), Fe(II,III), Co(II)) complexes are in detail investigated in our works [8, 39].

Research of the mechanism of formation of products of oxidation is a necessary condition of modeling effective catalytic systems for hydrocarbons oxidation in target products. In this work the routes of formation of the basic products of radical-chain toluene, cumene (and also ethylbenzene) oxidations by molecular oxygen, catalyzed with system {Ni(II) $(acac)_2(3\cdot10^{-3}$ mol/l)+L^2} (L^2=electron donating exo ligand-modifier N-methylpirrolidon-2, MP) has been researched for the first time with kinetic method.

It has been established that the route of formation of products for toluene (and ethylbenzene} varies depending on a parity of components catalytic systems and is defined by the nature of the active complexes formed in situ during oxidation: active primary $Ni(II)(acac)_2\cdot nMP$ (n = 1,2) and heteroligand complexes, $Ni_2(OAc)_3(acac)\cdot MP$ ("A"), the intermediate products of oxygenation of $Ni(II)(acac)_2\cdot nMP$.

In nickel catalyzed toluene oxidation the route of formation major oxidation products, benzaldehyde (BAL), benzyl alcohol (BZA) and benzoic acid (BA) is changed: from successive (catalysis by coordinated unsaturated complexes $Ni(II)(acac)_2\cdot MP$ on parallel (catalysis by complexes $Ni_2(OAc)_3(acac)\cdot MP$ ("A").

The increase in selectivity of toluene (and ethylbenzene) oxidation to ROOH, catalyzed with complexes $Ni(II)(acac)_2\cdot 2MP$ and $Ni_2(OAc)_3(acac)\cdot MP$ ("A"), as compared with catalysis by complexes $Ni(II)(acac)_2\cdot MP$ seems to be connected with as one of the reasons the change of a direction of formation of products BAL, BZA (toluene), AP, MPC (ethylbenzene), from consecutive *via* homolytic ROOH decomposition → the parallel, in steps of chain propagation with the assistance of the catalyst ($Ct + RO_2^{\cdot} \rightarrow$), and chain termination, and also with acceleration of in chain initiation (O_2 activation).

In the case of cumene oxidation changes in the mechanism of formation of products it was not observed. So as well as in uncatalyzed oxidation or at catalysis by $Ni(II)(acac)_2$, and by complexes $Ni(II)(acac)_2\cdot 2MP$ or $Ni_2(OAc)_3(acac)\cdot MP$, ketone AP and hydroperoxide CH are formed parallel each other, but alcohol DMPC is product of CH decomposition during of all reaction of oxidation. The growth of selectivity S_{CH} in this case seems to be connected with acceleration of formation ROOH for the account of participation of catalyst in a step of chain initiation (O_2 activation).

KEYWORDS

- **Benzaldehyde**
- ***Benzoic acid***
- **Cumene oxidation**
- **Ethylbenzene**
- **Heteroligand**
- **Monodentate ligands**
- **Uncatalyzed oxidation**

REFERENCES

1. Shul'pin, G. B. Metal-catalyzed hydrocarbon oxygenations in solutions: the dramatic role of additives; a review, *J. Mol. Catal. A: Chem.*, **2002**, *189*, 39.
2. In *The Activation ofDioxygen and Homogeneous Catalytic's Oxidation,* Barton, D. H. R.; Martell, A. E.; Sawyer, D. T., Eds.; Plenum: New York, 1993.
3. Mansuy, D. New Model Systems for Oxygenases, In *The Activation ofDioxygen and Homogeneous Catalyti'c Oxidation,* Barton, D. H. R.; Martell, A. E.; Sawyer, D. T., Eds.; Plenum: New York, 1993; p. 347.
4. Karasevich, E. I. Liquid-phase oxidation of alkanes in the presence of metals compounds, In *Reactions and Properties of Monomers and Polymers,* D'Amore, A.; Zaikov, G., Eds.; Nova Science Publ.: New York, 2007; p. 43.
5. Szaleniec, M.; Hagel, C.; Menke, M.; Novak, P.; Witko, M.; Heider, J. Kinetics and Mechanism of Oxygen-Independent Hydrocarbon Hydroxylation by Ethylbenzene *Biochemistry*, **2007**, *46,* 7637.
6. Weissermel, K.; Arpe, H-J. In *Industrial Organic Chemistry*, 3rd Ed.; VCH: New York, 1997.
7. Menage, S.; Galey, J.-B.; Dumats, J.; Hussler, G.; Seite, M.; Luneau, I. G.; Chottard, G.; Fontecave, M. O_2 Activation and Aromatic Hydroxylation Performed by Diiron Complexes, *J. Am. Chem. Soc.*, **1998**, *120*, 13370–13382.
8. Matienko, L. I. Solution of the problem of selective oxidation of alkylarenes by molecular oxygen to corresponding hydroperoxides. Catalysis initiated by Ni(II), Co(II), and Fe(III) complexes activated by additives of electron-donor mono- or multidentate extra-ligands, In *Reactions and Properties of Monomers and Polymers,* D'Amore, A.; Zaikov, G., Eds.; Nova Sience Publ. Inc.: New York, 2007; p. 21.
9. Mosolova, L. A.; Matienko, L. I.; Skibida, I. P. Composition catalysts of ethylbenzene oxidation based on bis(acetylacetonato) nickel (II) and phase transfer catalysts as ligands. 1. Macrocyclic polyethers, *Izv. Akad. Nauk, Ser.Khim.,* ***1994a***, 1406.
10. Matienko, L. I.; Mosolova, L. A. The mechanism of selective ethylbenzene oxidation with molecular O_2, catalyzed by complexes of nickel (II), cobalt (II) and iron (III) with macrocyclic polyethers, In *Uspekhi v Oblasti Geterogennogo Kataliza i Geterotsiklov* (Progress in Heterogeneous Catalysis and Heterocycles), Rakhmankulov, D. L., Ed.; Khimiya: Moscow, 2006; p. 235.
11. Matienko, L. I.; Mosolova, L. A.; Skibida, I. P. Composition catalysts of ethylbenzene oxidation based on bis(acetylacetonato) nickel (II) and phase transfer catalysts as ligands. 2. Quarternary ammonium salts, *Izv. Akad. Nauk, Ser. Khim.,* **1994**, 1412.
12. Matienko, L. I.; Mosolova, L. A. Selective oxidation of ethylbenzene with dioxygen into α-phenylethylhydroperoxide. Modification of catalyst activity of Ni(II) and Fe(III) complexes upon addition of quaternary ammonium salts as exoligands, In *New Aspects of Biochemical Physics. Pure and Applied Sciences,* Varfolomeev, S. D.; Burlakova, E. B.; Popov, A. A.; Zaikov, G. E., Eds.; Nova Science Publ.: New York, 2007, p. 95.
13. Matienko, L. I.; Mosolova, L. A. Selective Oxidation of Ethylbenzene with Dioxygen into α-Phenylethylhydroperoxide. Modification of Catalyst Activity of Ni(II) and Fe(III) Complexes upon Addition of Quaternary Ammonium Salts as Exoligands, In

Biochemical Physics. Research Trends. Varfolomeev, S. D.; Burlakova, E. B.; Popov, A. A.; Zaikov, G. E., Eds.; Nova Science Publ., Inc.: New York, 2009; p. 37.
14. Matienko, L. I.; Mosolova, L. A. Influence of Hexamethylphosphoric Triamide and Dimethylformamide Admixtures on the mechanism of Tris(acetylacetonato) iron(III)-Catalyzed Ethylbenzene Oxidation with Molecular Oxygen, *Neftekhimiya,* **2007**, *47*, 42 (*in Russian*).
15. Denisov, E. T.; Emanuel, N. M. *Usp. Khim., Russ. Chem. Rev.,* **1960**, *29,* 645.
16. Emanuel, N. M.; Gall, D. *Okislenie Etilbenzola. Model'naya Reaktsiya* Oxidation of Ethylbenzene. Model Reaction, Nauka: Moscow, 1984.
17. Mosolova, L. A.; Matienko, L. I.; Skibida, I. P. Activation of Bis(acetylacetonato)nickel by N-methyl 2-pyrrolidone in the Oxidation of Monoalkyl-substituted Benzenes to Hydroperoxides, *Kinetika i kataliz*, **1988**, *29*, 1078–1083 (*in Russian*).
18. Blanksby, S. J.; Ellison, G. B. Bond Dissociation Energies of Organic Molecules, *Acc. Chem. Res*., **2003**, *36*, 255.
19. Howard, J. A.; Schwalm, W. J.; Ingold, K. U. Absolute Rate Constants for Hydrocarbons autoxidation. VII. The Reactivities of Peroxy Radicals towards Hydrocarbons and Hydroperoxides, In *Preprints of the International Oxidation Simposium. San Francisco, California*. 1967; *1*, 3.
20. Partenheimer, W. Metodology and scope of metal/bromide autoxidation of hydrocarbon, *Catalysis Today*, **1991**, *23*, 69.
21. Lucas, H. R.; Li, L.; Sarjeant, A. A. N.; Vance, M. A.; Solomon, E. I.; Karlin, K. D. Toluene and Ethylbenzene Aliphatic C-H Bond Oxidations Initiated by a Dicopper(II)-μ-1,2-Peroxo Complex, *J. Am. Chem. Soc.*, **2009**, *131*, 3230.
22. Hanotier, J.; Hanotier-Bridoux, M.; de Radzitzky, P. Effect of Strong Acids of the Oxidation of Alkylarens by Manganic and Cobaltic Acetates in Acetic Acid, *J. Chem. Soc., Perk. Trans. II,* **1973**, 381.
23. Sergeev, P. G.; Phedorova, V. V.The liquid oxidation of fat aromatic hydrocarbons with gaseous oxygen. The hydroperoxide of toluene, *Dokl. AN SSSR*, **1956**, *109*, 796 (*in Russian*).
24. Reddy, T.R.; Murthy, G. S. S.; Jagannadham, V. Oxidation of lactic and mandelic acids by nickel(III) ion in H_2SO_4 medium via addition/elimination, a kinetic study, *Orient. J. Chem.,* **1986**, *2*, 92.
25. Reddy, T. R.; Murthy, G. S. S.; Jagannadham, V. Kinetiks and mechanism of oxidation of benzaldegyde and substituted benzaldegydes by Ni(III) ion in acedic acid – water mixtures, *Oxid. Commun*., **1986**, *9*, 83.
26. Russel, G. A. The Rates of Oxidation of Aralkyl Hydrocarbons. Polar Effects in Free Radical Reaction, *J. Am. Chem. Soc*., **1956**, *78*, 1047.
27. Sheldon, R. A. New catalytic methods for selective oxidation, *J. Mol. Catal.*, **1983**, *20*, 1.
28. Mlodnicka, T. Metalloporphyrin-Catalyzed Oxidation of Hydrocarbon with Dioxygen, In *Metalloporphyrins in Catalytic Oxidation,* Sheldon, R. A., Ed.; Marcel Dekker, Inc.: New York, Basel, Hong Kong, 1994; p. 261.
29. Boča, R. Molecular orbital study of coordinated dioxygen. V. Catalytic oxidation of toluel on cobalt-dioxygen complexes, *J. Mol. Catal*., **1981**, *12*, 351.
30. Golodov, V. A. The Synergistic phenomena in catalysis, *Ross. Khim. Zh.*, **2000**, *44*, 45 (*in Russian*).

31. Antonovsky, V. L.; Khursan, S. L. *Fizicheskaja khimia organicheskich peroksidov*, Physic chemistry of organic peroxides, Academbook: Moscow, 2003.
32. Suyrkin, Ja.K.; Moiseev, I. I. Mechanisms of some reactions with participation of peroxides, *Usp. Khim.*, **1960**, *29*, 425.
33. Boozer, C. B.; Ponder, P. W. Trisler, J. C. Wzightman, C. E. Deuterium isotope effects of the air oxidation of cumene, *J. Am. Chem. Soc*., **1956**, *78*, 1506.
34. Russel, G. A. Deuterium-isotope effects in the autoxidation of aralkylhydrocarbons. Mechanism of the interaction of peroxy radicals, *J. Am. Chem. Soc*., **1957**, *79*, 3871.
35. Weerarathas, S.; Hronec, M.; Malik, L.; Vesely, V. Selectivity of metal polyphthalocyanine catalyzed oxidation of cumene, *React. Kinet. Catal. Lett*., **1983**, *22*, 7.
36. Hsu, Y. F.; Yen, M. H.; Cheng, Ch. P. *J. Mol. Catal. A: Chem.*, Autoxidation of cumene catalyzed by transition metal compounds on polymeric supports. **1996**, *105*, 1377.
37. Kozlov, S. K.; Tovstokhat'ko, Ph. I.; Potekhin, V. M. Isopropyl oxidation catalysts, activated with N-heterocyclic ligand – 1,10-phenantrolin, *Zh. Prikl. Khim*., **1986**, *59*, 217.
38. Jones, R. D.; Summerville, D. A.; Basolo, F. Synthetic oxygen carriers related to biological systems, *Chem. Rev*., **1979**, *79*, 139.
39. Matienko, L. I.; Mosolova, L. A.; Zaikov, G. E. The Modeling of Transition Metal Complex Catalysts in the Selective Alkylarens Oxidations with Dioxygen. The Role of Hydrogen – Bonding Interactions, *Oxid. Commun.*, **2009**, *32,* 731.
40. Dai, Y.; Pochapsky, Th. C.; Abeles, R. H. Mechanistic Studies of Two Dioxygenases in the Methionine Salvage Pathway of *Klebsiella pneumoniae*, *Biochemistry*, **2001**, *40,* 6379.
41. Gopal, B.; Madan, L. L.; Betz, S. F.; Kossiakoff, A. A. The Crystal Structure of a Quercetin 2,3-Dioxygenase from *Bacillus subtilis* Suggests Modulation of Enzyme Activity by a Change in the Metal Ion at the Active Site(s), *Biochemistry,* **2005**, *44,* 193.
42. Balogh-Hergovich, E.; Kaizer, J.; Speier, G. Kinetics and mechanism of the Cu(I) and Cu(II) flavonolate-catalyzed oxygenation of flavonols, Functional quercetin 2,3-dioxygenase models, *J. Mol. Catal. A: Chem*., **2000**, *159,* 215.
43. Antonovsky, V. L.; Makalets, B. I. About sequence of formation of products at liquid cumene oxidation, *Dokl. AN SSSR,* **1961**, *140*, 1070 (*in Russian*).
44. Kim, C.; Dong, Y. H.; Que, L. Modeling Nonheme Diiron Enzymes: Hydrocarbon Hydroxylation and Desaturation by a High-Valent Fe_2O_2 Diamond Core, *J. Am. Chem. Soc.* **1997**, *119*, 3635.

CHAPTER 3

POLYMER-BASED SOLAR CELLS

A. D'AMORE, V. MOTTAGHITALAB, and A. K. HAGHI

CONTENTS

3.1 INTRODUCTION

In recent years, renewable energies attract considerable attention due to the inevitable end of fossil fuels and due to global warming and other environmental problems. Photovoltaic solar energy is being widely studied as one of the renewable energy sources with key significance potentials and a real alternate to fossil fuels. Solar cells are in general packed between weighty, brittle and rigid glass plates. Therefore, increasing attention is being paid to the construction of lighter, portable, robust, multipurpose and flexible substrates for solar cells. Textiles substrates are fabricated by a wide variety of processes, such as weaving, knitting, braiding and felting. These fabrication techniques offer enormous versatility for allowing a fabric to conform to even complex shapes. Textile fabrics not only can be rolled up for storage and then unrolled on site but also they can also be readily installed into structures with complex geometries.

Textiles are engaging as flexible substrates in that they have a enormous variety of uses, ranging from clothing and household articles to highly sophisticated technical applications. Last innovations on photovoltaic technology have allowed obtaining flexible solar cells, which offer a wide range of possibilities, mainly in wearable applications that need independent systems. Nowadays, entertainment, voice and data communication, health monitoring, emergency, and surveillance functions, all of which rely on wireless protocols and services and sustainable energy supply in order to overcome the urgent needs to regular battery with finite power. Because of their steadily decreasing power demand, many portable devices can harvest enough energy from clothing-integrated solar modules with a maximum installed power of 1–5 W.[1]

Increasingly textile architecture is becoming progressively of a feature as permanent or semi-permanent constructions. Tents, such as those used by the military and campers, are the best known textile constructions, as are sun shelter, but currently big textile constructions are used extensively for exhibition halls, sports complexes and leisure and recreation centers. Although all these structures provide protection from the weather, including exposure to the sun, but solar concept offers an additional precious use for providing power. Many of these large textile architectural constructions cover huge areas, sufficient to supply several kilowatts of power. Even the fabric used to construct a small tent is enough to provide a few hundred watts. In addition to textile architecture, panels made from robust solar textile fabrics could be positioned on the roofs of existing buildings. Compared to conventional and improper solar panels for roof structures lightweight and flexible solar textile panels is able to tolerate load-bearing weight without shattering.

Moreover, natural disaster extensively introduces the huge potential needs the formulation of unusual energy package based on natural source. Over the past 5 yr, more than 13 million *people have lost their home* and possessions because of earthquake, bush fire, flooding or other natural disaster. The victims of these disasters are commonly housed in tents until they are able to rebuild their homes. Whether they stay in tented accommodation for a short or long time, tents constructed from solar textile fabrics could provide a source of much needed power. This power could be stored in daytime and used at night, when the outdoor temperature can often fall. There are also a number of other important potential applications. The military would benefit from tents and field hospitals, especially those in remote areas, where electricity could be generated as soon as the structure is assembled.

3.2 THE BASIC CONCEPT OF SOLAR CELL

In 1839 French scientist, Edmond Becquerel found out photovoltaic effect when he observed increasing of electricity generation while light exposure to the two metal electrodes immersed in electrolytic solution [2], light is composed of energy packages known as photons. Typically, when a matter exposed to the light, electrons are excited to a higher level within material, but they return to their initial state quickly. When electrons take sufficient energy more than a certain threshold (band gap), move from the valance band to the conduction band holes with positive charge will be created. In the photovoltaic effect electron-hole pairs are separated and excited electrons are pulled and fed to an external circuit to buildup electricity [3] (Fig. 1).

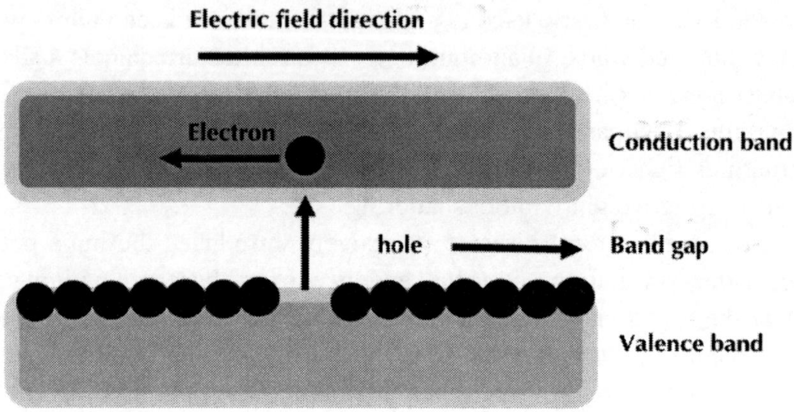

FIGURE 1 Electron excitation from valence band to conduction band.

An effective solar cell generally comprises an opaque material that absorbs the incoming light, an electric field that arises from the difference in composition between the semiconducting layers comprising the absorber, and two electrodes to carry the positive and negative charges to the electrical load. Designs of solar cells differ in detail but all must include the above features (Fig. 2).

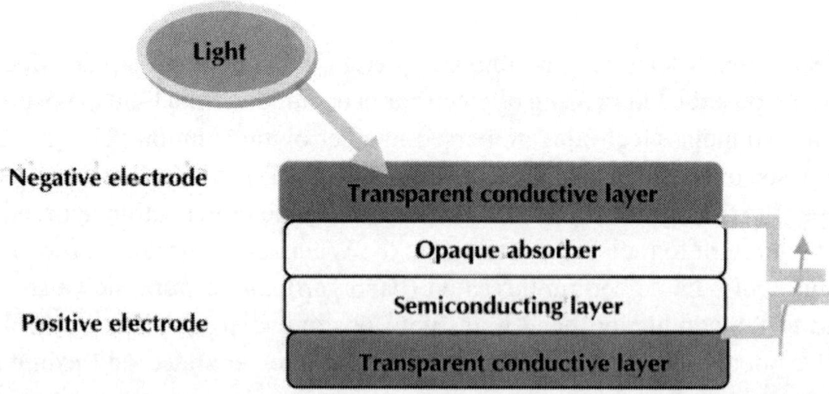

FIGURE 2 The general structure of a Solar cell.

Generally, Solar cells categorized into three main groups consisting of inorganic, organic and hybrid solar cells (Dye sensitized solar cells). Inorganic solar cells cover more than 95% of commercial products in solar cells industry.

3.2.1 ELECTRICAL MEASUREMENT BACKGROUND

All current–voltage characteristics of the photovoltaic devices were measured with a source measure unit in the dark and under simulated solar simulator source was calibrated using a standard crystalline silicon diode. The current-voltage characteristics of Photovoltaic devices are generally characterized by the short-circuit current (I_{sc}), the open-circuit voltage (V_{oc}), and the fill factor (*FF*). The photovoltaic power conversion efficiency (η) of a solar cell is defined as the ratio between the maximum electrical power (P_{max}) and the incident optical power and is determined by Eq. (1) [4].

$$\eta = \frac{I_{sc} \times V_{oc} \times FF}{P_{in}} \quad (1)$$

In Eq. (1), I_{sc} is the maximum current that can run through the cell. The open circuit voltage (V_{oc}) depends on the highest occupied molecular orbital (homo)level of the donor (*p*-type semiconductor quasi Fermi level) and the lowest unoccupied molecular orbital(lumo) level of the acceptor (*n*-type semiconductor quasi Fermi level), linearly. *P* in is the incident light power density. FF, the fill-factor, is calculated by dividing P_{max} by the multiplication of I_{sc} and V_{oc} and this can be explained by the following Eq. (2):

$$FF = \frac{I_{mpp} \times V_{mpp}}{I_{sc} \times V_{oc}} \quad (2)$$

In the Eq. (2), V_{mpp} and I_{mpp} represent, respectively, the voltage and the current at the maximum power point (MPP), where the product of the voltage and current is maximized [4].

3.2.2 INORGANIC SOLAR CELL

Inorganic solar cells based on semiconducting layer architecture can be divided into four main categories including P-N homo junction, hetrojunction either P-I-N or N-I-P and multi junction.

The P-N homojunction is the basis of inorganic solar cells in which two different doped semiconductors (*n*- type and *p*-type) are in contact to make solar cells (Fig. 3.a). *P*-type semiconductors are atoms and compounds with fewer electrons in their outer shell, which could create holes for the electrons within the lattice of *p*-type semiconductor. Unlike *p*-types semiconductors, the *n*-type have more electrons in their outer shell and sometimes there are exceed amount of electron on *n*-type lattice result lots of negative charges [5].

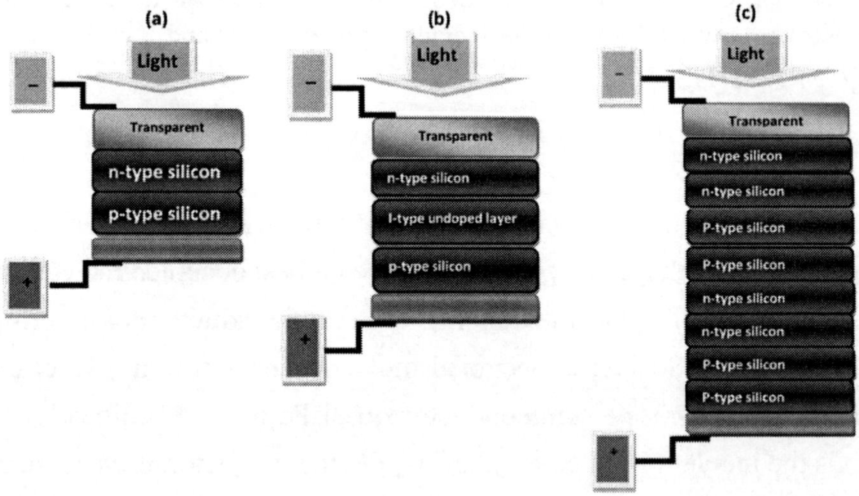

FIGURE 3 A general scheme of inorganic solar cell (a) P-N homojunction (b) P-I-N hetrojunction (c) multijunction.

Compared to homojunction structure amorphous silicon thin-film cells use a P-I-N hetrojunction structure, whereas cadmium telluride (CdTe) cells utilize a N-I-P arrangement. The overall picture embraces a three-layer sandwich with a middle intrinsic (*i*-type or undoped) layer between an N-type layer and a P-type layer (Fig. 3b). Multiple junction cells have several different semiconductor layers stacked together to absorb different wavebands in a range of spectrum, producing a greater voltage per cell than from a single junction cell, which most of the solar spectrum to electricity lies in the red (Fig. 3c).

Variety of semiconducting material such as single and poly crystal silicon, amorphous silicon, Cadmium-Telluride (CdTe), Copper Indium/Gallium Di Selenide (CIGS) have been employed to form inorganic solar cell based on layers configuration to enhance absorption efficiency, conversion efficiency, production and maintenance cost.

3.2.3 ORGANIC SOLAR CELLS (OSCS)

Photoconversion mechanism in organic or excitonic solar cells is differing from conventional inorganic solar cells in which exited mobile state are made by light absorption in electron donor. While, light absorption creates free electron-hole pairs in inorganic solar cells [5]. It is due to law dielectric constant of organic materials and weak non-covalent interaction between organic molecules. Consequently, exciton dissociation of electron-hole pairs occurs at the interface between electron donor and electron acceptor components [6]. Electron donor and acceptor act as semiconductor *p-n* junction in inorganic solar cells and should be blended together to prevent electron-hole recombination (Fig. 4).

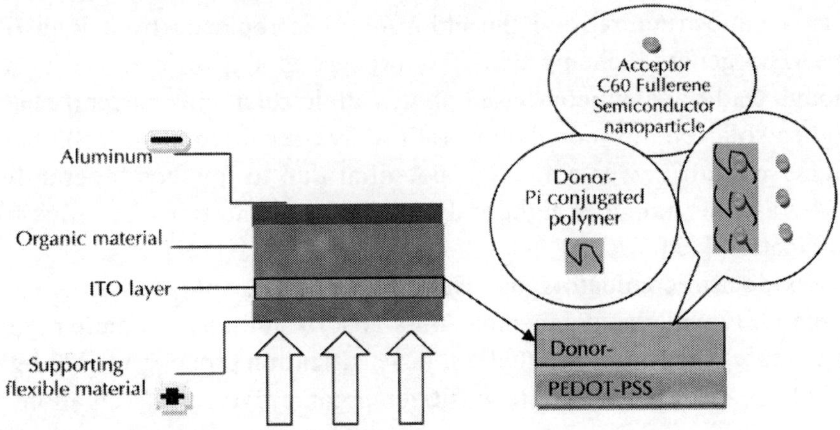

FIGURE 4 Schematic of organic solar cell.

There are two main types of PSCs including: bilayer heterojunction and bulk-heterojunction [7]. Bulk-heterojunction PSCs are more attractive due to their high surface area junction that increases conversion efficiency. This type of polymer solar cell consists of Glass, ITO, PEDOT: PSS, active layer, calcium and aluminum in which conjugated polymer are used as active layer [8]. The organic solar cells with maximum conversion efficiency about 6% still are at the beginning of development and have a long way to go to compete with inorganic solar cells. Indeed, the advantages of polymers including low-cost deposition in large areas, low weight on flexible substrates and sufficient efficiency are promising advent of new type of solar cells [9]. Conjugated small molecule attracted as an alternative approach of organic solar cells. Development of small molecule for OSCs interested because of their properties such as well-defined molecular structure, definite molecular weight, high purity, easy purification, easy mass-scale production, and good batch-to-batch reproducibility [10–12].

3.2.4 DYE-SENSITIZED SOLAR CELLS

DSSC use a variety of photosensitive dyes and common, flexible materials that can be incorporated into architectural elements such as windowpanes, building paints, or textiles. DSSC technology mimic photosynthesis process whereby the leaf structure is replaced by a porous titania nanostructure, and the chlorophyll is replaced by a long-life dye. The general scheme of DSSC process is shown in figure 5. Although traditional silicon-based photovoltaic solar cells currently have higher solar energy conversion ratios, dye-sensitive solar cells have higher overall power collection potential due to low-cost operability under a wider range of light and temperature conditions, and flexible application [13].

Oxide semiconductors materials such as TiO_2, ZnO_2 and SnO_2 have a relatively wide band gap and cannot absorb sunlight in visible region and create electron. Nevertheless, in sensitization process, visible light could be absorbed by photosensitizer organic dye results creation of

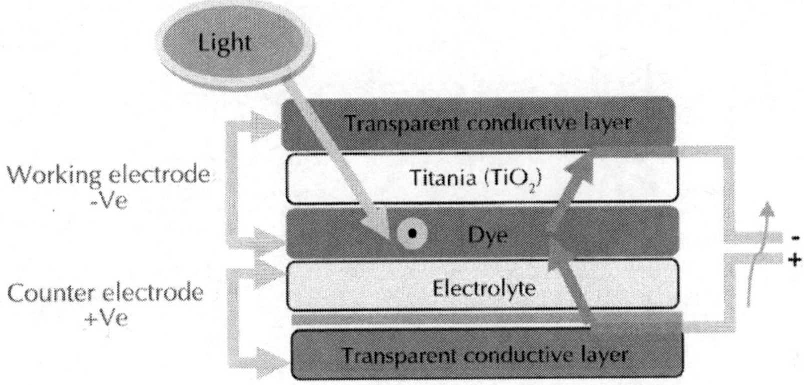

FIGURE 5 Schematic of Dye Senetesized Solar Cell (DSSC).

electron. Consequently, excited electrons are penetrated into the semiconductor conduction band. Generally, DSSC structures consist of a photoelectrode, photosensitizer dye, a redox electrolyte, and a counter electrode. Photoelectrodes could be made of materials such as metal oxide semiconductors. Indeed, oxide semiconductor materials, particularly TiO_2, are choosing due to their good chemical stability under visible irradiation, nontoxicity and cheapness. Typically, TiO_2 thin film photoelectrode prepared via coating the colloidal solution or paste of TiO_2 and then sintering at 450°C to 500°C on the surface of substrate, which led to increase of dye absorption drastically by TiO_2 [14]. The substrate must have high transparency and low ohmic resistance to high performance of cell could be achieved. Recently many researches focused on the both organic and inorganic dyes as sensitizer regarding to their extinction coefficients and performance. Among them, B4 (N3): RuL2(NCS)2 :L=(2,2'-bipyridyl-4,4'-dicarboxylic acid) and B2(N719): {cis-bis (thiocyanato)-bis(2,20-bipyridyl-4,40-dicarboxylato)-ruthenium(II) bis-tetrabutylammonium} due to its outstanding performance was interested.(Scheme 1)

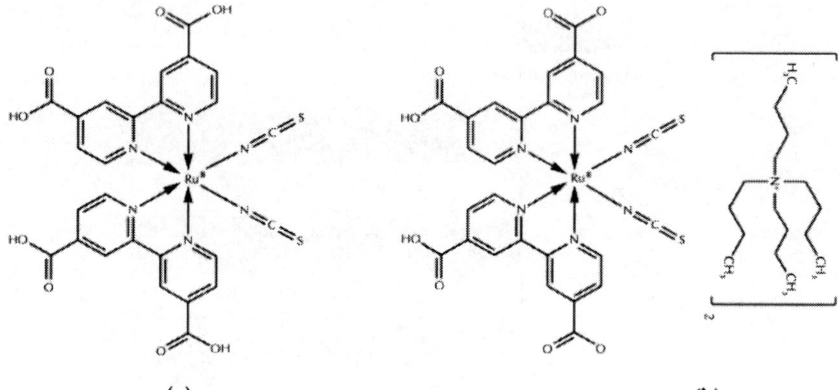

SCHEME 1 The chemical structure of (a) B4 (N3): RuL2(NCS)2 L(2,2'-bipyridyl-4,4'-dicarboxylic acid), (b) B2(N719) {cis-bis (thiocyanato)-bis(2,20-bipyridyl-4,40-dicarboxylato)-ruthenium(II) bis-tetrabutylammonium.

B2 is the most common high performance dye and a modified form of B4 to increase cell voltage. Up to now different methods have been performed to develop of new dyes with high molar extinction coefficients in the visible and near-IR in order to outperform N719 as sensitizers in a DSSC [15]. DNH2 is a hydrophobic dye, which very efficiently sensitizes wide band-gap oxide semiconductors, like titanium dioxide. DBL (otherwise known as "black dye") is designed for the widest range spectral sensitization of wide band-gap oxide semiconductors, like titanium dioxide up to wavelengths beyond 800 nm (Scheme 2).

In order to continuous electron movement through the cell, the oxidized dye should be reduced by electron replacement. The role of redox electrolyte in the DSSCs is to mediate electrons between the photoelectrode and the counter electrode. Common electrolyte used in the DSSC is based on I^-/I^-_3 redox ions [16]. The mechanism of photon to current has been summarized in the following equations:

Dye + light → Dye* + e^- (a)

Dye* + e^- + TiO_2 →e^-(TiO2) + oxidized Dye (b)

Oxidized Dye + 3/2I^- → Dye + 1/2I^{-3} (c)

1/2I^-_3 + e^-(counter electrode) → 3/2I^- (d)

Despite of many advantages of DSSCs, still lower efficiency compared to commercialized inorganic solar cells is a challenging area. Recently, one-dimensional nanomaterials, such as nanorods, nanotubes and nanofibers, have been proposed to replace the nanoparticles in DSSCs because of their ability to improve the electron transport leading to enhanced electron collection efficiencies in DSSCs.

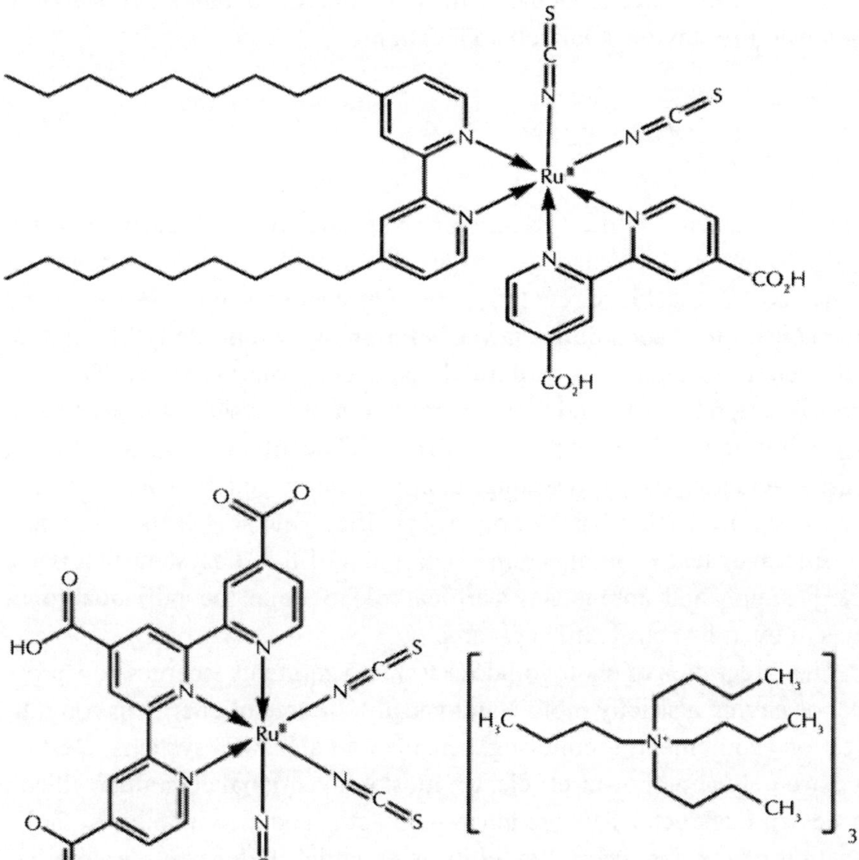

SCHEME 2 The chemical structure of (a) DNH2 (Z907) RuLL'(NCS)2 , L=2,2'-bipyridyl-4,4'-dicarboxylic acid ,L'= 4,4'-dinonyl-2,2'-bipyridine (b) DBL (N749) [RuL(NCS)3]: 3 TBA
L= 2,2':6',2"-terpyridyl-4,4',4"-tricarboxylic acid TBA=tetra-*n*-butylammonium.

Subsequent sections attempts to provide fundamental knowledge to general concept of textile solar cells and their recent progress based on. Of particular interest are electrospun TiO_2 nanofibers playing the role as a key material in DSSCs and other organic solar cell, which have been shown to improve the electron transport efficiency and to enhance the light harvesting efficiency by scattering more light in the red part of the solar spectrum. A detailed review on cell material selection and their effect on energy conversion are considered to elucidate the potential role of nanofiber in energy conversion for textile solar cell applications.

3.3 TEXTILE SOLAR CELLS

Clothing materials either for general or specific use are passive and the ability to integrate electronics into textiles provides great opportunity as smart textiles to achieve revolutionary improvements in performance and the realization of capabilities never before imagined on daily life or special circumstances such as battlefield. In general, smart textiles address diverse function to withstand an interactive wearable system. Development, incorporation and interconnection of flexible electronic devices including sensor, actuator, data processing, communication, internal network and energy supply beside basic garment specifications sketch the road map toward smart textile architecture. Regardless of the subsystem functions, energy supply and storage play a critical role to propel the individual functions in overall smart textile systems.

The integration of photovoltaic (PVs) into garments emerges new prospect of having a strictly mobile and versatile source of energy in communications equipment, monitoring, sensing and actuating systems. Despite of extremely good power efficiency, most conventional crystalline silicon based semiconductor PVs are intrinsically stiff and incompatible with the function of textiles where flexibility is essential. Extensive research has been conducted to introduce the novel potential candidates for shaping the textile solar cell (TSC) puzzle. In particular, polymer-based organic solar cell materials have the advantages of low price and ease of operation in comparison with silicon-based solar cells. Organic semiconductors, such as conductive polymers, dyes, pigments, and liquid crystals, can be manufactured cheaply and used in organic solar cell constructions easily. In the manufacturing process of organic solar cells, thin films are prepared utilizing specific techniques, such as vacuum evaporation, solution processing, printing [17, 18], or nanofiber formation [19] and electrospinning [20] at room temperatures. Dipping, spin coating, doctor blading, and printing techniques are mostly utilized for manufacturing organic solar cells based on conjugated polymers [17]. Recent TSC studies revealed two distinctive strategies for developing flexible textile solar cell and its sophisticated integration.

1. The first strategy involved the simple incorporation of a polymer PV on a flexible substrate Such as poly thyleneterphthalate (PET) directly into the clothing as a structural power source element.
2. The second strategy was more complicated and involved the lamination of a thin anti reflective layer onto a suitably transparent textile material followed by plasma, thermal or chemical treatment. The next successive step focuses on application of a photoanode electrode onto the textile material. Subsequent procedure led to the deposition of the active material and finally evaporation of the cathode electrode complete the device as a textile PV composed of organic, inorganic and also their composites.

Regardless of many gaps need to be bridged before large-scale application of this technology, the TSC fabrication based on second strategy may be envisaged through two routes to solve pertinent issues of efficiency and stability. The solar cell architecture in first approach is founded based on knitted or woven textile substrate, however second alternative follows a roadmap to develop a wholly PV fiber for further knitting or weaving process that may form energy-harvesting textile structures in any shape and structure.

Irrespective to fabric or fiber shaped of the photovoltaic unit, the light penetration and scattering in photo anode layer needs a waveguide layer. This basic requirement naturally mimicked by polar bear hair. The optical functions of the polar bear hair are scattering of incident light into the hair, luminescence wave shift and wave-guide properties due to total reflection. The hair has an opaque, rough-surfaced core, called the medulla, which scatters incident light. The simulated synthetic core-shell fiber can be manufactured through spinning of a core fiber or with sufficient wave-guiding properties followed by finishing with an optically active, i.e. fluorescing, coating as a shell to achieve a polar bear hair' effect [21]. As described in Tributsch's original work [22], a high-energy conversion can be expected from high frequency shifts as difference between the frequencies of absorbed and emitted light. According to Tributsch et al. [22] for the polar bear hair, the frequency shift is of the order of 2×10^{14} Hz.

In principle, the refractive index varies over the fiber diameter and this would provide a certain wave-guiding property of the fiber. In charge of wave-guiding property most specifically With regard to manufacturing on a larger scale, fiber morphology, crystalinity, alignment, diameter and geometry can be altered for preferred optical performance as solar energy transducer. The focus on both approached of second strategy and illuminates their opportunities and challenges are main area of interest in next parts.

3.3.1 RECENT PROGRESS OF PV FIBERS

Fiber shaped organic solar cells has been subject of a few patent, project and research papers. Polymers, small molecules and their combinations were used as light-absorbing layers in previous studies. A range of synthetic substrate with various level of flexibility including optical [23], polyimide [24] and poly propylene (PP)[25] subjected to processes in which functional electrodes and light absorbing layer continuously forms on fiber scaffold. In recent PV fiber studies, conducting polymers such as P3HT in combination to small molecules nanostructure materials such as branched fullerene plays main role as photoactive material. For instance, Reference 23 introduce a light absorbing layer on optical fiber composed of poly (3-hexylthiophene) (P3HT): phenyl-C61-butyric acid methyl ester (P3HT:PCBM) (Fig. 6) . While the light was travelling through the optical fiber and generating hole–electron pairs, a 100 nm top metal electrode (which does not let the light transmit from outside) was used to collect the electrons [23].

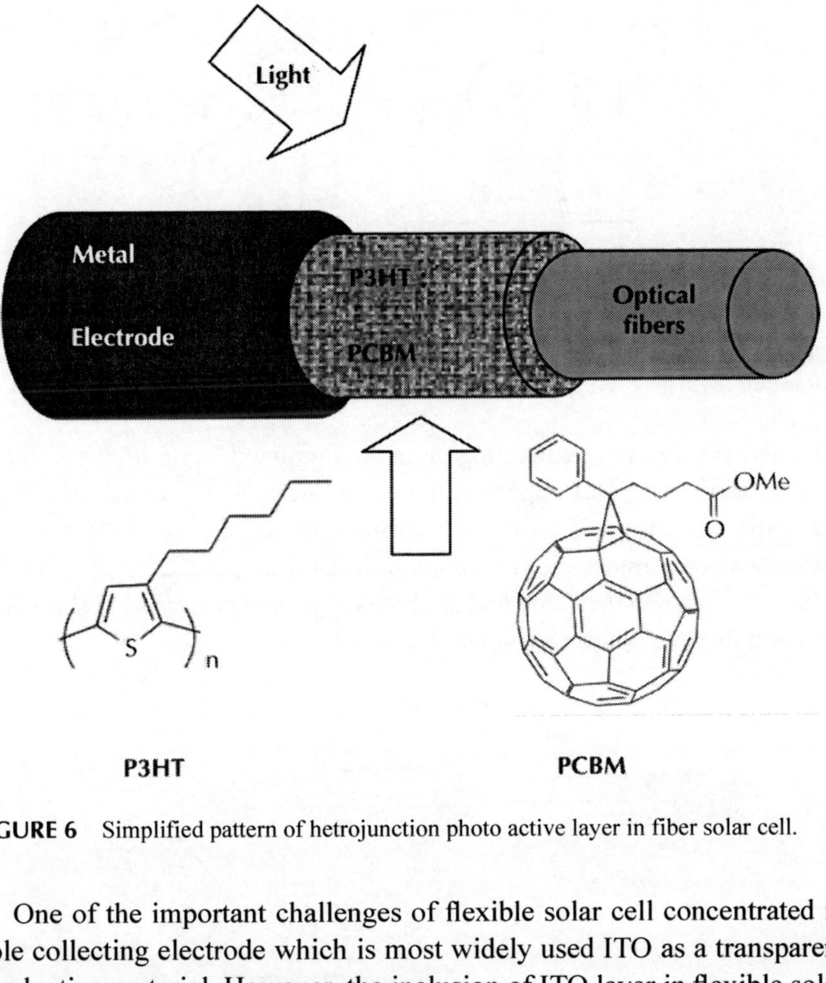

FIGURE 6 Simplified pattern of hetrojunction photo active layer in fiber solar cell.

One of the important challenges of flexible solar cell concentrated in hole collecting electrode which is most widely used ITO as a transparent conducting material. However, the inclusion of ITO layer in flexible solar cell could not be applicable. The restrictions are mostly due to the low availability and expense of indium, employment of expensive vacuum deposition techniques and providing high temperatures to guarantee highly conductive transparent layers. Accordingly, there are some ITO-free alternative approaches, such as using carbon nanotube (CNT) layers or different kinds of poly(3,4-ethylenedioxythiophene):poly(styrenesulfonate) (PEDOT:PSS) and its mixtures [26–28], or using a metallic layer [29] to perform as a hole-collecting electrode. (Scheme 3)

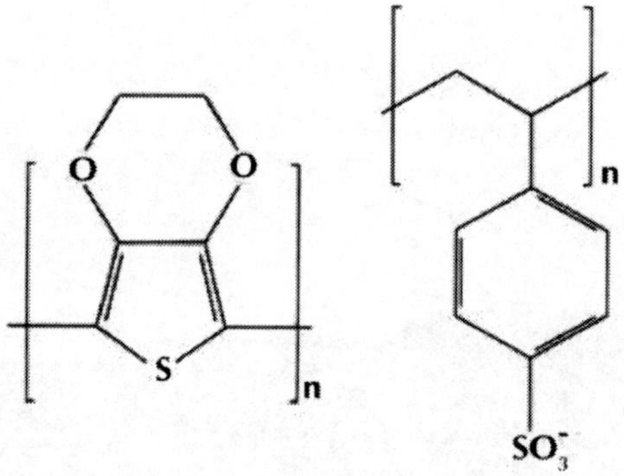

SCHEME 3 The chemical structure of poly (3,4-ethylenedioxythiophene):poly(styrene sulfonate).

The ITO free hole collecting layer was realized using highly conductive solution of PEDOT:PSS as a polymer anode that is more convenient for textile substrates in terms of flexibility, material cost, and fabrication processes compared with ITO material. Based on procedure described in reference 25 a sophisticated and simple design was presented to show how thin and flexible could be a solar cell panel.

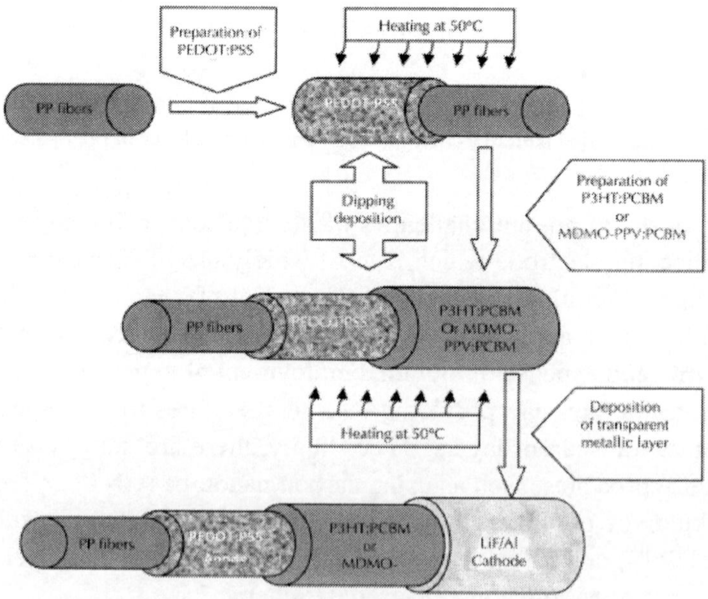

FIGURE 7 The Schematic of procedure for preparation of PV fiber.

Based on implemented pattern, the sunbeams entered into the photoactive layer with 4-10 mm^2 active area by passing through a 10 nm of lithium fluoride/aluminum (LiF/Al) layer as semi-transparent cathode outer electrode.

Table 1 gives the current density data versus voltage characteristics of the photovoltaic fibers consisting of P3HT: PCBM and MDMO-PPV: PCBM blends. Based on given results of open-circuit voltage , short-circuit current density and fill factor for two types of photoactive material, and also Eqs. (1) and (2) in Section 3.2.2, the power conversion efficiency of the MDMO-PPV: PCBM based photovoltaic fiber was higher than the P3HT: PCBM based photovoltaic fiber.

TABLE 1 Photoelectrical characteristics of photovoltaic fibers having different having different photoactive layers (P3HT: PCBM and MDMO-PPV:PCBM).

Solar cell pattern	V_{oc} (mV)	I_{sc} (mA/cm^2)	FF (%)	η (%)
PP|PEDOT:PSS| P3HT:PCBM|LiF/Al	360	0.11	24.5	0.010
PP|PEDOT:PSS| MDMO-PPV:PCBM| LiF/Al	300	0.27	26	0.021
ITO|PEDOT:PSS| MDMO-PPV:PCBM| LiF/Al	740	4.56	43.4	1.46

Comparing the solar cell characteristics of second and third pattern shows greater performance of ITO|PEDOT:PSS| MDMO-PPV:PCBM| LiF/Al rigid organic solar cell compared to PP|PEDOT:PSS| MDMO-PPV:PCBM| LiF/Al fiber solar cell. Since a same cathode, anode and photoactive material utilized in both pattern, the higher power conversion efficiency of ITO solar cell can be attributed to different wave guide property and transparency of cell pattern in sun's ray entrance angle(i.e. LiF/AL versus ITO glass).

Since using ITO is strictly restricted for PV fiber, enhancing the power conversion efficiency of needs to improving existing materials and techniques. In particular, the optical band gap of the polymers used as the active layer in organic solar cells is very important. Generally, the best bulk heterojunction devices based on widely studied P3HT: PCBM materials are active for wavelengths between 350 and 650 nm. Polymers with narrow band gaps can absorb more light at longer wavelengths, such as infrared or near-infra-red, and consequently enhance the device efficiency. Low band gap polymers (<1.8 eV) can be an alternative for better power efficiency in the future, if they are sufficiently flexible and efficient for textile applications [30, 31]. The variety of factors influence on polymer band gaps which can be categorized as intra-chain charge transfer, substituent effect, π-conjugation length.

Systematically the fused ring low band gap copolymer composes of a low energy level electron acceptor unit coupled with a high energy level electron donor unit. The band gap of the donor/acceptor copolymer is determined by the HOMO of the donor and LUMO of the acceptor, and therefore a high energy level of the HOMO of the donor and a low energy level of the LUMO of the acceptor results in a low band gap [32].

The substituent on the donor and acceptor units can affect the band gap. The energy level of the HOMO of the donor can be enhanced by attaching electron-donating groups (EDG), such as thiophene and pyrrole. Similarly, the energy level of the LUMO of the acceptor is lowered, when electron-withdrawing groups (EWG), such as nitrile, thiadiazole and pyrazine, are attached. This will result in improved donor and acceptor units, and hence, the band gap of the polymer is decreased [33].

3.3.2 PV INTEGRATION IN TEXTILE

Having a complete functional textile solar cell motivates the researchers to attempt an approach for direct incorporation of photovoltaic cell elements onto the textile. The textile substrates inherently scatter most part of the incident light outward. Therefore, it was found necessary to apply a layer of the very flexible polymer PE onto the textile substrate to have a surface compatible with a layered device. The textile-PE substrate was plasma treated before application of the transparent PEDOT electrode in order to obtain good adhesion of the PEDOT layer to the PE carrier. Then screen-printing was employed for the application of the active polymer poly[2-methoxy-5-(2'-ethylhexyloxy)-p-phenylene vinylene]. (MEH-PPV). [33]

The traditional solar cell geometry was re-invented in fractal forms that allow the building of structured modules by sewing the 25–40 cm cells realized. Figure 8 shows a step-by-step approach for fabrication textile solar cell pattern based on polymer photo absorbing layer.

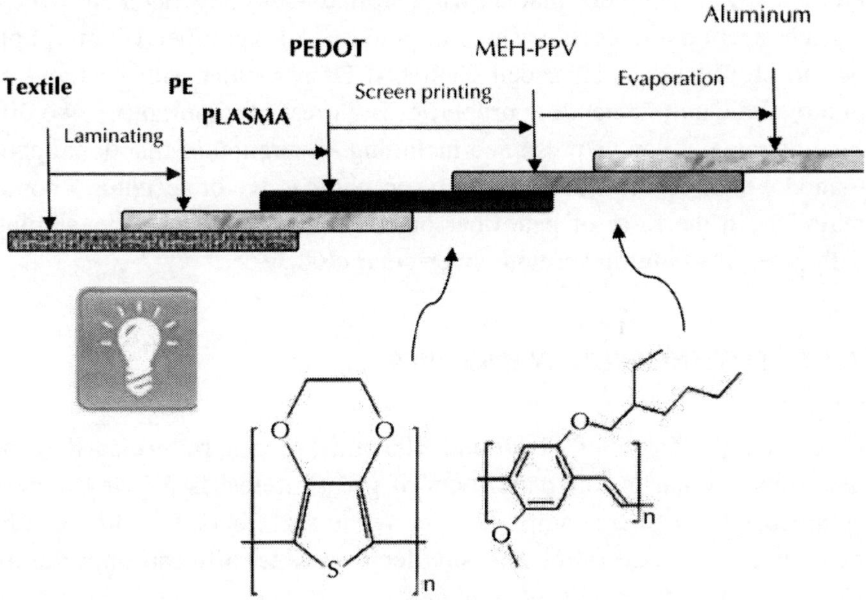

FIGURE 8 A typical fabrication procedure and key elements of textile solar cell.

The pattern designed was particularly challenging for application in solar cells and reduced the active area to 190 cm^2 (19% of the real area). The best module output power was found to be 0.27 mW with a Isc = 3:8 µA, Voc = 275 mV and a FF% of 25.7%. The pattern designed allows connections in different site of the cloth cell with reproducible performances within 5–10%.

3.4 NANOFIBERS AS A POTENTIAL KEY ELEMENT IN TEXTILE SOLAR CELLS

Previous sections present variety of solar cell structure and their corresponding elements and power conversion performance to indicate opportunities and challenges of producing of solar energy harvesting module based on a wholly flexible textile based photovoltaic unit. Current state of Textile Solar Cells is extremely far from commercial inorganic hetro junction solar cells that showing around 45% conversion efficiency. Current section addresses promising potential of nanofiber 1D morphology to be utilized as solar cell elements. Of particular, enhancement of photovoltaic unit demanding properties is a great of importance. Two different strategies can be presumed including integration of functional photoanode, photo cathode, scattering layer, photoactive or acceptor – donor materials in the form of nanofiber on to textile substrate or developing fully integrated multilayer nonwoven solar cloth.

3.4.1 ELECTROSPUN NANOFIBER

Fibers with a diameter of around 100 nm are generally classified as nanofibers. What makes nanofibers of great interest is their extremely small size. Nanofibers compared to conventional fibers, with higher surface area to volume ratios and smaller pore size, offer an opportunity for use in a wide variety of applications. To date, the most successful method of producing nanofibers is through the process of electrospinning. The electrospinning process uses high voltage to create an electric field between a droplet of polymer solution at the tip of a needle and a collector plate. When the electrostatic force overcomes the surface tension of the drop, a charged, continuous jet of polymer solution is ejected. As the solution moves away from the needle and toward the collector, the solvent evaporates and jet rapidly thins and dries. On the surface of the collector, a nonwoven web of randomly oriented solid nanofibers is deposited. Material properties such as melting temperature and glass transition temperature as well as structural characteristics of nanofiber webs such as fiber diameter distribution, pore size distribution and fiber orientation distribution determine the physical and mechanical properties of the webs. The surface of electrospun fibers is important when considering end-use applications. For example, the ability to introduce porous surface features of a known size is required if nanoparticles need to be deposited on the surface of the fiber.

The conventional setup for producing a nonwoven layer can be manipulated to fabricate diverse profile and morphology including oriented [34], Core-shell [35] and hollow [36] nanofiber. Figure 9 shows the latest nanofiber profiles and its corresponding electrospinning production instrument. The variety and propagation of nanofiber products opens new horizon for development of functional profile respect to demanding application. Amongst developed techniques, coaxial electrospinning forms core-shell and/or hollow nanofiber through combination of different materials in the core or shell side, novel properties and functionalities for nanoscale devices can be found.

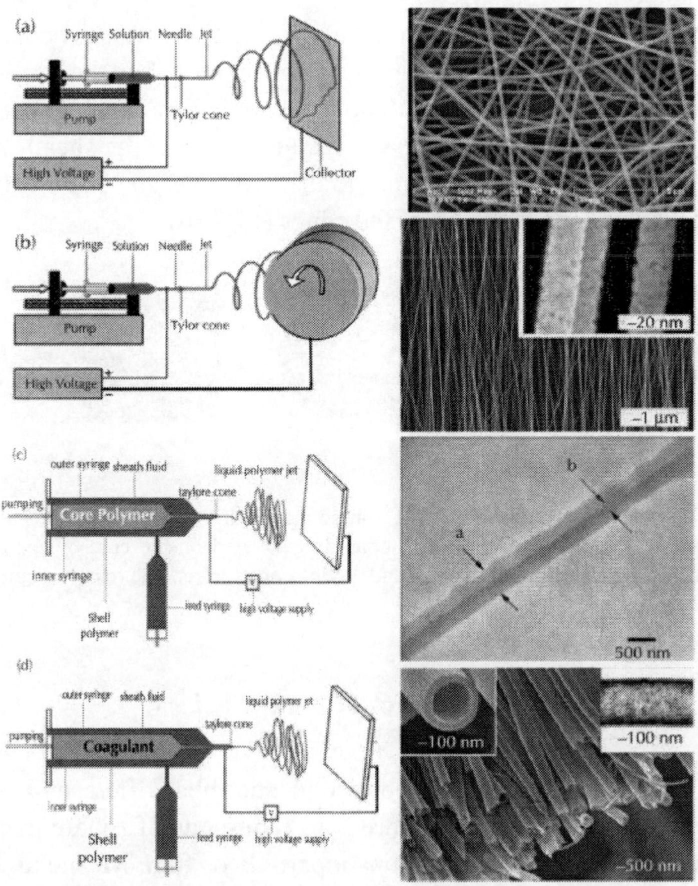

FIGURE 9 Electrospinning setup and its corresponding nanofiber profile (a) conventional nanofiber] (b) Oriented nanofiber (c) Core-shell nanofiber, (d) Hollow nanofiber.

Increasing demands for the manufacturing of bi-component structures, in which one is surrounded by the other or the particles of one are encapsulated in the matrix of the other, at the micro or nano level, show potential for a wide range of uses.

Application includes minimizing chances of decomposition of an unstable material, control releasing a substance to a particular receptor and improving mechanical properties of a core polymer by its reinforcing with another material. The electro-spinneret consists of concentric inner and outer syringe by witch two fluids are introduced to the spinneret, one in the core of the inner syringe and the other in the space between in the inner syringe and outer syringe.

The droplet of the sheath solution elongates and stretches due to the repulsing between charges and form a conical shape. When the applied voltage increases, the charge accumulation reaches a certain value so a thin jet extends from the cone. The stresses are generated in the sheath solution cause to the core liquid to deform into the conical shape and a compound co-axial jet develops at the tip of the cones (Fig. 10).

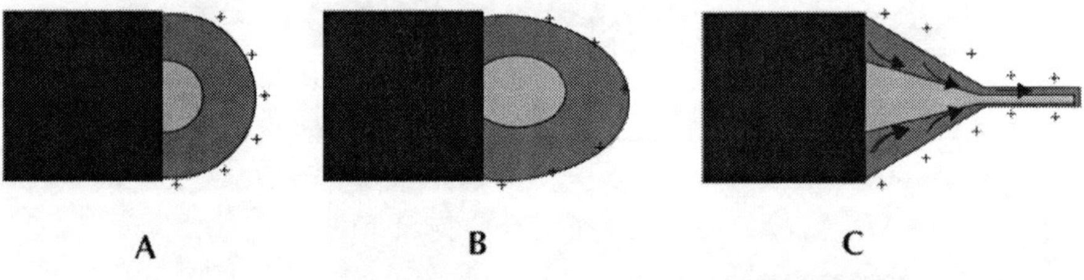

FIGURE 10 Schematic illustration of compound Taylor cone formation (A): Surface charges on the sheath solution, (B): viscous drag exerted on the core by the deformed sheath droplet, (C): Sheath-core compound Taylor cone formed due to continuous viscous drag).

3.4.2 NANOFIBERS IN DSSC SOLAR CELLS

It can be presumed that the electrospun nanofiber offers high specific surface areas (ranging from hundreds to thousands of square meters per gram) and bigger pore sizes than nanoparticle or film. Meanwhile, referring to Section 2.3, the particle-based titanium dioxide layers have low efficiencies due to the high density of grain boundaries, which exist between nanoparticles. The 1D morphology of metal oxide fibers attracts more interest because of the lower density of grain boundaries compared to those of sintered nanoparticles [37].

3.4.2.1 ELECTROSPUN SCATTERING LAYER

Ref. [38] compared the effect of TiO_2 nanofiber and nanoparticle as scattering layer and indicated the significant enhancement of all photovoltaic specefications. In another attempt ZNO nanofiber used instead of TiO_2 nanofiber to form photo anode [39].

TABLE 2 The effect of TiO_2 nanofiber on cell performance for DSSC solar cell [38–40].

Solar cell pattern	V_{oc} (mV)	I_{sc} (mA/cm^2)	FF (%)	η (%)	Ref
FTO|TiO_2 nanoparticle|N719|LiI/I2/TBT|Pt|FTO	630	11.3	54	3.85	[38]
FTO|TiO_2 nanofiber|N719|LiI/I2/TBT|Pt|FTO	660	14.3	53	4.9	[38]
FTO|ZNO nanofiber|N719|LiI/I2/TBT|Pt|FTO	690	2.87	44	0.88	[39]
FTO|TiO_2 nanofiber: Ag nano P|N719|LiI/I2/TBT|Pt|FTO	800	7.57	55	3.3	[40]

However, the measured photocurrent density–voltage shows poorer results compared to TiO_2. The influence of Ag nanoparticle was also studied showed nearly same fill factor but lower power conversion efficiency compared to neat TiO_2 nanofiber scattering layer [40]. Recently, a composite anatase TiO_2 nanofibers/nanoparticle electrode was fabricated through electrospinning [41]. This method avoided the mechanical grinding process, and offered a higher surface area, so conversion efficiencies of 8.14% and 10.3% for areas of 0.25 and 0.052cm^2, respectively, were reported Hybrid TiO_2 nanofibers with moderate multi walled carbon nanotubes (MWCNTs) content also can prolong electron recombination lifetimes [41]. Since MWCNTs can quickly transport charges generated during photocatalysis, the opportunity for charge recombination is reduced. Furthermore, MWCNTs decrease the agglomeration of TiO2 nanoparticles and increase the surface area of TiO_2. These advantages make this hybrid electrode a promising candidate for DSSCs.

3.4.2.2 NANOFIBER ENCAPSULATED ELECTROLYTE

Nanofiber can be considered as promising candidate for preparation o f solid or semi solid electrolyte. This is mostly because of the inherent long-term instability of electrolyte used in DSSCs usually consists of triiodide/iodide redox coupled in organic solvents [42]. Many solid or semi-solid viscous electrolytes with low level of penetration to TiO_2 layer such as ionic liquids [43], and gel electrolytes [44] utilized to triumph over these problems. However, nanofiber with may increase the penetration of viscous polymer gel electrolytes through large and controllable pore sizes.

A few research conducted to fabricate the Electrospun PVDF-HFP membrane by electrospinning process from a solution of poly(vinylidenefluoride-co-hexafluoropropylene) in a mixture of acetone/N,N-dimethylacetamide to encapsulate electrolyte solution [45, 46]. Although the solar energy-to electricity conversion efficiency of the quasi-solid-state solar cells with the electrospun PVDF-HFP membrane was slightly lower than the value obtained from the conventional liquid electrolyte solar cells, this cell exhibited better long-term durability because of the prevention of electrolyte solution leakage.

3.4.2.3 FLEXIBLE NANOFIBER AS COUNTER ELECTRODE

Efficient charge transfer from a counter electrode to an electrolyte is a key process during the operation of dye-sensitized solar cells. One of the greatest flexible counter electrode could be polyaniline (PAni) nanofibers on graphitized polyimide (GPi) carbon films for use in a tri-iodide reduction. These results are due to the high electrocatalytic activity of the PAni nanofibers and the high conductivity of the flexible GPi film. In combination with a dye-sensitized TiO_2 photoelectrode and electrolyte, the photovoltaic device with the PAni counter electrode shows an Current–voltage characteristics of the dye synthesized solar cells with various electrodes energy conversion efficiency of 6.85%. Short-term stability tests indicate that the photovoltaic device with the PAni counter electrode approximately preserves its initial performance [47] (Table 3). The major concern for the application of alternative counter electrodes to conventional platinized TCOs in DSSCs is long-term stability. Many publications indicate that during prolonged exposure in corrosive electrolyte, catalysts will detach from the substrate and deposit onto the surface of the semiconductor photoelectrode.

TABLE 3 Current–voltage characteristics of the dye synthesized solar cells with various electrodes.

Solar cell pattern	V_{oc} (mV)	I_{sc} (mA/cm²)	FF (%)	η (%)
FTO|TiO2 nanoparticle|N719|PMII/I2/TBP |Pt|FTO	820	12.61	62.3	6.44
FTO|TiO2 nanoparticle|N719|PMII/I2/TBP |PAni|FTO	831	12.22	62.1	6.31
FTO|TiO2 nanoparticle|N719|PMII/I2/TBP |PAni	856	11.59	58.7	5.82
FTO|TiO2 nanoparticle|N719|PMII/I2/TBP |PAni|GPi	901	9.68	28.3	2.49

3.4.2.4 A PROPOSED MODEL FOR DSSC TEXTILE SOLAR CELL USING NANOFIBERS

Choosing proper material and structure for DSSC textile solar cell using previously mentioned nanofiber propose potential candidates for designing of an integrated photovoltaic unit. As can be seen in Fig. 11 a multilayer textile DSSC solar cell composed of a complicated pattern while nanofiber is dominant in step-by-step fabrication process.

The major concern regarding the DSSC textile solar is TiO_2 nanofiber that needs to subject to high temperature for being scattering layer. This strategy is not compatible with other textile element and it is believed that the usage of Anatase TiO_2 spinning solution provide the possibility to avoid high temperature treatment. The proposed strategy is under intensive investigation in our laboratory and future results probably mostly illuminate the opportunities and challenges.

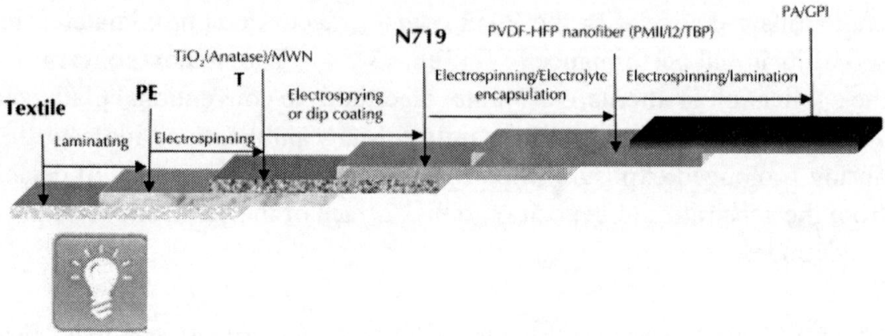

FIGURE 11 A proposed model for DSSC textile solar cell using nanofibers in successive layer.

3.4.3 ELECTROSPUN NONWOVEN ORGANIC SOLAR CELL

The idea of generating non-woven photovoltaic (PV) cloths using organic conducting polymers by electrospinning is quite new and has not been intensively investigated. Based on previously reported PV fiber composed of conjugated hole conducting polymer (see Section 3-2), a novel methodology was reported to generate a non-woven organic solar cloth. The fabrication of core–shell nanofibers has been achieved by co-electrospinning of two components such as poly(3-hexyl thiophene) (P3HT) (a conducting polymer) or P3HT/PCBM as the core and poly(vinyl pyrrolidone) (PVP) as the shell using a coaxial electrospinning set up [see Section 4.1] [48].

Initial measurements of the current density vs. voltage of the P3HT/PCBM solar cloth were carried out and showed current density (I_{sc}), open circuit voltage (V_{oc}) and fill factor (FF) of the fiber cloth around 3.2×10^{-6} mA/cm^2 0.12 V and 22.1%, respectively. In addition, a six order of magnitude lower photo conversion efficiency of the fiber cloth around 8.7×10^{-8} was observed that might sound disappointing. The low PV parameters of the fiber cloth could be attributed to the following factors:

a) The fiber cloth processing steps including electrospinning as well as ethanol washing were carried out under ambient conditions
(b) The thickness of the fiber cloth was ~5 μm compared to the diffusion length of the charge carriers in organic solar cells, which is only several nm.

Therefore, most of the charge carriers were lost in the fiber matrix itself. Drop casted films of the same thickness also showed similar PV parameters.

3.4.4 FUTURE PROSPECTS OF ORGANIC NANOFIBER TEXTILE SOLAR

Nanofiber revolutionizes the future trend of material selection to enhance the characteristics of textile solar cell. Latest experience regarding to polymer solar cell and low band gap material needs to be considered for ambitious plans with nanofiber morphology. Figure 12 shows a schematic for layer-by-layer hetrojunction textile solar cell according to previously mentioned concerns and promising potential solar absorbing and photoactive material. The usage of anti reflective and protective materials is extremely crucial for industrial scale production. A protective layer will save the organic material from moisture and oxygen. The anti-reflective layer in solar cells can obstruct the reflection of light and also contribute to the device performance. This is an important point that should be overcome in the case of large-scale production of textile solar cell. The PEDOT:PSS combination with CNTs forms first P-N junction in the form of bi layer nanofiber. The second layer composed of core- shell or general bi-layer nanofiber of MDMO/PPV:PCBM which absorb visible spectrum. The proposed plan can be realized in large-scale production through a needles electrospinning setup. The continuity of process also is not beyond expectation and a range of materials as nanofiber has been already provided on given substrates.

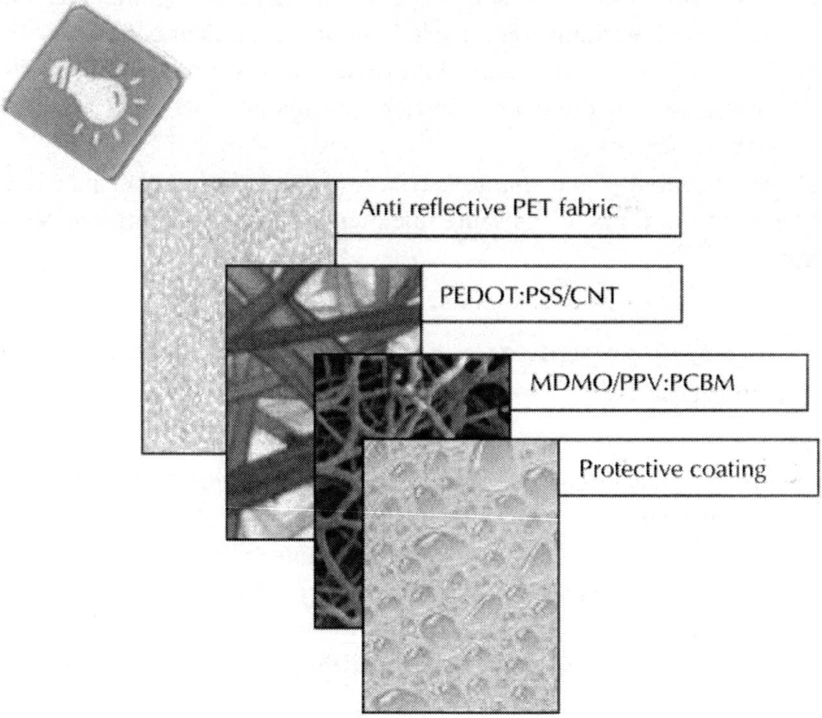

FIGURE 12 A general scheme of hetrojunction Organic nanofiber textile solar.

3.5 CONCLUDING REMARK

The multilayer solar cell energy conversion unit although obey simple theory of light scattering, absorption, electron excitation, charge transfer and its compensation but each layer for specific prescribed functions needs to be intensively investigated for being applicable in diverse circumstances. The commercial solar cell products including silicon either homo or hetrojunctions, dye synthesized solar cell and organic solar cell subjected to demanding research and reached to high level of maturity. The huge experience in silicon solar cell and other photovoltaic system should be reviewed for developing new generation of flexible solar cell. Therefore, current work in first part has a quick glance on variety of solar cells including inorganic, organic and hybrid structures. In overall, regardless of type of solar cell and its corresponding elements, following points needs to be addressed to find fascinating performance:

- The wider absorption wavebands in a range of spectrum.
- The thinner the solar cell.
- The lowering the band gap.
- The higher surface area per unit mass.
- The lowering the cathode thicknesses.
- Using anti reflective coating.
- Using protective coating.
- The multi junction cell to cover range of spectrum.
- Using semi-transparent wave-guide material.
- The lifetime enhancement.

Incorporation of organic solar cells into textiles has been realized reaching encouraging performances. Stability issues need to be solved before future commercialization can be envisaged. The mechanical stability of the devices was not limiting the function of the devices prepared. It would seem that low power conversion efficiency much more pertinent than the mechanical stability on the timescale of commercial.

KEYWORDS

- **Counter electrode**
- **Hetrojunction**
- **Homojunction**
- **Photovoltaic solar energy**

REFERENCES

1. Schubert, M. B. In *Conf. Rec. 31st IEEE Photovolt. Specialists Conf.*, IEEE, New York, 2005; 1488.
2. Nelson, J. In *The Physics of Solar Cells*. Imperial College Press, 2003; 1–16.
3. Miles, R.W.; Forbes, H. I.; Photovoltaic solar cells: An overview of state-of-the-art cell development and environmental issues. *Prog. Cryst. Growth Ch.*, **2005**, *51*, 1–42.
4. Günes, S.; Beugebauer, H.; Sariciftci, N. S. Conjugated Polymer-based Organic Solar Cells, *Chem. Rev.*, **2007**, *107*, 1324–1338.

5. Castafier, T. M. Solar Cells. In *Practical Handbook of Photovoltaics: Fundamentals and Applications*. Castafier, T. M. A. L. Ed.; Elsevier, 2003; 71–95.
6. Thompson, B. C. Polymer–Fullerene Composite Solar Cells. *Angew. Chem. Int. Ed.*, **2008**, *47*, 58–77.
7. Shrotriya, V.; Yao, Y.; Moriarty, T.; Emery, K.; Yang, Y. Accurate Measurement and Characterization of Organic Solar Cells. *Adv. Funct. Mater.*, **2006**, *16*, 2016–2023.
8. Krebs, F. C. In *Polymer Photovoltaics A Practical Approach*, Krebs, F. C., Ed.; SPIE, **2008**, 1–10.
9. Cai ,W.; Cao, Y. Polymer solar cells: Recent development and possible routes for improvement in the performance. *Sol. Energ. Mat. Sol. C.,* **2010**, *94,* 114–127.
10. Dutta, P.; Eom, S. H.; Lee, S. H. Synthesis and characterization of triphenylamine flanked thiazole-based small molecules for high performance solution processed organic solar cells. *Org. Electron.*, **2012**, *13(2)*, 273–282.
11. Soa, S.; Koa, H. M.; Kima, C.; Paeka, S.; Choa, N.; Songb, K.; Leec, J. K.; Koa, J. Novel unsymmetrical push–pull squaraine chromophores for solution processed small molecule bulk heterojunction solar cells. *Sol. Energ. Mat. Sol. C.,* **2012**, *98*, 224–232.
12. Lina, Y.; Liua, Y.; Shia, Q.; Hua, W.; Lia, Y.; Zhan, X. Small molecules based on bithiazole for solution-processed organic solar cells. *Org. Electron.*, **2012**, *13(4)*, 673–680.
13. Gratzel, M. *Nature*, **2001**, *414*, 338–344.
14. Ginger, D. S.; N. C. G. Electrical Properties of Semiconductor Nanocrystals, In *Semiconductor and Metal Nanocrystals*, Klimov, V. I., Ed.; Marcel Dekker, Inc., 2004; 236–285

14a. Ferrazza, F. Large size multicrystalline silicon ingots. Proc. E-MRS 2001 Spring Meeting, Symposium E on Crystalline Silicon Solar Cells. *Sol. Energ. Mater. Sol. C.*, **2002**, *72*, 77–81.

15. Kisserwan, H. Enhancement of photovoltaic performance of a novel dye, "T18", with ketene thioacetal groups as electron donors for high efficiency dye-sensitized solar cells. *Inorg. Chimi. Acta*, **2010**, *363*, 2409–2415.
16. Wei, D. Dye Sensitized Solar Cells, *Int. J. Mol. Sci.*, **2010**, *11*, 1103–1113.
17. Günes, S.; Beugebauer, H.; Sariciftci, N. S. Conjugated Polymer-based Organic Solar Cells, *Chem. Rev.*, **2007**, *107*, 1324–1338.
18. Brabec, C. J.; Dyakonov, V.; Parisi, J.; Sariciftci, N. S. In *Organic Photovoltaics Concepts and Realization*, 1st edition; Springer: New York, 2003.
19. Berson, S.; de Bettignies, R.; Bailly, S.; Guillerez, S. Poly(3-hexylthiophene) Fibers for Photovoltaic Applications, *Adv. Funct. Mater.*, **2007**, *17*, 1377–1384.
20. Gonzalez, R.; Pinto, N. J. Electrospun Poly(3-hexylthiophene-2,5-diyl) Fiber Field Effect Transistor, *Synthetic Met.*, **2005**, *151*, 275–278.
21. Bahners, T.; Schlosser, U.; Gutmann, R.; Schollmeyer, E. Textile solar light collectors based on models for polar bear hair, *Sol. Energ. Mat. Sol. C.,* **2008**, *92*, 1661–1667.
22. Tributsch, H.; Goslowski, H.; Ku, U.; Wetzel, H. Light collection and solar sensing through the polar bear pelt, *Sol. Energ. Mater.,* **1990**, *21*, 219–236.
23. Liu, J.; Namboothiry, M. A. G.; Carroll, D. L. Fiber based Architectures for Organic Photovoltaics, *Appl. Phys. Lett.*, **2007**, *90*, 063501.
24. O'Connor, B.; Pipe, K. P.; Shtein, M. Fiber Based Organic Photovoltaic Devices, *Appl. Phys. Lett.,* **2008**, *92*, 193306.

25. Bedeloglu, A., Demir, A., Bozkurt, Y., Sariciftci, N. S., A Photovoltaic Fiber Design for Smart Textiles, *Text. Res. J.*, **2007**, *80(11)*, 1065–1074.
26. Ouyang, J.; Chu, C. W.; Chen, F.-C.; Xu, Q.; Yang, Y. High-conductivity Poly (3,4-ethylenedioxythiophene): Poly(styrene sulfonate) Film and its Application in Polymer Optoelectronic Devices, *Adv. Funct. Mater.*, **2005**, *15*, 203–208.
27. Kushto, G. P.; Kim, W.; Kafafi, Z. H. Flexible Organic Photovoltaics using Conducting Polymer Electrodes, *Appl. Phys. Lett.*, **2005**, *86*, 093502.
28. Huang, J.; Wang, X.; Kim, Y.; deMello, A. J.; Bradley, D. D.C.; deMello, J. C. High Efficiency Flexible ITO-free Polymer/fullerene Photodiodes, *Phys. Chem. Chem. Phys.*, **2006**, *8*, 3904–3908.
29. Tvingstedt, K.; Inganäs, O. Electrode Grids for ITO-free Organic Photovoltaic Devices, *Adv. Mater.*, **2007**, *19*, 2893–2897.
30. Perzon, E.; Wang, X.; Admassie, S.; Inganäs, O.; Andersson, M. R. An Alternating Low Band-gap Polyfluorene for Optoelectronic Devices, *Polymer*, **2006**, *47*, 4261–4268.
31. Campos, L. M.; Tontcheva, A.; Günes, S.; Sonmez, G.; Neugebauer, H.; Sariciftci, N. S.; Wudl, F. Extended Photocurrent Spectrum of a Low Band Gap Polymer in a Bulk eterojunction Solar Cell, *Chem. Mater.*, **2005**, *17*, 4031–4033.
32. Shaheen, S. E.; Brabec, C. J.; Sariciftci, N.; Padinger, F.; Fromherz, T.;. Hummelen, J. C. *Appl. Phys. Lett,*. **2001**, *78,* 841–843.
33. Scharber, M. S.; Mühlbacher, D.; Koppe, M.; Denk, P.; Waldauf, C.; Heeger, A. J.; Brabec, C. J. *Adv. Mater.*, **2006**, *18*, 789–794.
33. Krebs, F. C.; Biancardo, M.; Jensen, B. W.; Spanggard, H.; Alstrup, J. Strategies for incorporation of polymer photovoltaics into garments and textiles, *Sol. Energ. Mater. Sol. C.*, **2006**, *90*, 1058–1067.
34. Li, D.; Wang, Y.; Xia, Y. Electrospinning Nanofibers as Uniaxially Aligned Arrays and Layer-by-Layer Stacked Films, *Adv. Mater.*, **2004**, *16(14)*, 361–366.
35. Yu, J. H.; Fridrikh, S. V.; Rutledge, G. C. Production of Submicrometer Diameter Fibers by Two-Fluid Electrospinning, *Adv. Mater.*, **2004**, *16(17)*, 1562–1566.
36. Li, D.; Xia, Y. Direct Fabrication of Composite and Ceramic Hollow Nanofibers by Electrospinning, *Nano Lett.*, **2004**, *4(5)*, 933–938.
37. Chuangchote, S.; Sagawa, T.; Yoshikawa, S. *Appl. Phys. Lett.*, **2008**, *93*, 033310.
38. Zhao, X.; Lin, H.; Li, X.; Li, J. The application of freestanding titanate nanofiber paper for scattering layers in dye-sensitized solar cells, *Mater. Lett.*, **2011**, *65*, 1157–1160.
39. Li, S.; Zhang, X.; Jiao, X.; Lin, H. One-step large-scale synthesis of porous ZnO nanofibers and their application in dye-sensitized solar cells, *Mater. Lett.*, **2011**, *65*, 2975–2978.
40. Li, J.; Chen, X.; Ai, N.; Hao, J.; Chen, Q.; Strauf, S.; Shi, Y. Silver nanoparticle doped TiO_2 nanofiber dye sensitized solar cells, *Chem. Phy. Lett.*, **2011**, *514*, 141–145.
41. Hu, G. J.; Meng, X. F.; Feng, X. Y.; Ding, Y. F.; Zhang, S. M.; Yang, M. S. *J. Mater. Sci.*, **2007**, *42*, 7162–7170.
42. Kubo, W.; Kitamura, T.; Hanabusa, K.; Wada, Y.; Yanagida, S. *Chem. Commun.*, **2002**, 374–375.
43. Wang, P.; Zakeeruddin, S. M.; Comte, P.; Exnar, I.; Gratzel, M. *J. Am. Chem. Soc.*, **2003**, *125*, 1166–1167.

44. Wang, P.; Zakeeruddin, S. M.; Moser, J. E.; Nazeeruddin, M. E.; Sekiguchi, T.; Gratzel, M. *Nat. Mater.,* **2003**, *2*, 402–407.
45. Park, S. H.; Kim, J. U.; Lee, S. Y.; Lee, W. K.; Lee, J. K.; Kim, M. R. *J. Nanosci. Nanotechnol.,* **2008**, *8*, 4889–4894.
46. Kim, J. U.; Park, S. H.; Choi, H. J.; Lee, W. K.; Lee, J. K.; Kim, M. R. *Sol. Energ. Mater. Sol. C.,* **2009**, *93*, 803–807.
47. Chen, J.; Lia, B.; Zheng, J.; Zhao, J.; Jing, Zhu, Z. Polyaniline nanofiber/carbon film as flexible counter electrodes in platinum-free dye-sensitized solar cells, *Electrochimi. Acta.,* **2011**, *56*, 4624–4630.
48. Sundarrajan, S.; Murugan, R.; Nair, A. S.; Ramakrishna, S. Fabrication of P3HT/PCBM solar cloth by electrospinning technique, *Mater. Lett.,* **2010**, *64*, 2369–2372.

CHAPTER 4

FORMATION OF METAL ORGANIC FRAMEWORKS

ALBERTO D'AMORE and A. K. HAGHI

CONTENTS

4.1 INTRODUCTION

Recently the application of nanostructured materials has garnered attention, due to their interesting chemical and physical properties. Application of nanostructured materials on the solid substrate such as fibers brings new properties to the final textile product [1]. Metal-organic frameworks (MOFs) are one of the most recognized nanoporous materials, which can be widely used for modification of fibers. These relatively crystalline materials consist of metal ions or clusters (named secondary building units, SBUs) interconnected by organic molecules called linkers, which can possess one, two or three dimensional structures [2–10]. They have received a great deal of attention, and the increase in the number of publications related to MOFs in the past decade is remarkable (Fig. 1).

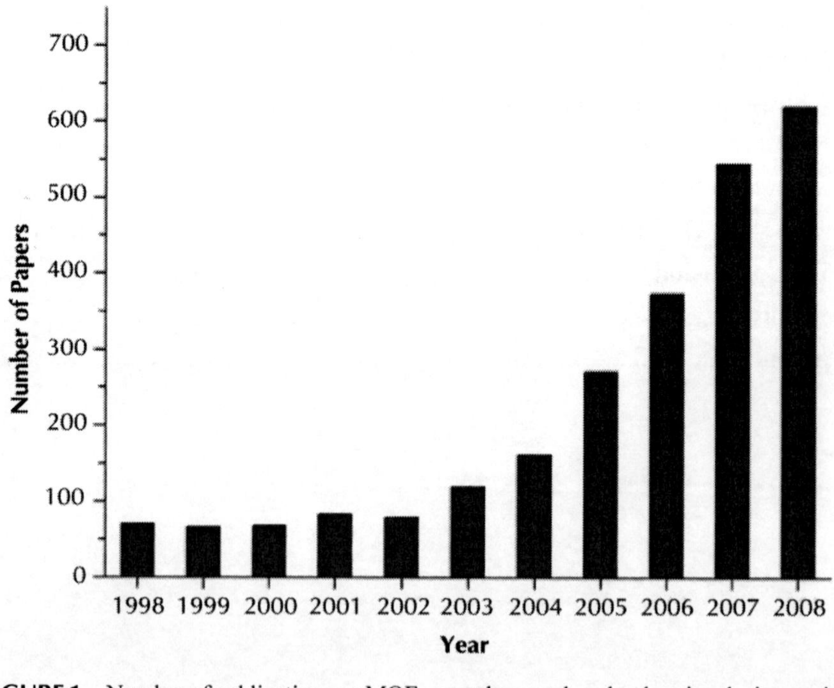

FIGURE 1 Number of publications on MOFs over the past decade, showing the increasing research interest in this topic.

These materials possess a wide array of potential applications in many scientific and industrial fields, including gas storage [11, 12], molecular separation [13], catalysis [14], drug delivery [15], sensing [16], and others. This is due to the unique combination of high porosity, very large surface areas, accessible pore volume, wide range of pore sizes and topologies, chemical stability, and infinite number of possible structures [17, 18].

Although other well-known solid materials such as zeolites and active carbon also show large surface area and nanoporosity, MOFs have some new and distinct advantages. The most basic difference of MOFs and their inorganic counterparts (e.g., zeolites) is the chemical composition and absence of an inaccessible volume (called dead volume) in MOFs [10]. This feature offers the highest value of surface area and porosities in MOFs materials [19]. Another difference between MOFs and other well-known nanoporous materials such as zeolites and carbon nanotubes is the ability to tune the structure and functionality of MOFs directly during synthesis [17].

The first report of MOFs dates back to 1990, when Robson introduced a design concept to the construction of 3D MOFs using appropriate molecular building blocks and metal ions. Following the seminal work, several experiments were developed in this field such as work from Yaghi and O'Keeffe [20].

In this review, synthesis and structural properties of MOFs are summarized and some of the key advances that have been made in the application of these nanoporous materials in textile fibers are highlighted.

4.2 SYNTHESIS OF MOFs

MOFs are typically synthesized under mild temperature (up to 200°C) by combination of organic linkers and metal ions (Fig. 2) in solvothermal reaction [2, 21].

Recent studies have shown that the character of the MOF depends on many parameters including characteristics of the ligand (bond angles, ligand length, bulkiness, chirality, etc.), solubility of the reactants in the solvent, concentration of organic link and metal salt, solvent polarity, the pH of solution, ionic strength of the medium, temperature and pressure [2, 21].

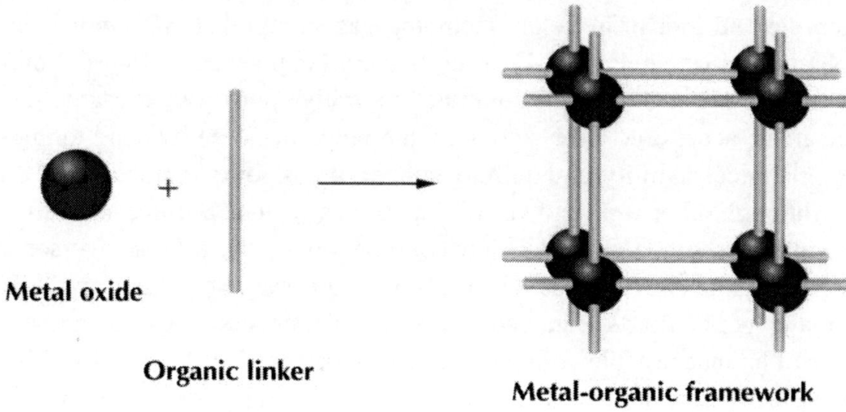

FIGURE 2 Formation of metal organic frameworks.

In addition to this synthesis method, several different methodologies are described in the literature such as ball-milling technique, microwave irradiation, and ultrasonic approach [22].

Post-synthetic modification (PSM) of MOFs opens up further chemical reactions to decorate the frameworks with molecules or functional groups that might not be achieved by conventional synthesis. In situations that presence of a certain functional group on a ligand prevents the formation of the targeted MOF, it is necessary to first form a MOF with the desired topology, and then add the functional group to the framework [2].

4.3 STRUCTURE AND PROPERTIES OF MOFs

When considering the structure of MOFs, it is useful to recognize the secondary building units (SBUs), for understanding and predicting topologies of structures [3]. Figure 3 shows the examples of some SBUs that are commonly occurring in metal carboxylate MOFs. Figure 3(a–c) illustrates inorganic SBUs include the square paddlewheel, the octahedral basic zinc acetate cluster, and the trigonal prismatic oxo-centered trimer, respectively. These SBUs are usually reticulated into MOFs by linking the carboxylate carbons with organic units [3]. Examples of organic SBUs are also shown in Fig. 3(d–f).

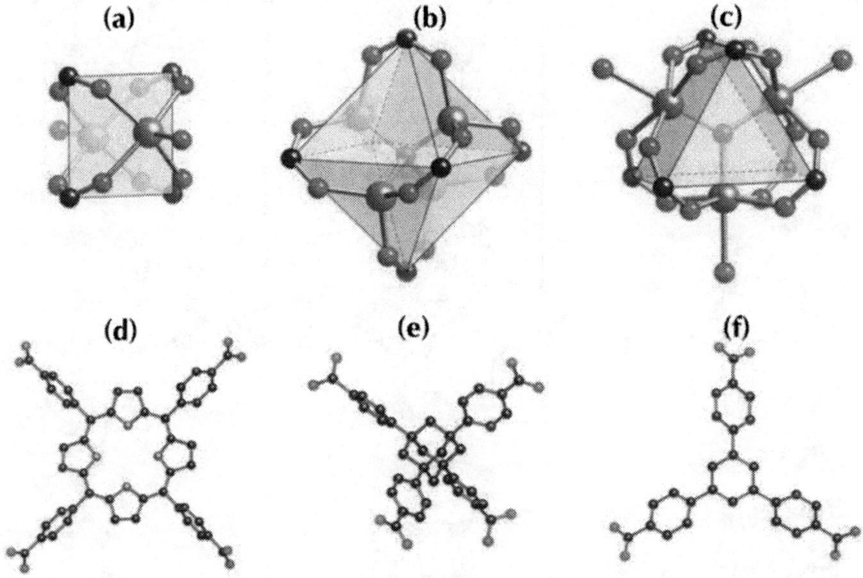

FIGURE 3 Structural representations of some SBUs, including (a–c) inorganic, and (b–f) organic SBUs. (Metals are shown as blue spheres, carbon as black spheres, oxygen as red spheres, nitrogen as green spheres).

It should be noted that the geometry of the SBU is dependent on not only the structure of the ligand and type of metal utilized, but also the metal to ligand ratio, the solvent, and the source of anions to balance the charge of the metal ion [2].

A large number of MOFs have been synthesized and reported by researchers to date. Isoreticular metal-organic frameworks (IRMOFs) denoted as IRMOF-*n* (n = 1 through 7, 8, 10, 12, 14, and 16) are one of the most widely studied MOFs in the literature. These compounds possess cubic framework structures in which each member shares the same cubic topology [3, 21]. Figure 4 shows the structure of IRMOF-1(MOF-5) as simplest member of IRMOF series.

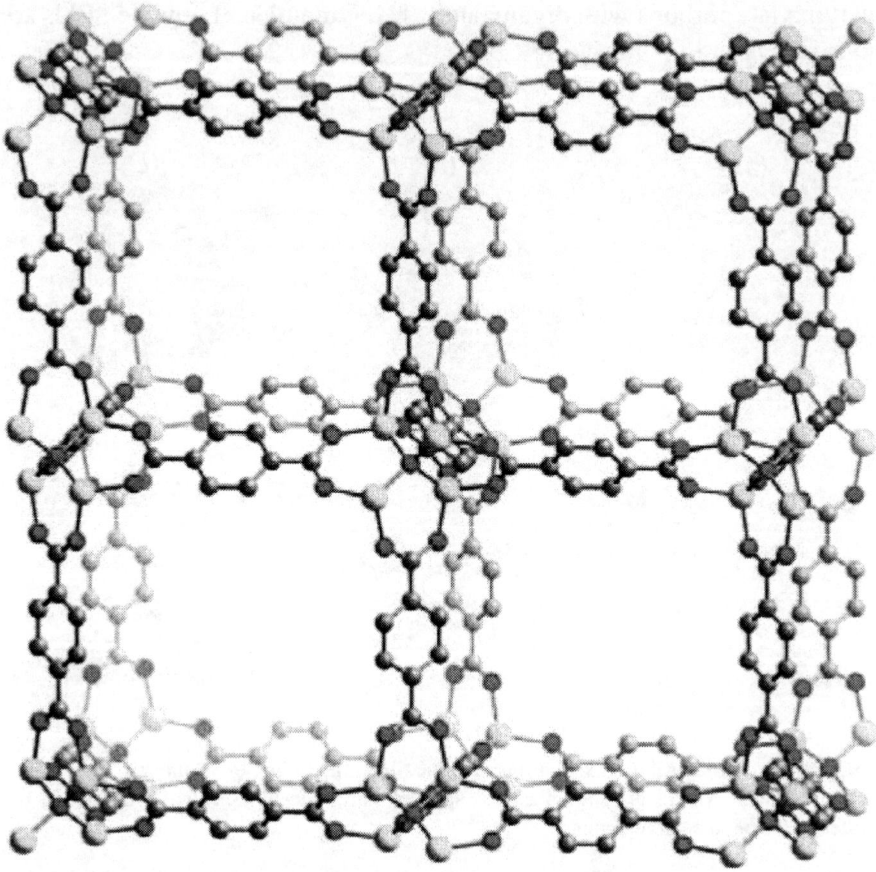

FIGURE 4 Structural representation of IRMOF-1. (Yellow, gray, and red spheres represent Zn, C, and O atoms, respectively).

4.4 APPLICATION OF MOFs IN TEXTILES

4.4.1 INTRODUCTION

There are many methods of surface modification, among which nanostructure based modifications have created a new approach for many applications in recent years. Although MOFs are one of the most promising nanostructured materials for modification of textile fibers, only a few examples have been reported to data. In this section, the first part focuses on application of MOFs in nanofibers and the second part is concerned with modifications of ordinary textile fiber with these nanoporous materials.

4.4.2 NANOFIBERS

Nanofibrous materials can be made by using the electrospinning process. Electrospinning process involves three main components including syringe filled with a polymer solution, a high voltage supplier to provide the required electric force for stretching the liquid jet, and a grounded collection plate to hold the nanofiber mat. The charged polymer solution forms a liquid jet that is drawn towards a grounded collection plate. During the jet movement to the collector, the solvent evaporates and dry fibers deposited as randomly oriented structure on the surface of a collector [23–28]. The schematic illustration of conventional electrospinning setup is shown in Fig. 5.

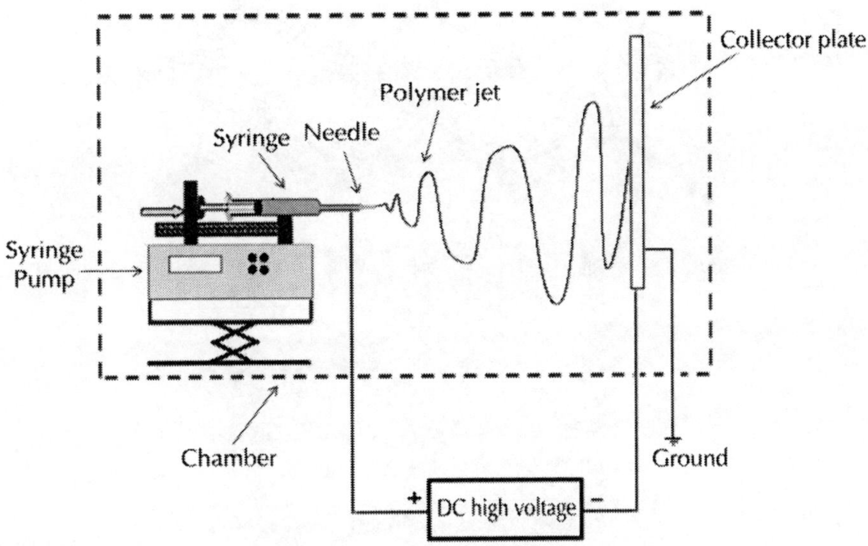

FIGURE 5 Schematic illustration of electrospinning set up.

At the present time, synthesis and fabrication of functional nanofibers represent one of the most interesting fields of nanoresearch. Combining the advanced structural features of metal-organic frameworks with the fabrication technique may generate new functionalized nanofibers for more multiple purposes.

While there has been great interest in the preparation of nanofibers, the studies on metal-organic polymers are rare. In the most recent investigation in this field, the growth of MOF (MIL-47) on electrospun polyacrylonitrile (PAN) mat was studied using in situ microwave irradiation [18]. MIL-47 consists of vanadium cations associated to six oxygen atoms, forming chains connected by terephthalate linkers (Fig. 6).

It should be mentioned that the conversion of nitrile to carboxylic acid groups is necessary for the MOF growth on the PAN nanofibers surface.

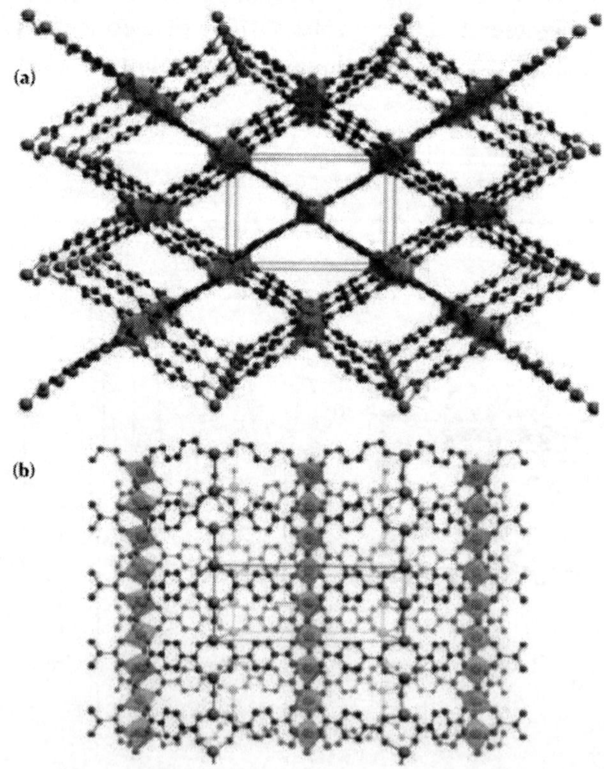

FIGURE 6 MIL-47 metal-organic framework structure: view along the b axis (a) and along the c axis (b).

The crystal morphology of MIL-47 grown on the electrospun fibers illustrated that after only 5 s, the polymer surface was partially covered with small agglomerates of MOF particles. With increasing irradiation time, the agglomerates grew as elongated anisotropic structures (Fig. 7) [18].

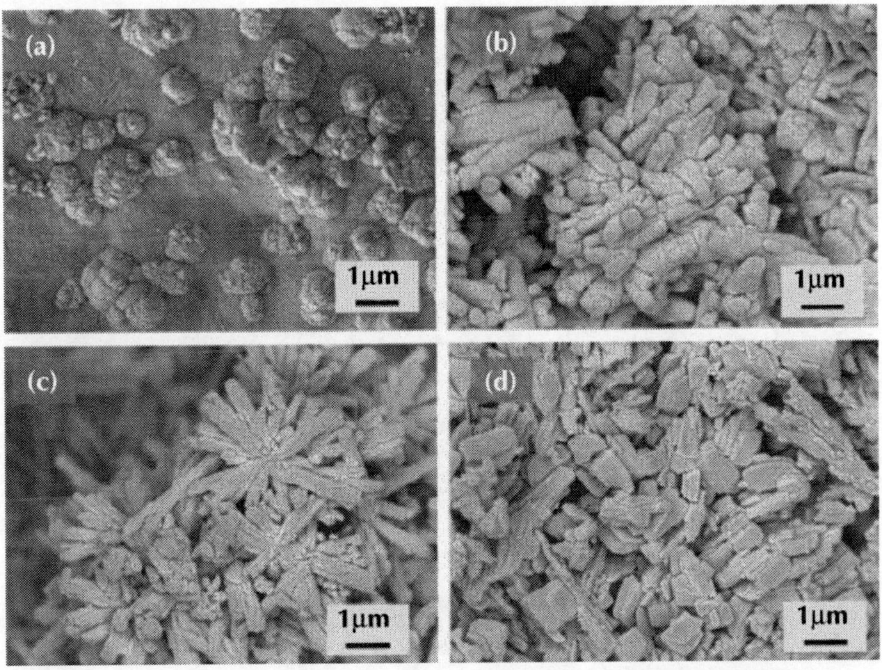

FIGURE 7 SEM micrograph of MIL-47 coated PAN substrate prepared from electrospun nanofibers as a function of irradiation time: (a) 5 s, (b) 30 s, (c) 3 min, and (d) 6 min.

It is known that the synthesis of desirable metal-organic polymers is one of the most important factors for the success of the fabrication of metal-organic nanofibers [29]. Among several novel microporous metal-organic polymers, only a few of them have been fabricated into metal-organic fibers.

For example, new acentric metal-organic framework was synthesized and fabricated into nanofibers using electrospinning process [29]. The two dimensional network structure of synthesized MOF is shown in Fig. 8. For this purpose, MOF was dissolved in water or DMF and saturated

MOF solution was used for electrospinning. They studied the diameter and morphology of the nanofibers using an optical microscope and a scanning electron microscope (Fig. 9). This fiber display diameters range from 60 nm to 4 μm.

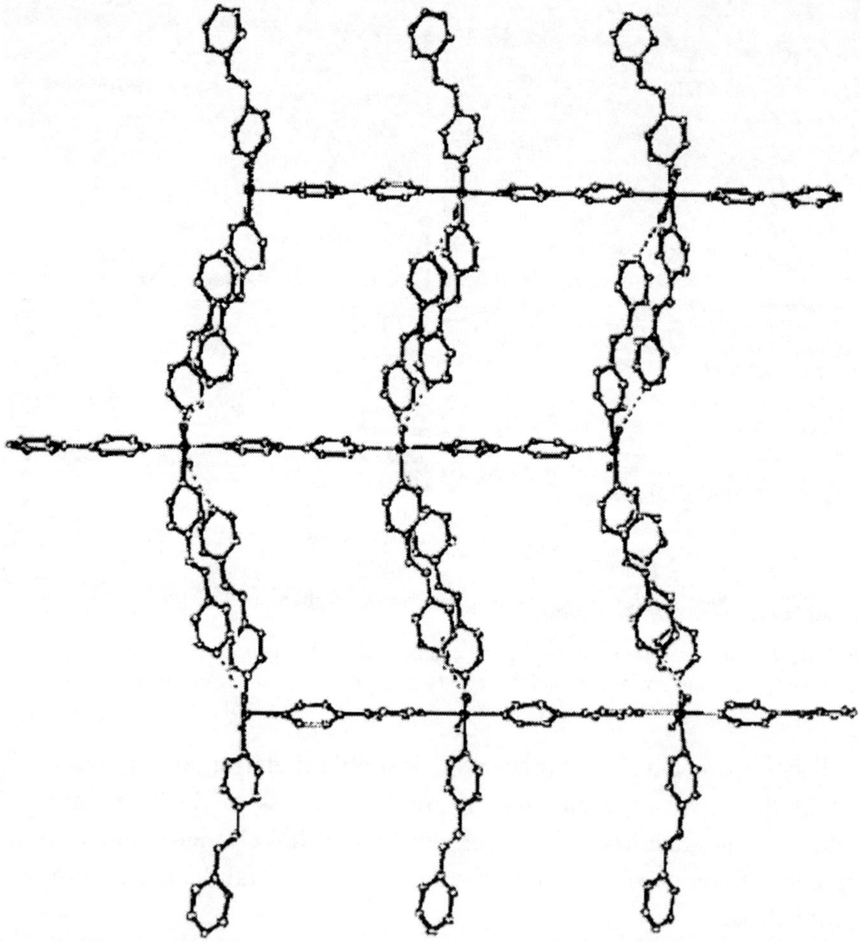

FIGURE 8 Representation of polymer chains and network structure of MOF.

FIGURE 9 SEM micrograph of electrospun nanofiber.

In 2011, Kaskel et al. [30], reported the use of electrospinning process for the immobilization of MOF particles in fibers. They used HKUST-1 and MIL-100(Fe) as MOF particles, which are stable during the electrospinning process from a suspension. Electrospun polymer fibers containing up to 80 wt% MOF particles were achieved and exhibit a total accessible inner surface area. It was found that HKUST-1/PAN gives a spider web-like network of the fibers with MOF particles like trapped flies in it, while HKUST-1/PS results in a pearl necklace-like alignment of the crystallites on the fibers with relatively low loadings.

4.4.3 ORDINARY TEXTILE FIBERS

Some examples of modification of fibers with metal-organic frameworks have verified successful. For instance, in the study on the growth of $Cu_3(BTC)_2$ (also known as HKUST-1, BTC=1,3,5-benzenetricarboxylate) MOF nanostructure on silk fiber under ultrasound irradiation, it was demonstrated that the silk fibers containing $Cu_3(BTC)_2$ MOF exhibited high antibacterial activity against the gram-negative bacterial strain *E. coli* and the gram-positive strain *S. aureus* [1]. The structure and SEM micrograph of $Cu_3(BTC)_2$ MOF is shown in Fig. 10.

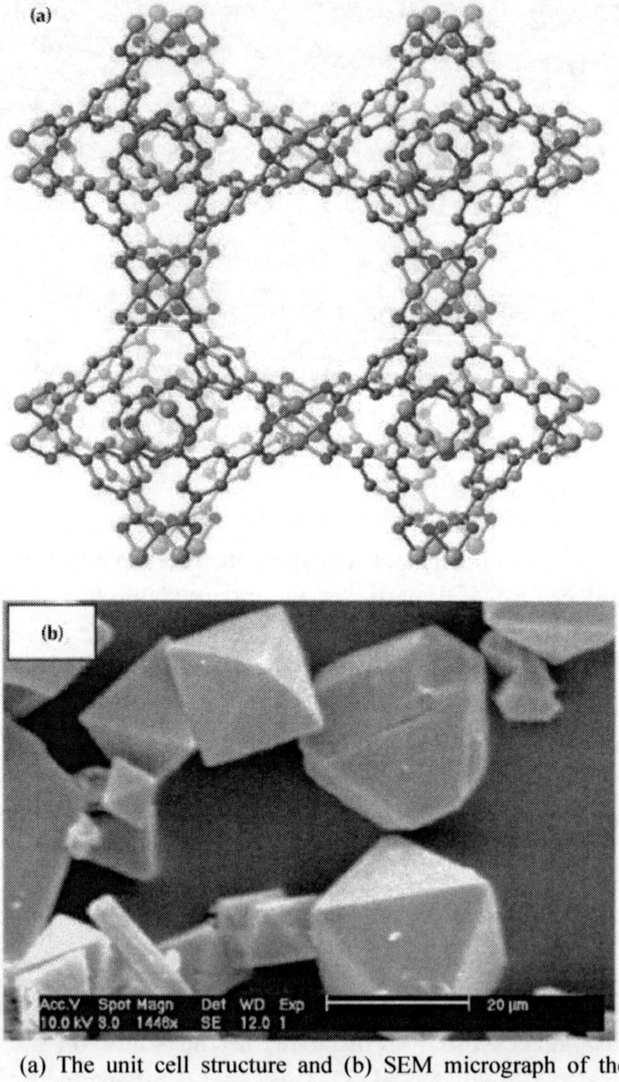

FIGURE 10 (a) The unit cell structure and (b) SEM micrograph of the $Cu_3(BTC)_2$ metal-organic framework. (Green, gray, and red spheres represent Cu, C, and O atoms, respectively).

$Cu_3(BTC)_2$ MOF has a large pore volume, between 62% and 72% of the total volume, and a cubic structure consists of three mutually perpendicular channels [32].

The formation mechanism of $Cu_3(BTC)_2$ nanoparticles upon silk fiber is illustrated in Fig. 11. It is found that formation of $Cu_3(BTC)_2$ MOF on silk fiber surface was increased in presence of ultrasound irradiation. In addition, increasing the concentration cause an increase in antimicrobial activity [1]. Figure 12 shows the SEM micrograph of $Cu_3(BTC)_2$ MOF on silk surface.

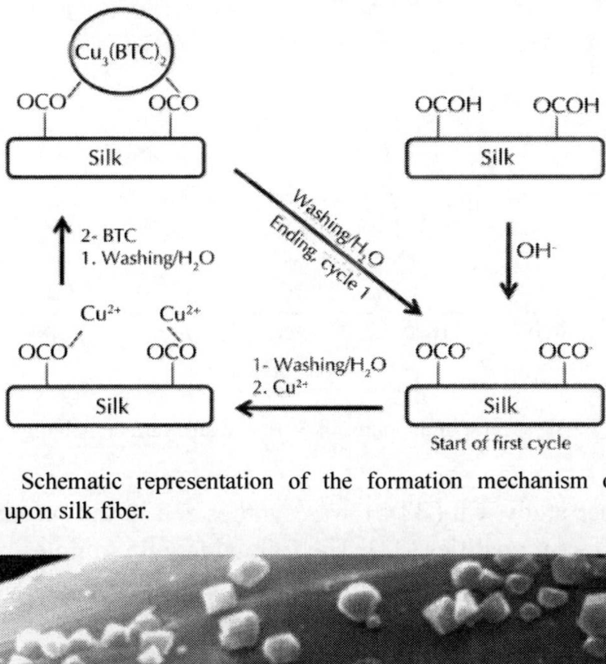

FIGURE 11 Schematic representation of the formation mechanism of $Cu_3(BTC)_2$ nanoparticles upon silk fiber.

FIGURE 12 SEM micrograph of $Cu_3(BTC)_2$ crystals on silk fibers.

The FT-IR spectra of the pure silk yarn and silk yarn containing MOF (CuBTC-Silk) are shown in Fig. 13. Owing to the reduction of the C=O bond, which is caused by the coordination of oxygen to the Cu^{2+} metal center (Fig. 11), the stretching frequency of the C=O bond was shifted to lower wavenumbers (1654 cm^{-1}) in comparison with the free silk (1664 cm^{-1}) after chelation [1].

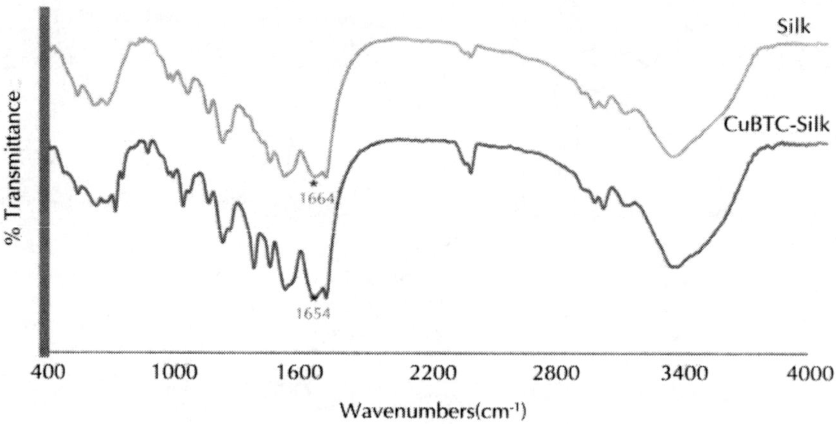

FIGURE 13 FT-IR spectra of the pure silk yarn and silk yarn containing $Cu_3(BTC)_2$.

In another study, $Cu_3(BTC)_2$ was synthesized in the presence of pulp fibers of different qualities [33]. The following pulp samples were used: a bleached and an unbleached kraft pulp, and chemithermomechanical pulp (CTMP).

All three samples differed in their residual lignin content. Indeed, owing to the different chemical composition of samples, different results regarding the degree of coverage were expected. The content of $Cu_3(BTC)_2$ in pulp samples, *k*-number, and single point BET surface area are shown in Table 1. *k*-number of pulp samples, which is indicates the lignin content indirectly, was determined by consumption of a sulfuric permanganate solution of the selected pulp sample [33].

TABLE 1 Some characteristics of the pulp samples.

Pulp sample	MOF content[a] (wt.%)	*k*-number[b]	Surface area[c] ($m^2 g^{-1}$)
CTMP	19.95	114.5	314
Unbleached kraft pulp	10.69	27.6	165
Bleached kraft pulp	0	0.3	10

[a]Determined by thermogravimetric analysis.
[b]Determined according to ISO 302.
[c]Single point BET surface area calculated at p/p_0=0.3 bar.

It is found that CTMP fibers showed the highest lignin residue and largest BET surface area. As shown in the SEM micrograph (Fig. 14), the crystals are regularly distributed on the fiber surface. The unbleached kraft pulp sample provides a slightly lower content of MOF crystals and BET surface area with 165 $m^2 g^{-1}$. Moreover, no crystals adhered to the bleached kraft pulp, which was almost free of any lignin.

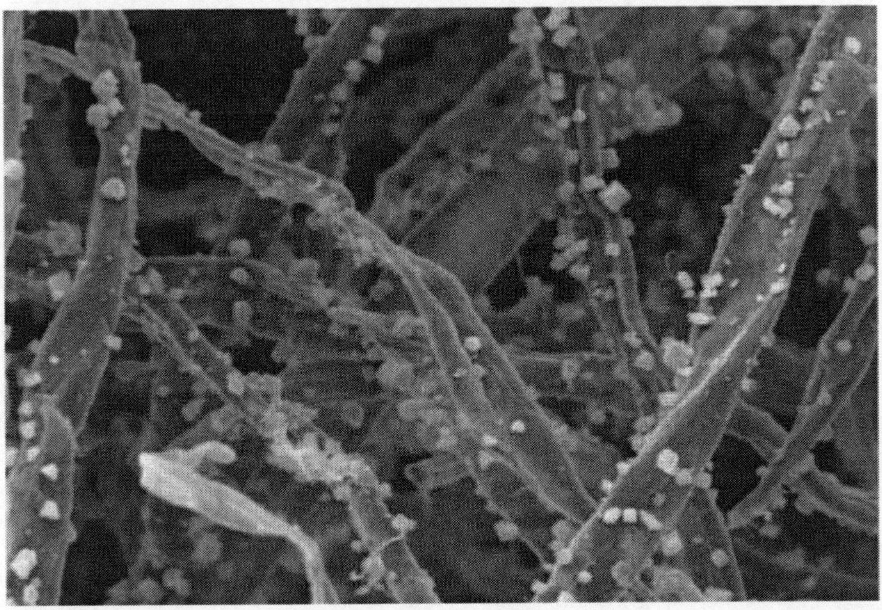

FIGURE 14 SEM micrograph of $Cu_3(BTC)_2$ crystals on the CTMP fibers.

4.5 CONCLUSION

New review on feasibility and application of several kinds of metal-organic frameworks on different substrate including nanofiber and ordinary fiber was investigated. Based on the researcher's results, the following conclusions can be drawn:

1. Metal-organic frameworks, as new class of nanoporous materials, can be used for modification of textile fibers.
2. These nanostructured materials have many exciting characteristics such as large pore sizes, high porosity, high surface areas, and wide range of pore sizes and topologies.
3. Although tremendous progress has been made in the potential applications of MOFs during past decade, only a few investigations have reported in textile engineering fields.
4. Morphological properties of the MOF/fiber composites were defined; the most advantageous, particle size distribution was shown.
5. It is concluded that the MOFs/fiber composite would be good candidates for many technological applications, such as gas separation, hydrogen storage, sensor, and others.

KEYWORDS

- **Ball-milling technique**
- ***E. coli***
- **Microwave irradiation**
- **Nanoporous materials**
- ***S. aureus***
- **Ultrasonic approach**

REFERENCES

1. Abbasi, A. R.; Akhbari, K.; Morsali, A. Dense coating of surface mounted CuBTC metal-organic framework nanostructures on silk fibers, prepared by layer-by-layer

method under ultrasound irradiation with antibacterial activity. *Ultrason. Sonochem.*, **2012**, *19*, 846–852.
2. Kuppler, R. J.; Timmons, D. J.; Fang, Q.-R.; Li, J.-R.; Makal, T. A.; Young, M. D.; Yuan, D.; Zhao, D.; Zhuang, W.; Zhou, H.-C. Potential applications of metal-organic frameworks. *Coordin. Chem. Rev.*, **2009**, *253*, 3042–3066.
3. Rowsell, J. L. C.; Yaghi, O. M. Metal-organic frameworks: A new class of porous materials. *Micropor. Mesopor. Mat.*, **2004**, *73*, 3–14.
4. An, J.; Farha, O. K.; Hupp, J. T.; Pohl, E.; Yeh, J. I.; Rosi, N. L. Metal-adeninate vertices for the construction of an exceptionally porous metal-organic framework. *Nature Commun.*, DOI: 10.1038/ncomms1618, 2012.
5. Morris, W.; Taylor, R. E.; Dybowski, C.; Yaghi, O. M.; Garcia-Garibay, M. A. Framework mobility in the metal-organic framework crystal IRMOF-3: Evidence for aromatic ring and amine rotation. *J. Mol. Struct.*, **2011**, *1004*, 94–101.
6. Kepert, C. J. Metal-organic framework materials. In *Porous Materials,* Bruce, D. W.; O'Hare, D.; Walton, R. I., Eds.; John Wiley & Sons: Chichester, 2011.
7. Rowsell, J. L. C.; Yaghi, O. M. Effects of functionalization, catenation, and variation of the metal oxide and organic linking units on the low-pressure hydrogen adsorption properties of metal-organic frameworks. *J. Am. Chem. Soc.*, **2006**, *128*, 1304–1315.
8. Rowsell, J. L. C.; Yaghi, O. M. Strategies for hydrogen storage in metal-organic frameworks. *Angew. Chemie Inl. Edit.*, **2005**, *44*, 4670–4679.
9. Farha, O. K.; Mulfort, K. L.; Thorsness, A. M.; Hupp, J. T. Separating solids: purification of metal-organic framework materials. *J. Am. Chem. Soc.*, **2008**, *130*, 8598–8599.
10. Khoshaman, A. H. In *Application of Electrospun Thin Films for Supra-Molecule Based Gas Sensing.* M.Sc. Thesis, Simon Fraser University, 2011.
11. Murray, L. J.; Dinca, M.; Long, J. R. Hydrogen storage in metal-organic frameworks. *Chem. Soc. Rev.*, **2009**, *38*, 1294–1314.
12. Collins, D. J.; Zhou, H.-C. Hydrogen storage in metal-organic frameworks. *J. Mat. Chem.*, **2007**, *17*, 3154–3160.
13. Chen, B.; Liang, C.; Yang, J.; Contreras, D. S.; Clancy, Y. L.; Lobkovsky, E. B.; Yaghi, O. M.; Dai S. A microporous metal-organic framework for gas-chromatographic separation of alkanes. *Angew. Chemie Inl. Edit.*, **2006**, *45*, 1390–1393.
14. Lee, J. Y.; Farha, O. K.; Roberts, J.; Scheidt, K. A.; Nguyen, S. T.; Hupp, J. T. Metal-organic framework materials as catalysts. *Chem. Soc. Rev.*, **2009**, *38*, 1450–1459.
15. Huxford, R. C.; Rocca, J. D.; Lin, W. Metal-organic frameworks as potential drug carriers. *Curr. Opin. Chem. Biol.*, **2010**, *14*, 262–268.
16. Suh, M. P.; Cheon, Y. E.; Lee, E. Y. Syntheses and functions of porous metallosupramolecular networks. *Coordin. Chem. Rev.*, **2008**, *252*, 1007–1026.
17. Keskin, S.; Kızılel, S. Biomedical applications of metal organic frameworks. *Ind. Eng. Chem. Res.*, **2011**, *50*, 1799–1812.
18. Centrone, A.; Yang, Y.; Speakman, S.; Bromberg, L.; Rutledge, G. C.; Hatton, T. A. Growth of metal-organic frameworks on polymer surfaces. *J. Am. Chem. Soc.*, **2010**, *132*, 15687–15691.
19. Wong-Foy, A. G.; Matzger, A. J.; Yaghi, O. M. Exceptional H_2 saturation uptake in microporous metal-organic frameworks. *J.Am. Chem. Soc.*, **2006**, *128*, 3494–3495.
20. Farrusseng, D. Metal-organic frameworks: Applications from Catalysis to Gas Storage. Wiley-VCH, Weinheim, 2011.

21. Rosi, N. L.; Eddaoudi, M.; Kim, J.; O'Keeffe, M.; Yaghi, O. M. Advances in the chemistry of metal-organic frameworks. *Cryst. Eng. Comm.*, **2002**, *4*, 401–404.
22. Zou, R.; Abdel-Fattah, A. I.; Xu, H., Zhao, Y.; Hickmott, D. D. Storage and separation applications of nanoporous metal-organic frameworks, *Cryst. Eng. Comm.*, **2010**, *12*, 1337–1353.
23. Reneker, D. H.; Chun, I. Nanometer diameter fibers of polymer, produced by electrospinning, *Nanotechnology*, **1996**, *7*, 216–223.
24. Shin, Y. M.; Hohman, M. M.; Brenner, M. P.; Rutledge, G. C. Experimental characterization of electrospinning: The electrically forced jet and instabilities. *Polymer,* **2001**, *42*, 9955–9967.
25. Reneker, D. H.; Yarin, A. L.; Fong, H.; Koombhongse S. Bending instability of electrically charged liquid jets of polymer solutions in electrospinning, *J. Appl. Phy.*, **2000**, *87*, 4531–4547.
26. Zhang, S.; Shim, W. S.; Kim, J. Design of ultra-fine nonwovens via electrospinning of Nylon 6: Spinning parameters and filtration efficiency, *Mater. Design,* **2009**, *30*, 3659–3666.
27. Yördem, O. S.; Papila, M.; Menceloğlu, Y. Z. Effects of electrospinning parameters on polyacrylonitrile nanofiber diameter: An investigation by response surface methodology. *Mater. Design,* **2008**, *29*, 34–44.
28. Chronakis, I. S. Novel nanocomposites and nanoceramics based on polymer nanofibers using electrospinning process—A review. *J. Mater. Proc. Tech.*, **2005**, *167*, 283–293.
29. Lu, J. Y.; Runnels, K. A.; Norman, C. A new metal-organic polymer with large grid acentric structure created by unbalanced inclusion species and its electrospun nanofibers. *Inorgan. Chem.*, **2001**, *40*, 4516–4517.
30. Rose, M.; Böhringer, B.; Jolly, M.; Fischer, R.; Kaskel, S. MOF processing by electrospinning for functional textiles. *Adv. Eng. Mater.*, **2011**, *13*, 356–360.
31. Basu, S.; Maes, M.; Cano-Odena, A.; Alaerts, L.; De Vos, D.E.; Vankelecom, I. F. J. Solvent resistant nanofiltration (SRNF) membranes based on metal-organic frameworks. *J. Membrane Sci.*, **2009**, *344*, 190–198.
32. Hopkins, J. B. In *Infrared Spectroscopy of H_2 Trapped in Metal Organic Frameworks.* B.A. Thesis, Oberlin College Honors, 2009.
33. Küsgens, P.; Siegle, S.; Kaskel, S. Crystal growth of the metal-organic framework $Cu_3(BTC)_2$ on the surface of pulp fibers. *Adv. Eng. Mater.*, **2009**, *11*, 93–95.

CHAPTER 5

QUANTUM AND WAVE CHARACTERISTICS OF SPATIAL-ENERGY INTERACTIONS

G. A. KORABLEV and G. E. ZAIKOV

CONTENTS

5.1 INTRODUCTION

Quantum conceptualizations on the composition of atoms and molecules make the foundation of modern natural science theories. Thus, the electronic angular momentum in stationary condition equals the integral multiple from Planck's constant. This main quantum number and three other combined explicitly characterize the state of any atom. The repetition factors of atomic quantum characteristics are also expressed in spectral data for simple and complex structures.

It is known that any periodic processes of complex shape can be shown as separate simple harmonic waves. "By Fourier theory, oscillations of any shape with period T can be shown as the total of harmonic oscillations with periods T_1, T_2, T_3, T_4, etc. Knowing the periodic function shape, we can calculate the amplitude and phases of sinusoids, with this function as their total" [1].

Therefore many regularities in intermolecular interactions, complex formation and nanothermodynamics are explained with the application of functional divisible quantum or wave energy characteristics of structural interactions.

In this research we tried to apply the conceptualizations on spatial-energy parameter (P-parameter) for this.

5.2 ON TWO PRINCIPLES OF ADDING ENERGY CHARACTERISTICS OF INTERACTIONS

The analysis of the kinetics of various physic-chemical processes demonstrates that in many cases the reciprocals of velocities, kinetic or energy characteristics of the corresponding interactions are added.

Here are some examples: ambipolar diffusion, total rate of topochemical reaction, and change in the light velocity when transiting from vacuum into the given medium, effective permeability of biomembranes.

In particular, such assumption is confirmed by the formula of electron transport probability (W_∞) due to the overlapping of wave functions 1 and 2 (in stationary state) during electron-conformation interactions:

$$W_\infty = \frac{1}{2}\frac{W_1 W_2}{W_1 + W_2} \tag{1}$$

Equation (1) is applied when evaluating the characteristics of diffusion processes accompanied with non-radiating electron transport in proteins [2].

Also: "It is known from the traditional mechanics that the relative motion of two particles with the interaction energy U(r) is the same as the motion of a material point with the reduced mass μ:

$$\frac{1}{\mu} = \frac{1}{m_1} + \frac{1}{m_2} \tag{2}$$

in the field of central force U(r), and total translational motion – as the free motion of the material point with the mass:

$$m = m_1 + m_2 \tag{3}$$

Such situation can be also found in quantum mechanics" [3].

The problem of two-particle interactions flowing by the bond line was solved in the time of Newton and Lagrange:

$$\mathring{A} = \frac{m_1 v_1^2}{2} + \frac{m_2 v_2^2}{2} + U\left(\overline{r}_2 - \overline{r}_1\right), \tag{4}$$

where *E* – system total energy, first and second components – kinetic energies of the particles, third – potential energy between particles 1 and 2, vectors $\overline{r}_2$ and $\overline{r}_1$ characterize the distance between the particles in final and initial states.

For moving thermodynamic systems the first law of thermodynamics can be shown as follows [4]:

$$\delta\mathring{A} = d\left(U + \frac{mv^2}{2}\right) \pm \delta A, \tag{5}$$

where $\delta\mathring{A}$ – amount of energy transferred to the system; component $d\left(U + \frac{mv^2}{2}\right)$ characterizes changes in internal and kinetic energies of the system; $+\delta \grave{A}$ – work performed by the system; – $\delta \grave{A}$ – work performed on the system.

Since the work numerically equals the change in the potential energy, then:

$$+\delta \grave{A} = -\Delta U \text{ и } -\delta A = +\Delta U \tag{6,7}$$

Probably not only the value of potential energy but also its changes are important in thermodynamic and also in many other processes in the dynamics of interactions of moving particles. Therefore, by the analogy with Eq. (4) the following should be fulfilled for two-particle interactions:

$$\delta \mathring{A} = d\left(\frac{m_1 v_1^2}{2} + \frac{m_2 v_2^2}{2}\right) \pm \Delta U \tag{8}$$

Here,

$$\Delta U = U_2 - U_1, \tag{9}$$

where U_2 and U_1 – potential energies of the system in final and initial states.

At the same time, the total energy (E) and kinetic energy $\left(\frac{mv^2}{2}\right)$ can be calculated from their zero value. In this case only the last component is modified in the Eq. (4).

The character of the changes in the potential energy value (ΔU) was analyzed by its index for different potential fields as given in Table 1.

From the table it is seen that the values of – ΔU and consequently $+\delta A$ (positive work) correspond to the interactions taking place by the potential gradient, and ΔU and $-\delta A$ (negative work) take place during the interactions against the potential gradient.

The solution of two-particle problem of the interaction of two material points with masses m_1 and m_2 obtained under the condition of no external forces available corresponds to the interactions taking place by the gradient, the positive work is performed by the system (similar to attraction process in the gravitation field).

The solution for this equation through the reduced mass (μ) [5] is Lagrangian equation for the relative motion of the isolated system of two interacting material points with masses m_1 and m_2, in coordinate x it looks as follows:

$$\mu \cdot x'' = -\frac{\partial U}{\partial x}; \frac{1}{\mu} = \frac{1}{m_1} + \frac{1}{m_2}.$$

Here, U – mutual potential energy of material points; μ – reduced mass. At the same time $x'' = a$ (characteristic of system acceleration). For elementary regions of interactions Δx can be taken as follows:

$\frac{\partial U}{\partial x} \approx \frac{\Delta U}{\Delta x}$ i.e., $\mu a \Delta x = -\Delta U$. Then,

$$\frac{1}{1/(a\Delta x)} \frac{1}{(1/m_1 + 1/m_2)_1} \approx -\Delta U; \frac{1}{1/(m_1 a\Delta x) + 1/(m_2 a\Delta x)} \approx -\Delta U$$

or

$$\frac{1}{\Delta U} \approx \frac{1}{\Delta U_1} + \frac{1}{\Delta U_2} \tag{10}$$

where ΔU_1 and ΔU_2 – potential energies of material points on the elementary region of interactions, ΔU – resulting (mutual) potential energy of these interactions.

Therefore,

1. In systems in which the interaction takes place by the potential gradient (positive work), the resultant potential energy is found by the principle of adding the reciprocals of the corresponding energies of subsystems [6]. The reduced mass for the relative motion of isolated system of two particles is calculated in the same way.
2. In systems in which the interaction takes place against the potential gradient (negative work), their masses and corresponding energies of subsystems (similar to Hamiltonian) are added algebraically.

5.3 INITIAL CRITERIA

From the Eq. (10) it is seen that the resultant energy characteristic of the system of interaction of two material points is found by the principle of adding the reciprocals of initial energies of interacting subsystems.

TABLE 1 Directedness of interaction processes.

No.	Systems	Potential field type	Process	U	r_2/r_1 (x_2/x_1)	U_2/U_1	Index ΔU	Index δA	Process directedness in the potential field
1	Opposite electric charges	Electrostatic	Attraction	$-k\frac{q_1q_2}{r}$	$r_2 < r_1$	$U_2 > U_1$	–	+	By gradient
			Repulsion	$-k\frac{q_1q_2}{r}$	$r_2 > r_1$	$U_2 < U_1$	+	–	Against gradient
2	Same electric charges	Electrostatic	Attraction	$k\frac{q_1q_2}{r}$	$r_2 < r_1$	$U_2 > U_1$	+	–	Against gradient
			Repulsion	$k\frac{q_1q_2}{r}$	$r_2 > r_1$	$U_2 < U_1$	–	+	By gradient
3	Elementary masses m_2 and m_2	Gravitational	Attraction	$-\gamma\frac{m_1m_2}{r}$	$r_2 < r_1$	$U_2 > U_1$	–	+	By gradient
			Repulsion	$-\gamma\frac{m_1m_2}{r}$	$r_2 > r_1$	$U_2 < U_1$	+	–	Against gradient

TABLE 1 *(Continued)*

No.	Systems	Potential field type	Process	U	r_2/r_1 (x_2/x_1)	U_2/U_1	Index ΔU	Index δA	Process directedness in the potential field
4	Spring deformation	Field of spring forces	Compression	$k\frac{\Delta x^2}{2}$	$U_2 > U_1$	$U_2 > U_1$	+	–	Against gradient
			Stretching	$k\frac{\Delta x^2}{2}$	$x_2 > x_1$	$U_2 > U_1$	+	–	Against gradient
5	Photo effect	Electrostatic	Repulsion	$k\frac{q_1 q_2}{r}$	$U_2 < U_1$	$U_2 < U_1$	–	+	By gradient

"Electron with the mass m moving near the proton with the mass M is equivalent to the particle with the mass $m_r = \frac{mM}{m+M}$" [7].

Therefore, modifying the Eq. (10), we can assume that the energy of atom valence orbitals (responsible for interatomic interactions) can be calculated [6] by the principle of adding the reciprocals of some initial energy components based on the equations:

$$\frac{1}{q^2/r_i} + \frac{1}{W_i n_i} = \frac{1}{P_E} \text{ or } \frac{1}{P_0} = \frac{1}{q^2} + \frac{1}{(Wrn)_i}; P_E = P_0/r_i \qquad (11\text{–}13)$$

where, W_i – orbital energy of electrons [8]; r_i – orbital radius of i orbital [9]; $q=Z^*/n^*$ – by [10,11], n_i – number of electrons of the given orbital, $Z^*$and n^* – nucleus effective charge and effective main quantum number, r – bond dimensional characteristics.

P_O is called a spatial-energy parameter (SEP), and P_E – effective P-parameter (effective SEP). Effective SEP has a physical sense of some averaged energy of valence orbitals in the atom and is measured in energy units, e.g., in electron-volts (eV).

The values of P_0-parameter are tabulated constants for electrons of the given atom orbital.

For SEP dimensionality:

$$[P_0] = [q^2] = [E] \times [r] = [h] \times [\upsilon] = \frac{kgm^3}{s^2} = Jm$$

where $[E]$, $[h]$ and $[\upsilon]$ – dimensionalities of energy, Plank's constant and velocity.

The introduction of P-parameter should be considered as further development of quasi-classical concepts with quantum-mechanical data on atom structure to obtain the criteria of phase-formation energy conditions. For the systems of similarly charged (e.g., orbitals in the given atom) homogeneous systems the principle of algebraic addition of such parameters is preserved:

$$\sum P_E = \sum (P_0/r_i); \sum P_E = \frac{\sum P_0}{r} \qquad (14), (15)$$

or

$$\sum P_0 = P_0' + P_0'' + P_0''' + ...; r\sum P_E = \sum P_0 \qquad (16), (17)$$

Here *P*-parameters are summed up by all atom valence orbitals.

To calculate the values of P_E-parameter at the given distance from the nucleus either the atomic radius (*R*) or ionic radius (r_I) can be used instead of *r* depending on the bond type.

Let us briefly explain the reliability of such an approach. As the calculations demonstrated the values of P_E-parameters equal numerically (in the range of 2%) the total energy of valence electrons (*U*) by the atom statistic model. Using the known correlation between the electron density (β) and intra-atomic potential by the atom statistic model [12], we can obtain the direct dependence of P_E-parameter on the electron density at the distance r_i from the nucleus.

The rationality of such technique was proved by the calculation of electron density using wave functions by Clementi [13] and comparing it with the value of electron density calculated through the value of P_E-parameter.

5.4 WAVE EQUATION OF *P*-PARAMETER

To characterize atom spatial-energy properties two types of *P*-parameters are introduced. The bond between them is a simple one:

$$P_E = \frac{P_0}{R}$$

where *R* – atom dimensional characteristic. Taking into account additional quantum characteristics of sublevels in the atom, this equation can be written down in coordinate *x* as follows:

$$\Delta P_E \approx \frac{\Delta P_0}{\Delta \tilde{o}} \quad \text{or} \quad \partial P_E = \frac{\partial P_0}{\partial \tilde{o}}$$

where the value *ΔP* equals the difference between P_0-parameter of *i* orbital and P_{CD}–countdown parameter (parameter of main state at the given set of quantum numbers).

According to the established [6] rule of adding P-parameters of similarly charged or homogeneous systems for two orbitals in the given atom with different quantum characteristics and according to the energy conservation rule we have:

$$\Delta P_E'' - \Delta P_E' = P_{E,\lambda}$$

where $P_{E,\lambda}$ – spatial-energy parameter of quantum transition.

Taking for the dimensional characteristic of the interaction $\Delta\lambda = \Delta x$, we have:

$$\frac{\Delta P_0''}{\Delta\lambda} - \frac{\Delta P_0'}{\Delta\lambda} = \frac{P_0}{\Delta\lambda}$$

or

$$\frac{\Delta P_0'}{\Delta\lambda} - \frac{\Delta P_0''}{\Delta\lambda} = -\frac{P_0\lambda}{\Delta\lambda}$$

Let us again divide by $\Delta\lambda$ term by term: $\left(\frac{\Delta P_0'}{\Delta\lambda} - \frac{\Delta P_0''}{\Delta\lambda}\right) \Big/ \Delta\lambda = -\frac{P_0}{\Delta\lambda^2}$,

where:

$$\left(\frac{\Delta P_0'}{\Delta\lambda} - \frac{\Delta P_0''}{\Delta\lambda}\right) \Big/ \Delta\lambda \sim \frac{d^2 P_0}{d\lambda^2}, \text{ i.e.: } \frac{d^2 P_0}{d\lambda^2} + \frac{P_0}{\Delta\lambda^2} \approx 0$$

Taking into account only those interactions when $2\pi\Delta x = \Delta\lambda$ (closed oscillator), we have the following equation: $\frac{d^2 P_0}{dx^2} + 4\pi^2 \frac{P_0}{\Delta\lambda^2} \approx 0$

Since $\Delta\lambda = \frac{h}{mv}$, then: $\frac{d^2 P_0}{dx^2} + 4\pi^2 \frac{P_0}{h^2} m^2 v^2 \approx 0$

or

$$\frac{d^2 P_0}{dx^2} + \frac{8\pi^2 m}{h^2} P_0 E_k = 0 \qquad (18)$$

where $E_k = \frac{mV^2}{2}$ – electron kinetic energy.

Schrodinger equation for the stationery state in coordinate x:

$$\frac{d^2\psi}{dx^2} + \frac{8\pi^2 m}{h^2}\psi E_k = 0$$

When comparing these two equations we see that P_0-parameter numerically correlates with the value of Ψ-function: $P_0 \approx \Psi$, and is generally proportional to it: $P_0 \sim \Psi$. Taking into account the broad practical opportunities of applying the P-parameter methodology, we can consider this criterion as the materialized analog of Ψ-function [14, 15].

Since P_0-parameters like Ψ-function have wave properties, the superposition principles should be fulfilled for them, defining the linear character of the equations of adding and changing P-parameter.

5.5 QUANTUM PROPERTIES OF P-PARAMETER

According to Planck, the oscillator energy (E) can have only discrete values equaled to the whole number of energy elementary portions-quants:

$$nE = h\nu = hc/\lambda \tag{19}$$

where h – Planck's constant, ν – electromagnetic wave frequency, c – its velocity, λ – wavelength, $n = 0, 1, 2, 3\ldots$

Planck's equation also produces a strictly definite bond between the two ways of describing the nature phenomena – corpuscular and wave.

P_0-parameter as an initial energy characteristic of structural interactions, similarly to the Eq. (19), can have a simple dependence from the frequency of quantum transitions:

$$P_0 \sim \hbar(\lambda\nu_0) \tag{20}$$

where λ – quantum transition wavelength [16]; $\hbar = h/(2\pi)$; ν_0 – kayser, the unit of wave number equaled to $2.9979 \cdot 10^{10}$ Hz.

In accordance with Rydberg equation, the product of the right part of this equation by the value $(1/n^2 - 1/m^2)$, where n and m – main quantum numbers – should result in the constant.

Therefore the following equation should be fulfilled:

$$P_0(1/n^2_1 - 1/m^2_1) = N\hbar(\lambda v_0)(1/n^2 - 1/m^2) \quad (21)$$

where the constant N has a physical sense of wave number and for hydrogen atom equals $2 \cdot 10^2 Å^{-1}$.

The corresponding calculations are demonstrated in Table 2. There: r_i' = 0.5292 Å – orbital radius of 1S-orbital and r_i'' = =$2^2 \cdot 0.5292$ = 2.118 Å – the value approximately equaled to the orbital radius of 2S-orbital.

The value of P_0-parameter is obtained from the Eq. (12), e.g., for 1S-2P transition:

$$1/P_0 = 1/(13.595 \cdot 0.5292) + 1/14.394 \rightarrow P_0 = 4.7985 \text{ eVÅ}$$

The value q^2 is taken from Refs. [10, 11], for the electron in hydrogen atom it numerically equals the product of rest energy by the classical radius.

The accuracy of the correlations obtained is in the range of percentage error 0.06 (%), i.e., the Eq. (21) is in the accuracy range of the initial data.

In the Eq. (21) there is the link between the quantum characteristics of structural interactions of particles and frequencies of the corresponding electromagnetic waves.

But in this case there is the dependence between the spatial parameters distributed along the coordinate. Thus in P_0-parameter the effective energy is multiplied by the dimensional characteristic of interactions, and in the right part of the Eq. (21) the kayser value is multiplied by the wavelength of quantum transition.

In Table 2 you can see the possibility of applying the Eq. (21) and for electron Compton wavelength ($\lambda_к = 2.4261 \cdot 10^{-12}$ m), which in this case is as follows:

$$P_0 = 10^7\hbar(\lambda_к v_0) \quad (22)$$

(with the relative error of 0.25%).

Integral-valued decimal values are found when analyzing the correlations in the system "proton-electron" given in Table 3:

1. Proton in the nucleus, energies of three quarks 5 + 5 + 7 ≈ 17 (MeV) → $P_p \approx 17$ MeV·0.856·10^{-15} m ≈ 14.552·10^{-9} eVm. Similarly for the electron P_e = 0.511 (MeV)·2.8179·10^{-15} m (electron classic radius) → P_e = 1.440·10^{-9} eVm.

Therefore:
$$P_p \approx 10\, P_E \tag{23}$$

2. Free proton P_n = 938.3 (MeV)·0.856·10^{-15} (m) = 8.0318·10^{-7} eVm. For electron in the atom P_a = 0.511 (MeV)·0.5292·10^{-5}(m) = 2.7057·10^{-5} eVm.

Then,
$$3P_a \approx 10^2 P_n \tag{24}$$

The relative error of the calculations by these equations is found in the range of the accuracy of initial data for the proton (δ ≈ 1%).

From Tables 2 and 3 we can see that the wave number N is quantized by the decimal principle:

$$N = n10^Z,$$

where n and Z – whole numbers.

Other examples of electrodynamics equations should be pointed out in which there are integral-valued decimal functions, e.g., in the formula:

$$4\pi\varepsilon_0 c^2 = 10^7,$$

where ε_0 – electric constant.

In Ref. [17] the expression of the dependence of constants of electromagnetic interactions from the values of electron получено P_e-parameter was obtained:

$$k\mu_0 c = k/(\varepsilon_0 c) = P_e^{1/2} c^2 \approx 10/\alpha \tag{25}$$

Here, $k = 2\pi/\sqrt{3}$; μ_0 – magnetic constant; c – electromagnetic constant; α – fine structure constant.

All the above conclusions are based on the application of rather accurate formulas in the accuracy range of initial data.

TABLE 2 Quantum properties of hydrogen atom parameters.

Orbitals	W_i (eV)	r_i (Å)	q_i^2 (eVÅ)	P_0 (eVÅ)	$P_0(1/n_1^2 - 1/m_1^2)$ (eVÅ)	N (Å^{-1})	λ (Å)	Quantum transition	$N\hbar\lambda\nu_0$ (eVÅ)	$N\hbar\lambda\nu_0 \cdot (1/n^2 - 1/m^2)$ (eVÅ)
1S	13.595	0.5292	14.394	4.7985	3.5989	$2 \cdot 10^2$	1215	1S-2P	4.7951	3.5963
1S						$2 \cdot 10^2$	1025	1S-3P	4.0452	3.5954
1S						$2 \cdot 10^2$	912	1S-nP	3.5990	3.5990
2S	3.3988	2.118	14.394	4.7985	3.5990	$2 \cdot 10^2$	6562	2S-3P		3.5967
2S						$2 \cdot 10^2$	4861	2S-4P		3.5971
2S						$2 \cdot 10^2$	3646	2S-nP		3.5973
1S	13.595	0.5292	14.394	4.7985		10^7	$2.4263 \cdot 10^{-2}$	–	4.7878	

TABLE 3 Quantum ratios of proton and electron parameters.

Particle	E (eV)	r (Å)	$P = Er$ (eVÅ)	Ratio
Free proton	$938.3 \cdot 10^6$	$0.856 \cdot 10^{-5}$	$8.038 \cdot 10^3 = P_n$	$3P_a/P_n \approx 10^2$
Electron in an atom	$0.511 \cdot 10^6$	0.5292	$2.7042 \cdot 10^5 = P_a$	
Proton in atom nuclei	$(5 + 5 + 7) \cdot 10^6 = 17 \cdot 10^6$	$0.856 \cdot 10^{-5}$	$145.52 = P_p$	$P_p/P_e \approx 10$
Electron	$0.511 \cdot 10^6$	$2.8179 \cdot 10^{-5}$	$14.399 = P_e$	

5.6 CONCLUSIONS

1. Two principles of adding interaction energy characteristics are functionally defined by the direction of interaction by potential gradient (positive work) or against potential gradient (negative work).
2. Equation of the dependence of spatial-energy parameter on spectral and frequency characteristics in hydrogen atom has been obtained.

KEYWORDS

- **Against potential gradient**
- **Lagrangian equation**
- ***P*-parameter**
- **Planck's constant**
- **potential gradient**
- **Spatial-energy**

REFERENCES

1. Gribov, L. A.; Prokofyeva, N. I. In *Basics of Physics*. M.: Vysshaya shkola, 1992; 430 p.

2. Rubin, A. B. Biophysics. Book 1. In *Theoretical Biophysics*. M.: Vysshaya shkola, 1987; 319 p.
3. Blokhintsev, D. I. In *Basics of Quantum Mechanics*. M.: Vysshaya shkola, 1961; 512 p.
4. Yavorsky, B. M.; Detlaf, A. A. In *Reference-Book in Physics*. M.: Nauka, 1968; 939 p.
5. Christy R. W.; Pytte A. In *The Structure of Matter: An Introduction to Modern Physics*. Translated from English. M.: Nauka, 1969; 596 p.
6. Korablev, G. A. In *Spatial-Energy Principles of Complex Structures Formation*, Brill Academic Publishers and VSP: Netherlands, 2005; 426 p. (Monograph).
7. Eyring, G.; Walter, J.; Kimball, G. In *Quantum Chemistry*. M.: F.L., 1948; 528 p.
8. Fischer C. F. *Atomic Data*, **1972**, *4*, 301–399.
9. Waber, J. T.; Cromer, D. T. *J. Chem. Phys.*, **1965**, *42(12)*, 4116–4123.
10. Clementi, E.; Raimondi D. L. Atomic Screening constants from S.C.F. Functions, 1. *J. Chem. Phys.*, **1963**, *38(11)*, 2686–2689.
11. Clementi, E.; Raimondi, D. L. *J. Chem. Phys.*, **1967**, *47(4)*, 1300–1307.
12. Gombash, P. In *Statistic Theory of An Atom and Its Applications*. M.: I.L., 1951; 398 p.
13. Clementi, E. *J.B.M. S. Res. Develop. (Suppl.)*, **1965**, *9(2)*, 76.
14. Korablev, G. A.; Zaikov, G. E. *J. Appl. Poly. Sci., USA*, **2006**, *101(3)*, 2101–2107.
15. Korablev, G. A.; Zaikov, G. E. In *Progress on Chemistry and Biochemistry*, Nova Science Publishers, Inc.: New York, 2009; 355–376.
16. Allen, K. W. In *Astrophysical Values*. M.: Mir, 1977; 446 p.
17. Korablev, G. A. In *Exchange Spatial-Energy Interactions*. Izhevsk Publishing House "Udmurt University," 2010; 530 p. (Monograph).

CHAPTER 6

FRACTAL ANALYSIS IN POLYCONDENSATION PROCESS

G. V. KOZLOV, G. B. SHUSTOV, G. E. ZAIKOV, E. M. PEARCE, and G. KIRSHENBAUM

CONTENTS

6.1 INTRODUCTION

It is well known [1, 2], that depending on reactionary medium and synthesis carrying out conditions a polycondensation processes can be ceased at different temporal scales of the experiment that in the long run results to a considerable variation of the main parameters characterizing this process: conversion degree Q and molecular weight MW. Let's note, that the indicated reaction cessation is not defined by the exhaustion of the available reactive groups. At present there are quite enough partial explanations of this effect, applicable to individual combinations of reagents and solvents [1, 2] that does not allows developing general conception of this important effect both theoretically and practically. Therefore for the description of polycondensation process cessation we used much more general notions, namely, irreversible aggregation models. The application of the mentioned models is due to the fact, that any polymerization process in the most general treatment presents unification of small molecules (particles) in much larger macromolecule (cluster).

Irreversible aggregation models of different universality classes (particle-cluster, cluster–cluster) were developed for the description frequently occurring in nature aggregation processes like flocculation, coagulation, polymerization and so on [3, 4]. In the last years the applicability of these models for the description of all kind of polymerization processes within the framework of the very general physical notations [5–7]. The main feature of the indicated models is, that processes of aggregation described by them generate fractal objects as final aggregates [3, 4]. As it is known [8], macromolecular coils in solution are fractals, which fractal dimension D varies within the range ~ 1.5–2.2. This circumstance allows using for the description of polymerization processes in solutions irreversible diffusion-limited aggregation models cluster–cluster [3]. At present several types of such models were developed, allowing to take into consideration a number of specific features of polymerization processes, namely, simultaneous occurrence of aggregation and disaggregation [9], different probability of cluster joining [10] and so on. The purpose of the present paper is irreversible aggregation model used for the description of some features of different polymers polycondensation: aromatic copolyethersulfoneformals (APESF), diblockcopolyether of oligoformal 2,2-di-(4-oxiphenyl)-propane and oligosulfone of phenolphthalein (CP-OF-10/OS-10) and also polyarylate on the base of dichloroanhydride of terephthalic acid and phenolphthalein (F-2).

6.2 EXPERIMENTAL

APESF have the following chemical structure:

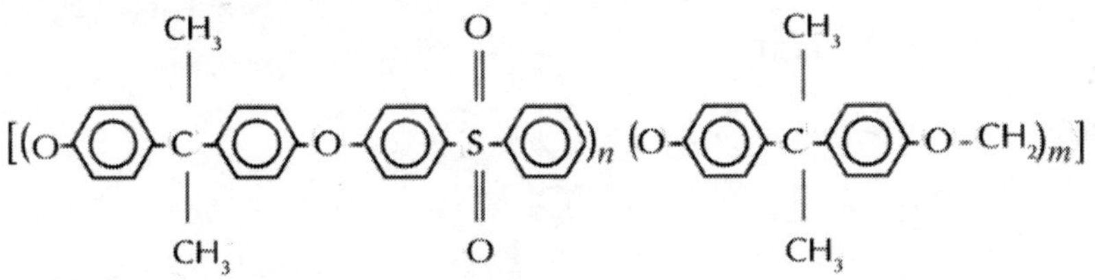

The APESF synthesis was carried out according to the following technique. In a three-neck retort of 750 ml capacity supplied with a stirrer, Din-Stark trap and cleaned by nitrogen, 40 ml of dimethyl sulfoxide and 7.8953 g (0.09285 mole) of sodium hydroxide in a state of concentrated aqueous solution, 10.5984 g (0.04642 mole) of 2,2-di-(4-oxiphenyl)-propane and 40 ml of toluene are placed. At stirring temperature was gradually raised and azeotropic mixture of water and toluene was distilled off. After cessation of water separation temperature was raised to 443 K and toluene was distilled off. To the solution of bis-phenolate cooled to 413 K 9.3322 g (0.032497 mole) of 4,4'-dichlorodiphenysulfone were added. The temperature was raised to 433 K and at this temperature the mixture was sustained for an hour. Then the mixture was cooled up to 303 K, Din-Stark trap was replaced by the opposite refrigerator and to 0.9 ml (0.01393 mole) of methylene chloride were added to the reactionary mixture. The temperature was gradually being raised to 423 K and the mixture was being sustained at this temperature for 2 hr. The obtained viscous solution was cooled off, diluted with dimethylsulfonoxide and poured into distilled water stirred intensively and oxidized by oxalic acid. After separation, purification and drying 16.3 g (92%) copolymer, containing 30 mol.% of methylene chloride residuals, were received. The reduced viscosity was equal to 0.78 dl/g (0.5%-th solution in chloroform at 298 K). The other APESF were synthesized by the similar way.

Diblockcopolymers CP-OF-10/OS-10 have the following chemical structure:

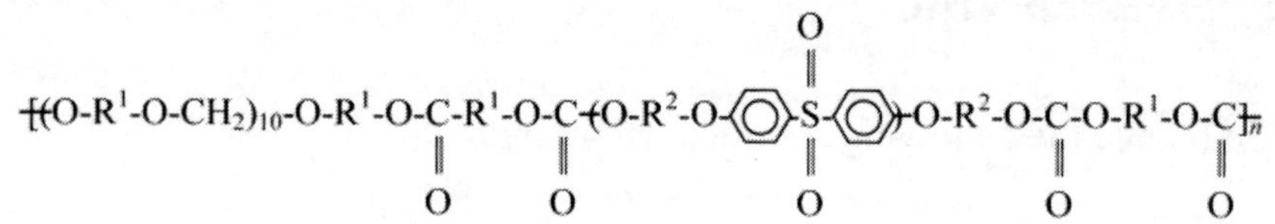

where

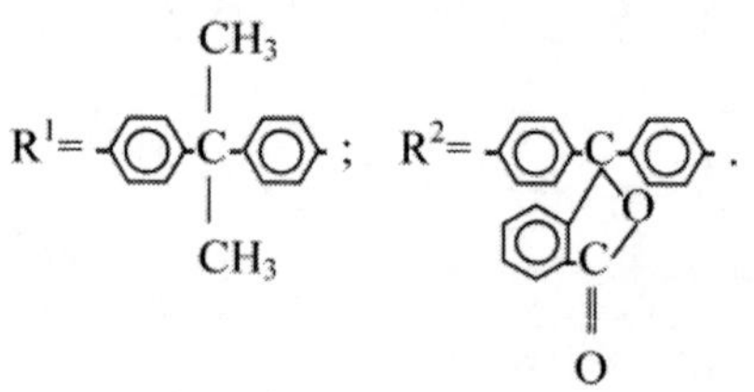

The synthesis of diblockcopolymers was made according to the following technique. In a three-neck retort of 250 ml capacity supplied with mechanical stirrer 1.365 g (0.000525 mole) of OF-10, 2.94 g (0.000525 mole) of OS-10 and 30 ml of methylene chloride are placed. After solution of the oligomers 0.3 ml (0.0021 mole) of triethylamine, then 0.3709 g (0.00105 mole) of bischloroformiate 2,2-di-(4 oxyphenyl)-propane were added. The reactionary mixture was being stirred during 2 h at room temperature, after that the viscous solution was diluted by methylene chloride and it was poured into intensively stirred iso-propanol spirit. After separation, purification and drying 4.215 g (92%) of diblockcopolymer containing 50 mol.% of OF-10 with reduced viscosity 0.52 dl/g (0.5%-th solution in chloroform at 298 K) were received. The other diblockcopolymers were synthesized by the similar way.

The glass transition temperatures T_g of APESF and CP-OF-10/OS-10 were determined by thermomechanical technique. They are within the range 353–493 K. The T_g value for F-2 was accepted equal to 503 K [11]. The reduced viscosity of copolymers in simm-tetrachloroethane (good solvent) and 1,4-dioxane (θ-solvent) was measured with the aid of Ubbelode viscosimeter at temperature 298 K. For the intrinsic viscosity [η] calculation the following relationship was used [12]:

$$[\eta] = \frac{\eta_{red} / c_p}{1 + K_{\eta}\eta_{red}}, \tag{1}$$

where η_{red} is reduced viscosity, c_p is polymer concentration in solution, K_η is coefficient, equal to 0.28.

The values of reduced viscosity η_{red} in various solvents for F-2 were accepted according to the data of paper [1]. The initial concentration c_0 is determined as general mass of loading reagents, reduced to solvent volume of 30 ml.

6.3 RESULTS AND DISCUSSION

6.3.1 *METHODICS OF THE MACROMOLECULAR COIL FRACTAL DIMENSION DETERMINATION*

As it is known [13], linear polymeric macromolecules can be found in solution depending on *MM*, solvent quality, temperature, concentration and other factors in various conformational and/or phase states. The most trivial of them are statistically twisted coil in ideal (θ) solvent, nontransparent coil in good solvent and transparent coil. In all listed states the macromolecular coil in solution is fractal, i.e., self-similar object with dimension *D*, which does not coincide with its topological dimension d_t. The dimension *D* is named the fractal (Hausdorf) one and it characterizes the volume distribution of coil elements in space [3]. Value *D* determination is the first stage of macromolecular coils study within the framework of fractal analysis and similar estimations are usually made by the exponents measurement in the equations type of Mark-Kuhn-Houwink, connecting [η], translational diffusion coefficient D_0 or high-speed sedimentation coefficient S_0 with *MW* of polymers [14]:

$$[\eta] \sim MW^{a_\eta}, \tag{2}$$

$$D_0 \sim MW^{-a_D}, \tag{3}$$

$$S_0 \sim MW^{a_S}. \tag{4}$$

Then D value can be calculated according to the relationship [14]:

$$D = \frac{3}{a_\eta + 1}, \tag{5}$$

$$D = \frac{1}{a_D}, \tag{6}$$

$$D = \frac{1}{1 - a_S}. \tag{7}$$

All considered methods require complicated enough and laborious measurements [14–16]. The simplest and not requiring complicated devices from the listed methods is [η] measurement, which can be practically fulfilled in any laboratory. Therefore, we will consider here a simple method of the value D estimation based on the same principles that were used at the Eqs. (5)–(7) derivation.

As it is known [17], the coefficient of swelling of a macromolecular coil α is determined as follows:

$$\alpha = \left(\frac{\langle \bar{h}^2 \rangle}{\langle \bar{h}_\theta^2 \rangle} \right)^{1/2}, \tag{8}$$

where $\langle \bar{h}^2 \rangle$ and $\langle \bar{h}_\theta^2 \rangle$ are the root-mean-square distances between the ends of the macromolecule in an arbitrary and ideal (θ) solvents, respectively.

In its turn, the value α is related with intrinsic viscosities [η] and $[\eta]_\theta$ of the polymer in arbitrary and ideal solvents, respectively, according to the Eq. [17]:

$$\alpha^3 = \frac{[\eta]}{[\eta]_\theta}. \tag{9}$$

The parameter ε of the bulk interactions (which cause a deviation of the coil shape from the ideal Gaussian one) is determined according to the relationship [17]:

$$\varepsilon = \frac{d \ln \alpha^2}{d \ln MM} = \frac{\alpha^2 - 1}{5\alpha^2 - 3}. \tag{10}$$

In its turn, both ε, and coil fractal dimension D depend on the exponent a_η value (further simply a) in Mark-Kuhn-Houwink Eq. (2) [16]:

$$\varepsilon = \frac{2a - 1}{3}, \tag{11}$$

and the dependence D on a is given by the Eq. (5).

The combination of the Eqs. (5) and (8)–(11) allows to obtain the following relationship for D determination only from intrinsic viscosities values $[\eta]$ and $[\eta]_\theta$:

$$D = \frac{5\left([\eta]/[\eta]_\theta\right)^{2/3} - 3}{3\left([\eta]/[\eta]_\theta\right)^{2/3} - 2} \tag{12}$$

The value $[\eta]_\theta$ can be estimated either directly from the experiment, as it is made in the present paper for APESF and CP-OF-10/OS-10, or from the relationship Eq. (2) under the condition a=0.5, which is true for the θ-point, if the constant in this equation is known, as it is made for F-2.

6.3.2 MOLECULAR WEIGHT ESTIMATION

MW estimation for all studied polymers was carried out with the aid of Mark-Kuhn-Houwink equation. For F-2 solution in simm-tetrachloroethane this equation has the form [18]:

$$[\eta] = 0.266 \times 10^{-4} MW^{0.938} \tag{13}$$

For APESF and CP-OF-10/OS-10 the coefficient K in Mark-Kuhn-Houwink equation can be calculated as follows [18]:

$$K = \frac{21}{m_e}\left(\frac{1}{2500 m_e}\right)^a, \tag{14}$$

where m_e is an average weight of polymer elementary link (without side substituents).

The value a was determined according to the Eq. (5). The values m_e for APESF homopolymers were calculated from their molecular weight MW_e with the aid of the relationship [19]:

$$m_e = \frac{MW_e}{N_A}, \tag{15}$$

where N_A is Avogadro number.

For CP-OF-10/OS-10 in a homopolymers case molecular weight of one of 10 fragments of oligomer in brackets was calculated and the tenth of molecular weights sum of oligomer link and chain lengthener remaing part were added to it. From total MW_e obtained in this way the value m_e was calculated according to the Eq. (15). The values m_e for copolymers were calculated in supposition of their additive change with composition.

The combination of the Eqs. (2), (5) and (14) allows to obtain Mark-Kuhn-Houwink equation fractal variant:

$$[\eta] = \frac{21}{m_e}\left[MW\left(\frac{1}{2500m_e}\right)\right]^{(3-D)/D} \tag{16}$$

6.3.3 THEORETICAL PRINCIPLES OF IRREVERSIBLE AGGREGATION MODELS APPLICATION

Let's consider irreversible aggregation models application for polycondensation process cessation criterion obtaining. The dependence of aggregation processes on aggregating particles concentration c (with initial concentration c_0) in some closed space has the direct relation to the considered problem. If during the growth of cluster (macromolecular coil) its density is reduced up to the density of the environment (solution with concentration c at the given moment), then the transition between universality classes diffusion-limited aggregation (DLA) a particle-cluster and Eden happens and further macromolecular coil ceases its growth, achieving a critical gyration radius R_c [3]. Let's note, that earlier one more transition occurs within the framework of DLA between universality classes cluster–cluster and particle-cluster. This strongly decreases the reaction rate, but, unlike the transition indicated earlier, does not cease it [20]. The scale of transition from a particle-cluster aggregation to Eden model depends on the initial concentration of particles [3]:

$$R_c \sim c_0^{-\frac{1}{d^D}} \tag{17}$$

where d is dimension of Euclidean space, in which fractal object was considered (obviously, in our case d=3).

It is obvious also, that the value c_0 can be expressed through weights of monomers loading during the reaction m_m and the particle (molecule of mono- or oligomer calculated according to the Eq. (15)) weight m_0:

$$\tilde{n}_0 = \frac{m_m}{m_0} \tag{18}$$

In its turn, value R_c is related to the largest possible degree of polymerization N_c as follows [3]:

$$R_c \sim N_c^{1/D} \tag{19}$$

The combination of the Eqs. (17)–(19) allows to obtain the interrelation between MW and molecular weight of the aggregating particles MW_0 (accounting equality MW=$N_c MW_0$) in the following form:

$$MW \sim \frac{MW_0^{d/(d-D)}}{m_m^{D/(d-D)}} \tag{20}$$

6.3.4 COMPARISON OF THEORETICAL AND EXPERIMENTAL RESULTS FOR COPOLYMERS

In Fig. 1 the comparison of the calculated according to Mark-Kuhn-Houwink equation and obtained within the framework of irreversible aggregation model values MW (the Eq. (20), the last ones were fitted up to the best correspondence with the first one) for copolymers APESF and CP-

OF-10/OS-10 is shown. Let's note, that for APESF theoretical data are arranged systematically lower than experimental dependence of *MW* on formal blocks contents C_{form} and for diblockcopolymers CP-OF-10/OS-10 the good correspondence of theory and experiment up to C_{form}≈30 mol.% is observed and then (at C_{for}≈50 and 70 mol.%) the calculated data are essentially lower than the experimental ones. To explain and remove this discrepancy let's consider the dependences $T_g(C_{form})$ for the indicated copolymers, shown in Fig. 2. From the adduced plots nonadditive dependence $T_g(C_{form})$ follows for both copolymers: for APESF T_g exceeds additive glass transition temperature T_g^{ad} at all compositions and for CP-OF-10/OS-10 for small C_{form} the value $T_g < T_g^{ad}$, at C_{form}≈30 mol.% $T_g \approx T_g^{ad}$ and at C_{form}>30 mol.% T_g of copolymers exceeds T_g^{ad}. This means, that for APESF at any composition the same type of copolymer is formed, whereas at CP-OF-10/OS-10 syntheses depending on the composition different copolymer types are formed. As it was known [21], at the analysis of copolymers composition influence on their T_g Gordon-Tailor-Wood equation is often used:

$$T_g = K_m\left[\left(T_{g_2} - T_g\right)\left(\frac{W}{1-W}\right)\right] + T_{g_1}, \tag{21}$$

where T_{g_1} and T_{g_2} are the glass transition temperatures of homopolymers, K_m is copolymer constant, which serves as quantitative characteristic of distribution sequence of links in copolymer chain [22], *W* is comonomer molar fraction.

The value K_m, which can be calculated according to the Eq. (21), characterizes copolymer type [22]. For purely statistical copolymers K_m=1, for copolymers, forming long sequences of the same links, K_m=0.5 and for strictly alternating blocks K_m=2.0. This classification assumes that in the APESF case copolymers with more or less regular alternation of links are synthesized irrespective of the composition and for CP-OF-10/OS-10 at small C_{form} copolymers with long sequences of the same blocks are synthesized, at C_{form}≈30 mol.% the statistical copolymer was obtained and at C_{form}>30 mol.% copolymers with more or less regular blocks alternation is formed.

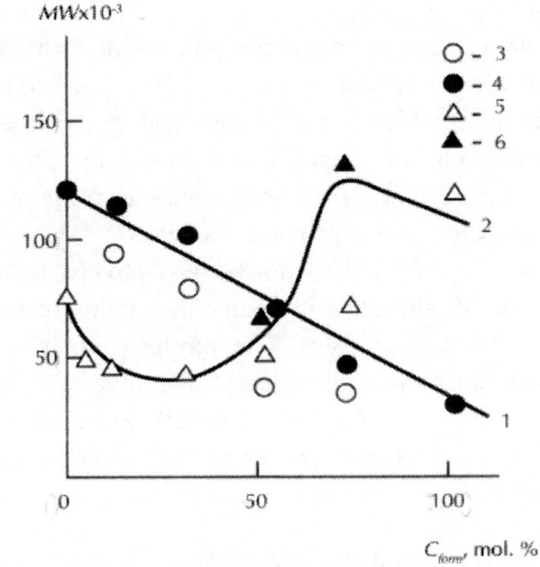

FIGURE 1 The dependences of molecular weight MW on formal block contents C_{form} for copolymers APESF (1, 3, 5) and CP-OF-10/OS-10 (2, 4, 6). 1, 2 – experimental data; 3, 4 – calculation according to the equation (20) with determination of MW_0 according to the Eq. (15); 5, 6 – the same, but with determination of MW_0^{ef} according to the Eq. (22).

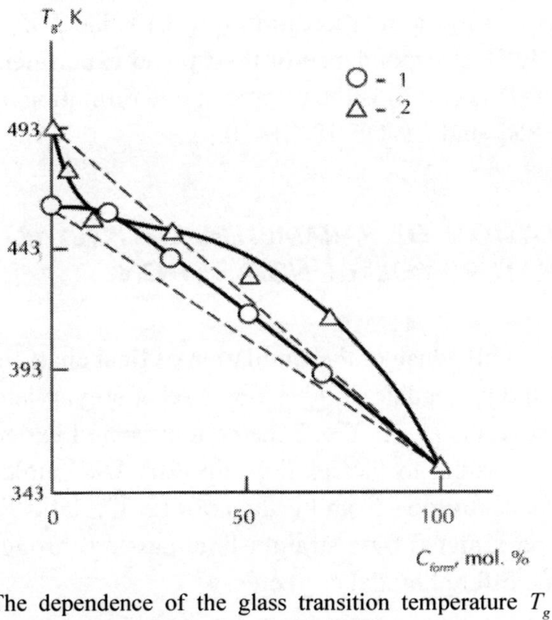

FIGURE 2 The dependence of the glass transition temperature T_g on formal block contents C_{form} for APESF (1) and CP-OF-10/OS-10 (2).

We assume that in the last case, i.e., at K_m>1, the intermediate aggregation process occurs in a polycondensation process the formation of regular sequences from the even number of oligomer fragments or from the even number of monomers, which being in such aspect are inserted into the growing macromolecule. For copolymer with strictly alternating blocks K_m=2.0 [22] that means, that all 100% of monomer (oligomer) molecules intermediate aggregates. We obtain the values K_m=1.3 and 1.8 for CP-OF-10/OS-10 with C_{form}=50 and 70 mol.%, respectively, and this assumes, that in them about 65 and 90% oligomer molecules respectively were formed by intermediate aggregates. The similar estimations were made for APESF. It is obvious, that in this treatment at invariable initial concentration of aggregating particles c_0 the value MW_0 increases proportionally to the intermediate aggregates fraction and that's why the effective value MW_0^{ef} can be expressed as follows:

$$MW_0^{ef} = K_m MW_0 \tag{22}$$

It is easy to see, that in the relationship Eq. (22) only values K_m>1 can be used, as K_m=1 dois not change the value MW_0^{ef} and K_m<1 means oligomer (monomer) molecules destruction, which is not confirmed experimentally.

The replacement in the relationship Eq. (20) value MW_0 on MW_0^{ef} allows obtaining full correspondence of theory and experiment, as follows from Fig. 1 data. This confirms the offered above formation mechanism of copolymers APESF and CP-OF-10/OS-10.

6.3.5 SIMULATION OF PARAMETERS OF POLYARYLATE F-2 LOW-TEMPERATURE POLYCONDENSATION

In this section we will consider the simulation of final characteristics (conversion degree and molecular weight), obtained at polyarylate F-2 synthesis in different solvents [1]. In Fig. 3 the comparison of experimental MW and estimated according to the relationship (20) MW^T molecular weight for F-2 is shown. As follows from the data of Fig. 3, relation between MW and MW^T is approximated by a straight line, passing through the origin, that confirms the offered model correctness.

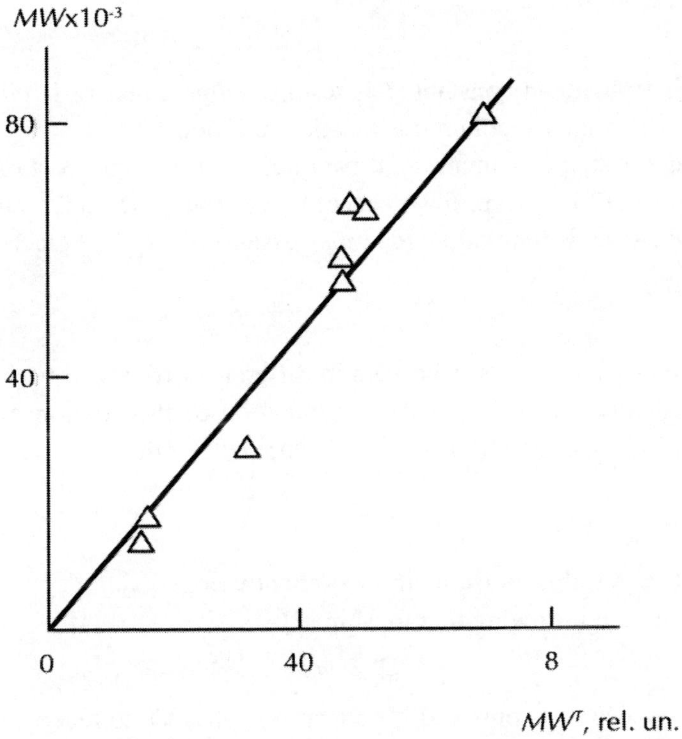

FIGURE 3 The relation between experimental *MW* and calculated according to the Eq. (20) MW^T molecular weight values for polyarylate F-2.

Let's note, that polymers synthesis reactions simulation within the framework of the offered model is possible in virtue of macromolecular coil fractality only, whose density ρ depends on its gyration radius R_g and, hence, on molecular weight [3]:

$$\rho \sim R_g^{D-d} \tag{23}$$

It is obvious, that in case of Euclidean object $D=d$ and ρ=const.

The gyration radius R_g of macromolecular coil (aggregate) in cluster–cluster model can be described by the following relationship [23]:

$$R_g \sim \left(\frac{4c_0 kT}{3\eta_0 m_0}\right)^{1/D} t^{1/D}, \tag{24}$$

where k is Boltzmann constant, T is testing temperature, η_0 is initial viscosity of reactionary medium, t is reaction duration.

If as the first approximation all parameters in the right part of the relationship Eq. (30), except for t, accept by constant, the under conditions $R_g=R_c$ and $t=t_c$ (t_c is time of the reaction cessation), we shall receive:

$$t_c \sim R_c^D \tag{25}$$

As for polyarylates F-2, synthesized in different solvents, m_0=const, then we shall finally receive from the combination of the relationships Eqs. (19) and (25) under conditions of $R_g=R_c$ and $MW=N_c m_0$:

$$t_c \sim MW \tag{26}$$

Further it is possible to estimate theoretically conversion degree Q (Q^T) according to the following relationship [7]:

$$Q^T \sim m_m \eta_0 t_c^{(3-D)/2} \tag{27}$$

If to suppose m_m=const and η_0=const, it is possible to receive Q^T from the more simple relationship (with allowance for the relationship Eq. (26)):

$$Q^T \sim MW^{(3-D)/2} \tag{28}$$

In Fig. 4 the comparison of experimental values Q [1] and Q^T for polyarylates F-2 is shown. As one can see, though the variation tendencies Q and Q^T are identical (the shaded line in Fig. 4), but the quantitative correspondence is poor. It is explained by the acceptance of the condition η_0=const in the relationship Eq. (27). As the experimental values η_0 are absent, we shall make rough, but simple estimation of this parameter from the following considerations. From the relationship Eq. (27) at $Q^T=Q$, m_m=const and t_c=const=60 min. (representative reaction duration [2]) we'll estimate η_0 and then calculate Q^T according to the same relationship with variable η_0. The results of comparison of Q and estimated by indicated method Q^T are shown, in Fig. 5 from which much better correspondence of theory and experiment follows in comparison with Fig. 4.

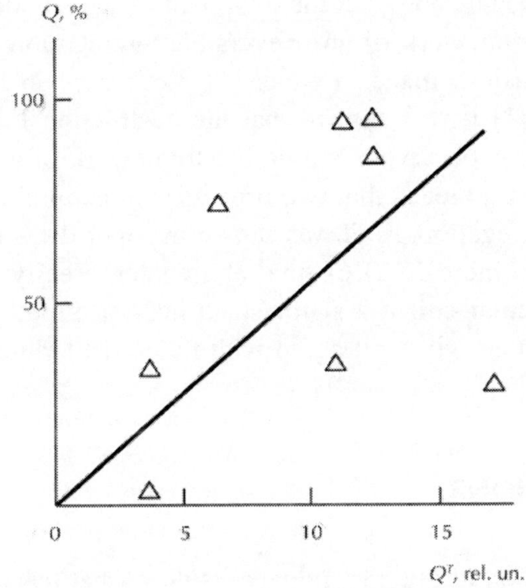

FIGURE 4 The relation between experimental Q and calculated according to the Eq. (28) Q^T polymer conversion degree values for polyarylate F-2.

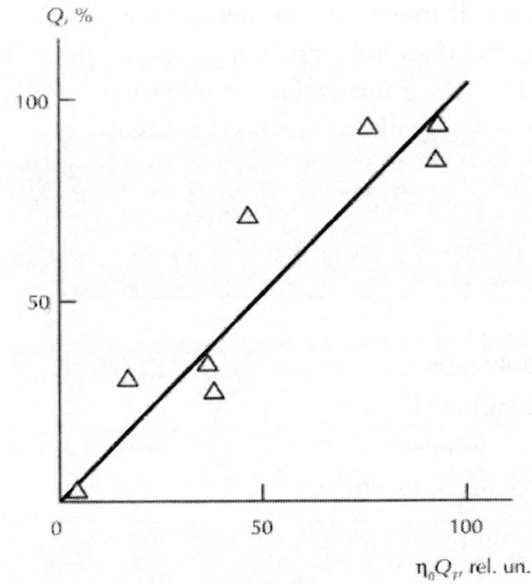

FIGURE 5 The relation between experimental values of conversion degree Q of polymer and parameter $\eta_0 Q^T$ for polyarylate F-2.

Let's note in conclusion, that the treatment of polycondensation process within the framework of an irreversible aggregation models does not except conclusions made in paper [1]. Let's explain it on one example. In paper [1] it was shown, that the destruction F-2 in N,N-dimethylformamide is observed. Within the framework of an irreversible aggregation model it means that two processes in parallel occur: aggregation and disaggregation. As it was shown in paper [9], such combination results to the increase D of final aggregates. Really, the value D for a macromolecular coil F-2 synthesized in N,N-dimethylformamide is equal to 2.03, that will well agree with theoretical value D for such aggregation type [9].

6.4 CONCLUSIONS

The main conclusion of the present paper is that, what reaction cessation in polycondensation process is limited by purely physical factor, namely, the achievement of macromolecular coil density of solution density in which reaction proceeds. The possibility of such treatment is due fully to macromolecular coil fractality. The indicated processes are simulated within the framework of an irreversible aggregation models over mechanism cluster–cluster. The critical values of molecular weight and reaction duration, above which a synthesis reaction ceases, exist.

KEYWORDS

- **Cluster**
- **Diblockcopolymers**
- **Mark-Kuhn-Houwink**
- **Polycondensation process**
- **Thermomechanical technique**

REFERENCES

1. Korshak, V. V.; Vinogradova, S. V.; Vasnev, V. A. *Vysokomol. Soed. A*, **1968**, *10(6)*, 1329–1335.
2. Korshak, V. V.; Vinogradova, S. V. Nonequilibrium Polycondensation. Nauka: Moscow, 1972; 695 p.
3. Kokorevich, A. G.; Gravitis, Ya. A.; Ozol-Kalnin, V. G. *Khimiya Drevesiny*, **1989**, *1*, 3–24.
4. Feder, F. In *Fractals*. Plenum Press: New York, 1989; 248 p.
5. Kaufman, J. H.; Melroy, O. R.; Abraham, F. F.; Nazzal, A. I. *Solid State Commun.*, **1986**, *60(9)*, 757–761.
6. Shiyan, A. A. *Vysokomol. Soed. B*, **1995**, *37(9)*, 1578–1580.
7. Shogenov, V. N.; Temiraev, K. B.; Kozlov, G. V. In *Physics and Chemistry of Perspective Materials*. Nal'chik, KBSU, 1997; 80–84.
8. Vilgis, T. A. *Physica A*, **1988**, *153(2)*, 341–354.
9. Botet, R.; Jullien, R. *Phys. Rev. Lett.*, **1985**, *55(19)*, 1943–1946.
10. Kolb, M.; Jullien, R. J. *Phys. Lett. (Paris)*, **1984**, *45(10)*, L977–L981.
11. Büller, K.-U. In *Heat- and Thermostable Polymers*. Khimiya: Moscow, 1984; 1056 p.
12. Brow, D.; Sherdron, G.; Kern, V. In *The Practical Guidance on Synthesis and Polymers Properties Study*. Zubkov, V. P., Ed.; Khimiya: Moscow, 1976; 256 p.
13. Baranov, V. G.; Frenkel, S. Ya.; Brestkin, Yu. V. *Doklady AN SSSR*, **1986**, *290(2)*, 369–372.
14. Karmanov, A. P.; Monakov, Yu. B. *Vysokomol. Soed. B*, **1995**, *37(2)*, 378–331.
15. Pavlov, G. M.; Korneeva, E. V.; Mikhailova, N. A.; Anan'eva, E. P. *Biofizika*, **1992**, *37(6)*, 1035–1040.
16. Pavlov, G. M.; Korneeva, E. V. *Biofizika*, **1995**, *40(6)*, 1227–1233.
17. Ptytsin, O. B.; Eizner, Yu. E. *Zhurnal Technicheskoi Fiziki*, **1959**, *29(9)*, 1117–1134.
18. Askadskii, A. A. In *Physics–Chemistry of Polyarylates*. Khimiya: Moscow, 1968; 216 p.
19. Kozlov, G. V.; Shogenov, V. N.; Kharaev, A. M.; Mikitaev, A. K. *Vysokomol. Soed. B*, **1987**, *29(4)*, 311–314.
20. Kolb, M. *Physica A*, **1986**, *140(2)*, 416–420.
21. Aliguliev, R. M.; Oganyan, V. A.; Yurkhanov, V. B.; Ibragimov, Kh. D. *Vysokomol. Soed. A*, **1987**, *29(3)*, 611–615.
22. In *Encyclopaedia of Polymers*, Kargin, V. A., Ed.; Sovetskaya Entsiklopediya: Moscow, 1972; *1*, 1223 p.
23. Weitz, D. A.; Huang, J. S.; Lin, M. Y.; Sung, J. *Phys. Rev. Lett.*, **1984**, *53(17)*, 1657–1660.

CHAPTER 7

SELECTION OF MEDICAL PREPARATIONS FOR TREATING LOWER PARTS OF THE URINARY SYSTEM

Z. G. KOZLOVA

CONTENTS

7.1 INTRODUCTION

Stones – a metabolism illness due to various endogenous or exogenous causes and often of a hereditary nature characterized by the urino-formation of stones in the urinary system.

There are people in all age groups who suffer from irretention of urine. Thirty percent of women suffer from this in one form or another, i.e., unable to regulate the functioning of the urinary bladder. This problem may be solved by strengthening the wall of the urinary bladder, decreasing the inflammatory process in the urinary tract and strengthening the connective tissue.

Antioxidants (AO) are nutrient substances (vitamins, microelements, etc.), which human organisms require constantly. They serve to maintain a balance between free-radicals and AO forces.

Modern medicine uses AO to improve people's health and as a prophylaxis. Therefore, in recent times, the AO properties of various compounds are being widely studied. The most prevalent source of AO is considered to be vegetative objects on the basis of which medicinal preparations and BAAs are prepared.

As a criterion for evaluating the quality of a medicinal preparation, we took its Antioxidant activity (AOA), i.e., concentration of natural AO in it, which was measured on a model chain reaction of liquid-phase oxidation of hydrogen by molecular oxygen.

Set task: Quantitatively measure AO content in investigated preparations since they constitute a vegetation composition and evaluate their effectiveness in improving the quality of life.

The following are medicinal preparations and BAAs investigated for treating illnesses of the urethra (cystitis, enuresis and urine stones): Cyston (India), Urotractin (Italy), Promena (USA), Spasmex (Germany), Urolyzin (Russia), Contrinol (USA), Tonurol (Anti-Enuresis) (Russia), Blemaren (Germany).

7.2 CHARACTERISTICS OF PREPARATIONS

CYSTON (India) – Complex therapy for stones in the bladder, crystallization, infection of the urethra, podagra.

Each tablet contains: Didymocarpus pedicellata R. Br., Saxifraga Ligulata Wall, Rubia cordifolia L.; Cyperus scariosus R.Br.; Achyranthes aspera L.; Onosma bracteatum Wall.; Vernonia cinerea (L.) Less. BAA.

CONTRINOL (USA) – A Mixture of eastern and western medicinal plants used for normal functioning of the urethra. It strengthens the connective tissue.

Each capsule contains: Horsetail Herb, White Poplar Bark, Dogwood Berry and Schizsandra Berry. BAA.

PROMENA (USA) – provides important support for the prostate, possesses anti-inflammation properties, normalizes functioning of the urethra and strengthens the immune system.

Each capsule contains: Vitamin E, Vitamin C, Zinc, Vitamin A, Parsley Leaf, Echinacea, Pumpkin Seed, Gravel Root, Corn silk, Bee Pollen. BAA.

SPASMEX (Germany), TROSPIYA CHLORED (PRO. MED. CS PRAHA a.s.) – quarter amine is safer to use because of its unique chemical structure. Lowers tone of the smooth muscles of the urinary bladder, reduces detrusion of the bladder.

Each capsule contains: Dry birch bark extract, Irish Moss, Origanum vulgare L.

UROLYZIN (Russia) – source of Arbutin and Flavonoids used as a diuretic and antiseptic.

Content: Extracts of Folium Betula, Polygonum avicular L., Orthosiphon stamineus Benth, Sprout vaccinium myrtillus L., Fructus Aronia melanocarpa Elliot, Fructus Sorbus aucuparia L., Burdock Root. BAA.

TONUROL (ANTI-ENURESIS) (Russia) – supports urethra organs, strengthens bladder wall, helping in case of incontinence and has antiseptic and anti-inflammation properties.

Capsule content: Equisetum arvense L., White Poplar Bark, Hypericum perforatum L. flowers, Dogwood Berry, Schisandra chinensis Baill (Turez), Matricaria recutita (L.) extract.

BLEMAREN (Germany) – for prophylactic and treatment of stones in the bladder. Burbling tablets.

7.3 METHOD OF EXPERIMENT

Chain reactions of oxidation can be used for quantitative characterization of the properties of inhibitors (antioxidants). The investigated samples were analyzed by means of a model chain reaction of initiating oxidation of cumene [1–4]. Initiated oxidation of cumene in the presence of studied AO proceeds in accordance with the following scheme:

Initiation of chain Origination of $RO_2^{\cdot}$ radicals, initiation rate W_i
Continuation of chain $RH+RO_2^{\cdot}$ ($_3$+O_2ROOH+$RO_2^{\cdot}$
Break of chain $2RO_2^{\cdot}$ ^{k6}; $\longrightarrow$ molecular products
$InH+RO_2^{\cdot}$ ^{k7}; $\longrightarrow$ $RO_2H+In^{\cdot}$
$RO_2^{\cdot}+In^{\cdot}$ ^{k8}; $\longrightarrow$ molecular products

(We use the widely accepted numeration of rate constants of elementary reaction of inhibited oxidation.)

In accordance with this scheme, each independent inhibiting group of AO breaks two chains of oxidation. This makes it possible by means of the formula:

$$\tau = \frac{2[InH]_0}{W_i} \tag{1}$$

to determine the initial concentration of the inhibitor (more exactly, the concentration of inhibiting groups) taken for the reaction from the experimentally determined value of the period of induction τ. In the right member of Eq. (1), we have: W_i, the standard given rate of initiation; f, the inhibiting coefficient, equal to 2; n, the number of inhibiting groups in an antioxidant molecule; $[InH]_0$, the initial AO concentration. The kinetic curve of oxygen absorption is described by the following equation:

$$\frac{\Delta O_2}{[RH]} = \frac{k_3}{k_7} \ln(1 - t/\tau) \tag{2}$$

The constant of inhibiting rate k_7, determining the anti-radical activity of AO and being its qualitative characteristic, is found from relation Eq. (2), using the known constant of chain continuation rate k_3, concentration

[RH] for hydrocarbon, experimentally determined period of induction τ and quantity of absorbed oxygen ΔO_2.

Cumene (isopropyl benzene) was used as oxidizing hydrocarbon and azo-bis-isobutyronitrile as initiator, which forms free-radicals upon thermal decay. Initiating rate was determined from the following formula:

$W_i = 6.8 \times \cdot 10^{-8}$ [AIBN] mol/l · s,

where [AIBN] (AZO-bis-ISOBUTYRONITRILE) is the initiator concentration in mg per ml of cumene.

The period of induction τ is determined by plotting the dependence of the quantity of absorbed oxygen ΔO_2 in the reaction against time t. The end of the kinetic curve is a linear portion representing non-inhibited reaction, i.e., the portion after expending AO. The AO expenditure time τ is determined graphically on the kinetic dependence of oxygen absorption by the point of intersection of two straight lines: one of the lines is the line that the kinetic curve assumes after AO is consumed and the other is a tangent to the kinetic curve, the tangent of the angle of inclination of which is one half the tangent of the angle of the first. The greater the amount of AO in the sample the greater the period of induction τ.

The method is direct and based on the use of the chain reaction of liquid-phase oxidation of hydrocarbon by molecular oxygen.

The method is functional, i.e., the braking of the oxidizing reaction is determined only by the presence of AO in the system being analyzed. Other possible components of the system (not AO) do not exert a significant effect on the oxidizing process, which enables to analyze AO in complex systems, avoiding a stage of separation.

The method is very sensitive, exact and informative.

The method is absolute, i.e., does not require calibration, is simple to apply and does not require complex equipment [1–5].

7.4 RESULTS AND DISCUSSION

The following medicinal preparations were taken for investigation: Cyston, Spasmex, Urotractin, Urolyzin, Contrinol, Promena, Tonurol (Anti-Enuresis), Blemaren. The AOA of these preparations was determined.

As an example, the figure shows the kinetic dependences of oxygen absorption in a model reaction of initiated cumene oxidation in the absence of antioxidant (straight line 1) and in the presence of Urotractin (curve 2), Spasmex (curve 3), Promena (curve 4), Contrinol (curve 5), Cyston (curve 6), Urolyzin (curve 7).

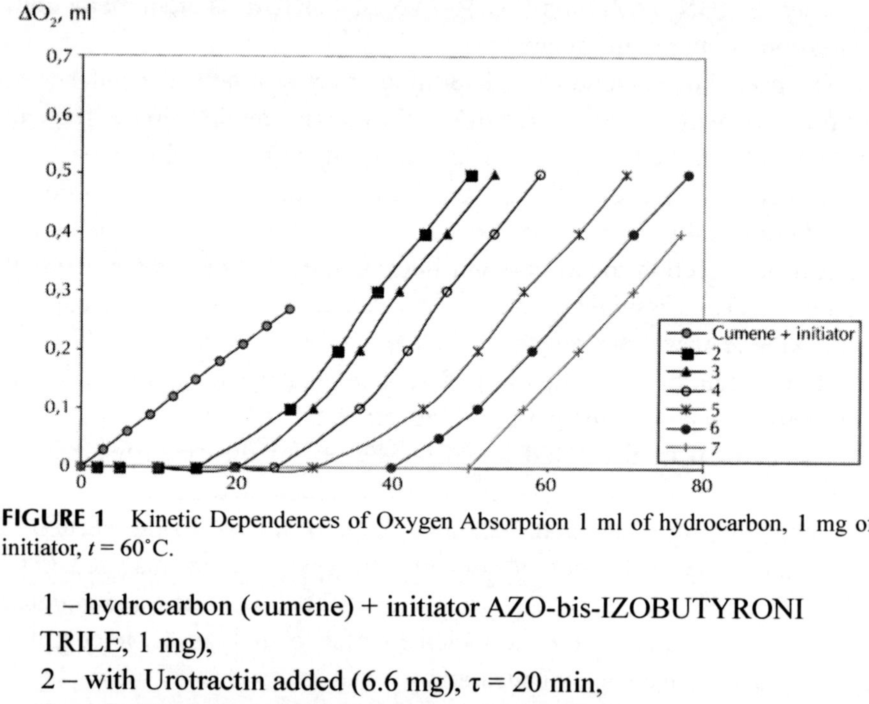

FIGURE 1 Kinetic Dependences of Oxygen Absorption 1 ml of hydrocarbon, 1 mg of initiator, $t = 60^{\circ}$C.

1 – hydrocarbon (cumene) + initiator AZO-bis-IZOBUTYRONI TRILE, 1 mg),
2 – with Urotractin added (6.6 mg), $\tau = 20$ min,
3 – with Spasmex added (6.0 mg), $\tau = 25$ min,
4 – with Promena added (7.5 mg), $\tau = 30$ min,
5 – with Contrinol added (20 mg), $\tau = 38$ min,
6 – with Cyston added (9.6 mg), $\tau = 45$ min,
7 – with Urolyzin added (14.6 mg), $\tau = 50$ min.

It can be seen from the figure that, in the absence of the additive, hydrocarbon oxidation proceeds at constant rate (straight line 1). When the preparation is added, the oxidation rate at the beginning is strongly retarded but begins to increase after a certain period of time. This is indicative of the presence of antioxidant in the additive.

The rise in reaction rate is due to the expenditure of antioxidant. When it is used up, the reaction proceeds at the constant rate of an uninhibited reaction. The time of antioxidant (τ) is determined graphically by the intersection of two straight lines on the kinetic curve. One of these is the straight-line portion of the kinetic curve after AO has been used up. The other is the tangent to the kinetic curve whose inclination angle is one- half the tangent angle of the first.

Data on the antioxidant content of investigated preparations are presented in the table. These data are illustrated by the diagram.

TABLE 1 Concentration of Antioxidants in Studied Preparations.

№	Preparations	Antioxidant Concentration (M/kg)
1	Cyston	9.6×10^{-3}
2	Spasmex	8.5×10^{-3}
3	Promena	8.2×10^{-3}
4	Urolyzin	7.0×10^{-3}
5	Urotractin	6.2×10^{-3}
6	Tonurol (Anti-Enuresis)	5.4×10^{-3}
7	Contrinol	3.9×10^{-3}
8	Blemaren	–

Diagram Illustrating Results in Table

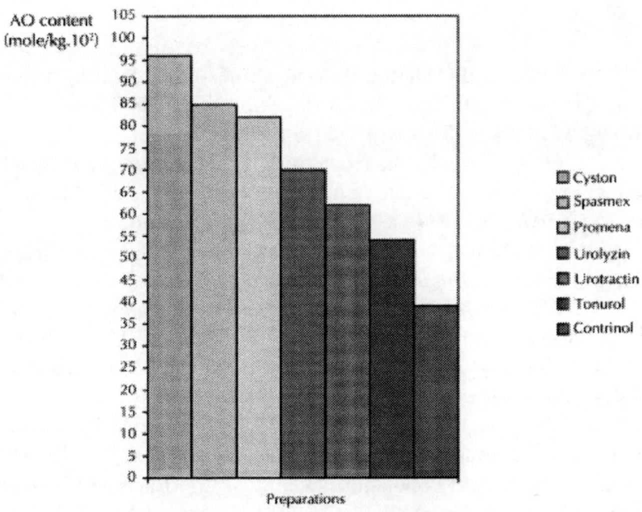

Results of analysis show that AOA is in the interval of values (3.9 – 9.6) × 10^{-3} M/kg, which correlates with the values determined earlier for AO in plants: Basilicum, Hypericum perforatum L. – 8.8×10^{-3} M/kg, Coriandrum sativum L. – 3.9 × 10^{-3} M/kg, Valeriana officinalis L. – 7.7 × 10^{-3} M/kg, Eleutherococcus Senticosus maxim. – 9.4 × 10^{-3} M/kg. Thus, the effectiveness of certain BAAs may be due to their AO content (more than 10^{-3} M/kg of dry substance) and their inclusion in complex therapy of corresponding illnesses may be justified. The close values of AO concentration between the studied preparations and medicinal plants gives basis for assuming their having a positive effect on the human organism [6].

KEYWORDS

- **Antioxidant activity**
- **Basilicum**
- **Coriandrum sativum L.**
- **Eleutherococcus Senticosus maxim**
- **Hypericum perforatum L.**
- **Prophylaxis**
- **Valeriana officinalis L.**

REFERENCES

1. Tsepalov, V. F. A Method of Quantitative Analysis of Antioxidants by means of a Model Reaction of Initiated Oxidation, In *Investigation of Synthetic and Natural Antioxidants In vitro and In vivo*, Moscow, 1992 (*in Russian*).
2. Kharitonova, A. A.; Kozlova, Z. G.; Tsepalov, V. F.; Gladyshev, G. P. Kinetic Analysis of Antioxidant Properties in Complex Compositions by means of a Model Chain Reaction, *Kinetika i Kataliz. J.*, **1979**, *20(3)*, 593–599 (*in Russian*).
3. Tsepalov, V. F.; Kharitonova, A. A.; Kozlova, Z. G. Physicochemical Characteristics of Natural Food Products, In *Bioantioxidant*, Moscow, 1998; p. 94 (*in Russian*).
4. Tsepalov, V. F.; Kharitonova, A. A.; Kozlova, Z. G.; Bulgakov, V. G. Natural Antioxidants: An Express Method of Analysis and Prospects of Use as Food Additives. Collection, In *Pishchevye Ingredienty*, Moscow, 2000; pp. 7–8 (*in Russian*).
5. Tsepalov, V. F. *Zavodskaya Lab.*, **1964**, *1*, 111 (*in Russian*).
6. Kozlova, Z. G.; Eliseyeva, L. G.; Nevolina, O. A.; Tsepalov, V. F. Content of Natural Antioxidants in Spice-aromatic and Medicinal Plants. Theses of Report at International Scientific Conference: Populations Quality of Life-Basis and Goal of Economic Stabilization and Growth, Oryol, 23.09–24.09.1999; 184–185 (*in Russian*).

CHAPTER 8

IMPROVEMENT OF THE FUNCTIONAL PROPERTIES OF LYSOZYME BY INTERACTION WITH 5-METHYLRESORCINOL

E. I. MARTIROSOVA and I. G. PLASHCHINA

CONTENTS

8.1 INTRODUCTION

Wide using of lysozyme in medicine for the treatment of various infections and in food and cosmetic industry to prevent bacterial contamination of the stuffs leads to production of the resistance of microorganisms to lysozyme action. In recent years, a great deal of attention has been devoted to the investigations aimed at studying lysozyme modifications, affecting lysozyme properties while retaining its enzymatic activity. One of the ways to solve this problem is modification of lysozyme structure with using weak nonspecific interaction with 5-methylresorcinol. It is known numeral effects of MR on the structure and functions of lysozyme. In particularity, in concentration range of 10^{-7}–10^{-3}M MR the specific and nonspecific enzymatic activities raise, its substratum specificity extends and temperature range of lysozyme catalysis expands [1].

Earlier the interrelation between the influence of MR concentration on nonspecific activity of lysozyme and the destabilizing effect of MR on its native conformation were established. It is shown, that activity and thermostability of lysozyme depend on the concentration of MR and the incubation time of their mixed solutions [2]. The MR ability to self-organization in a solution with micelle like structure formation was established via methods of mixing microcalorimetry and dynamic light scattering [3].

Due to diphylic character of MR and its ability of self-organization in a solution it can behave similarly to nonionic surfactants. Adsorption of MR was shown to be mixed-diffusion barrier controlled (pH 7.4). MR was also shown to change thermodynamic affinity of lysozyme to the solvent with the resulting the protein being slightly less or more surface active depending on MR concentration. Under used conditions MR and lysozyme can compete in adsorption process at air/water interface [4].

The aim of this work is to study the effect of MR on the surface activity of lysozyme and rheological properties of its adsorption layers at the air/solution (0.05 M phosphate buffer, pH 6.0) interface.

8.2 MATERIALS AND METHODS

A sample of hen egg white lysozyme (Sigma, USA) with activity 20,000 U/mg and molecular mass 14,445 Da was used.

Alkyl-substituted hydroxybenzenes, 5-Methylresorcinol (5-Methylbenzene-1,3-diol) (Sigma, USA) with molecular mass 124,14 g/mol (anhydr) was taken.

All reagents for phosphate buffer preparation in Milli-Q water were of analytical grade.

8.2.1 SURFACE TENSION MEASUREMENT

Dynamic surface tension was measured with an automatic drop Tracker tensiometer (ITC Concept, France), connected to thermostatic bath to maintain the temperature constant at 25°C during the measurements. The principle of tensiometer is to determine the surface tension of the studied solution from the axisymmetric shape of a rising bubble analysis [5]. Due to the active control loop, the instrument allows long-time experiments with a constant drop/bubble volume or surface area.

Surface tension, σ (mN/m), was measured in 7 ml samples at constant lysozyme (0.075 mg/ml) and varying 5-methylresorcinol (0.16–67.2 mM) concentrations and their mixed solutions in 0.05 M phosphate buffer, pH 6.0 at 25°C.

The mixed solutions LYS-MR were prepared with using of lysozyme and MR after 24 h incubated separately at 25°C in darkness. After mixing of these solutions 1:1 they were stored 3 h at 25°C away from the light. The dynamic surface tension was measured over 60 000 s to guarantee steady-state of the adsorption layer. A $t \rightarrow \infty$ asymptotic extrapolation was used to find the steady-state surface tension values. Standard deviations were always less, than 0.5 mN/m, and duplicate measurements were made for each MR concentration. Finally phosphate buffer were confirmed not to present surface activity by measuring separately the surface tension of a phosphate buffer solution, obtaining values practically equal to those of pure water.

From the kinetic curves (surface tension versus time) for the solutions of different composition, the steady-state surface tension isotherms (surface tension versus concentration) were obtained. As the concentration of the protein in the mixture was fixed, the abscissa in all graphs represents the concentration of the surfactant in the mixed solution. From surface tension data the MR critical concentration of self-organization (micelle-formation) in presence of lysozyme were estimated. For this purpose surface tension values were fitted with the logarithm of the MR concentration.

8.2.2 SURFACE DILATATIONAL PROPERTIES MEASUREMENT

The surface rheological parameters of adsorbed LYS-MR films at the air-water interface; the surface dilatational modulus (E) and the phase angle (θ) were measured as a function of time with constant amplitude ($\Delta A/A$) of 3% and angular frequency (ω). The range of angular frequencies used was 0.007–0.625 rad/s. Measurements of rheological properties were performed after formation of enough stable adsorption layer during of 60,000–70,000 s. The sinusoidal oscillation for surface dilatational measurement was made with 3–5 oscillation cycles followed by a time of 3–5 cycles without any oscillation. The average standard accuracy of the surface pressure is roughly 0.1 mN/m. The reproducibility of the results (for at least two measurements) was better than 0.5%.

8.3 RESULTS AND DISCUSSION

8.3.1 ADSORPTION OF METHYLRESORCINOL IN THE PRESENCE OF LYSOZYME

MR is weak non-ionic surfactant. In the case of protein/non-ionic surfactant mixtures, there is a competitive adsorption phenomenon, which is combined with weak hydrophobic interactions between protein and surfactant, resulting in modification of the protein molecule.

Earlier we have demonstrated the ability of MR to increase surface activity of lysozyme [4]. We used phosphate buffer with pH 7.4, $I = 0.05$ M as a solvent, that is corresponds to the conditions of lysozyme catalytic activity determination in the presence of MR [1]. However using dynamic light scattering method we have shown that there were associates (or aggregates) of some lysozyme molecules in solution at this pH value. To avoid this effect we changed the pH value from 7.4 to 6.0.

The dynamic surface tension measurements performed in phosphate buffer solutions (pH 6.0, $I = 0.05$ M) at constant lysozyme concentration, 5.1×10^{-6} M, and varying concentrations of MR, show that the LYS-MR mixture is characterized by more high rate of adsorption and more low quasi-equilibrium surface tension, than pure LYS (Fig. 1). As concerning the effect of MR on the surface activity of lysozyme, it is seen from Fig. 2 that MR enhances it at all using concentration-making lysozyme more hydrophobic. MR critical micelle concentration in lysozyme presence is 32.7 mM.

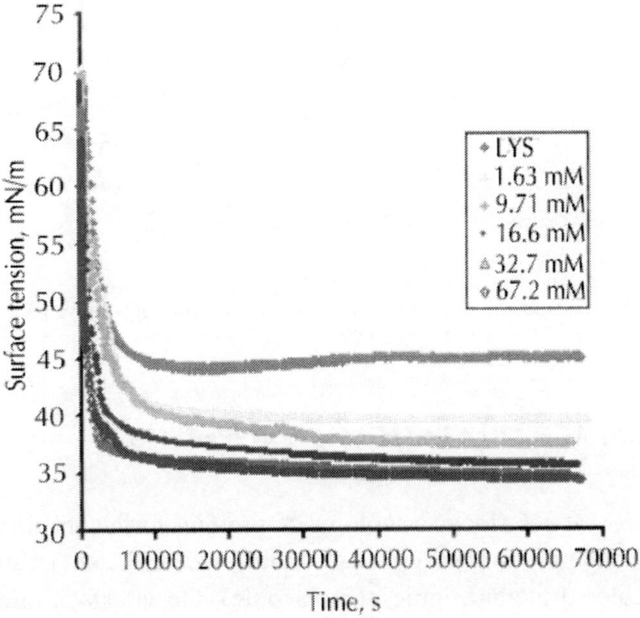

FIGURE 1 Dynamic curves of adsorption of MR in mixture with lysozyme at air/water interface at 25°C; MR concentrations 1.63-67.2 mM and constant protein concentration 5.1×10^{-6} M.

In literature the surface activity of phenols were investigated in [6]. The effect mainly depended on the nature of the phenolic compound and its physicochemical properties in relation to their affinity for the air/water interfaces [6]. For example, catechin was proven to be able to accumulate at the air/water interface, decreasing the surface tension values with increasing its concentration.

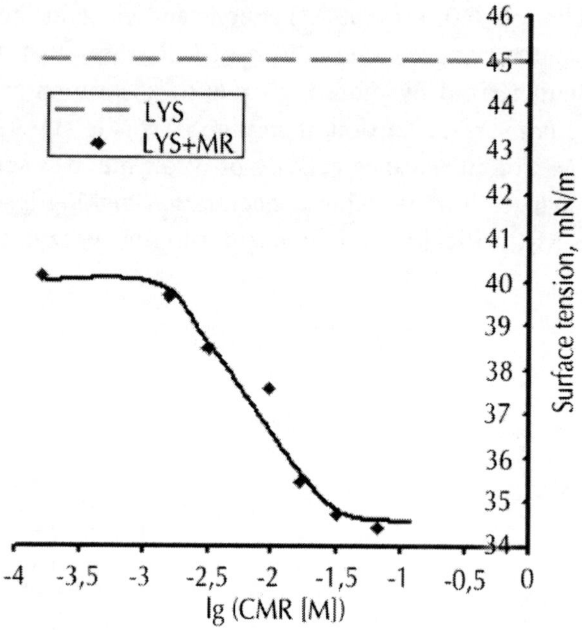

FIGURE 2 Surface tension isotherms for MR in mixed solution with lysozyme at constant protein concentration (5.1×10^{-6} M) at 25°C.

8.3.2 SURFACE DILATATIONAL PROPERTIES

It is known that interfacial rheology of protein–surfactant mixed layers depends on the protein (random or globular), the surfactant (water-soluble or oil-soluble surfactant, ionic or non-ionic), the interface (air–water or oil–water), the interfacial (protein/surfactant ratio) and bulk (i.e., pH, ionic strength, etc.) compositions, the method of formation of the interfacial layer (by spreading or adsorption, either sequentially or simultaneously), the interactions (hydrophobic and/or electrostatic), and the displacement of protein by surfactant [7].

The quasi-equilibrium adsorption layers (the formation time of 60,000–70,000 sec) were subjected to compressive/tensile deformation sinusoidally in the field of linear viscoelasticity. The dependences of the complex viscoelastic modulus of adsorption layers (**E**), as well as its elastic (real part,

$\mathbf{E_{rp}}$) and viscous (imaginary part, $\mathbf{E_{ip}}$) components from the frequency of the applied deformation were obtained (fig. 3).

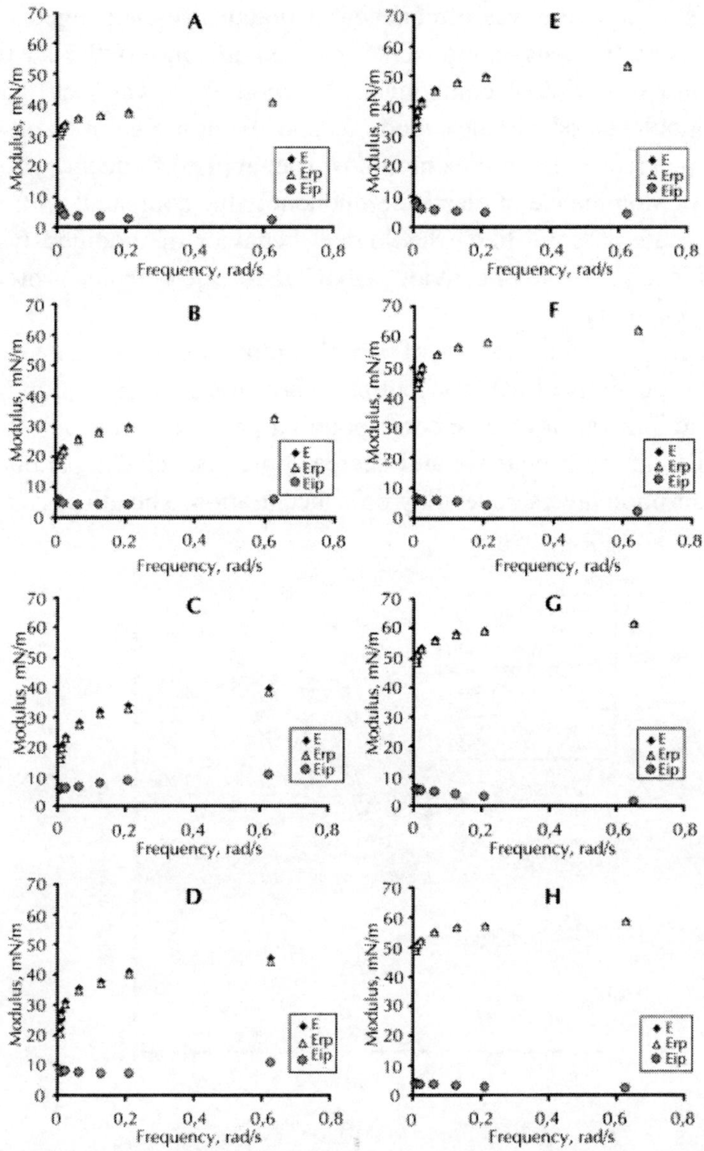

FIGURE 3 Dependence of elasticity modulus on frequency in mixtures of lysozyme and MR different concentration (A – native lysozyme, B – 0.16 mM, C – 1.63 mM, D – 3.21 mM, E – 9.71 mM, F – 16.6 mM, G – 32.7 mM, H – 67.2 mM).

It was found that the value of the complex viscoelastic modulus is almost equal to its elastic component in all range of frequencies used (0.007–0.625 rad/s) (Fig. 3). With increasing the frequency, the rise of the viscoelastic modulus was established at practically unchanged value of the imaginary (viscous) component. The predominance of the elastic component above of viscous component in several times was seen. The practical coincidence of complex modulus and its elastic component values, weak dependence of complex modulus from applied frequency of loading and the predominance of elastic component value compared with viscous ones indicate a "solid-like" rheological behavior of modified lysozyme adsorption layers. This behavior is typical for all globular proteins and LYS among them.

In Figs. 4 and 5 the results of complex modulus and phase angle measuring as functions of MR concentration are presented. The drop lines indicate the interval of change corresponding parameters for pure LYS. As seen, MR can both increase and decrease the viscoelastic parameters of LYS adsorption layers depending on concentration. The effect is more expressed at low frequencies.

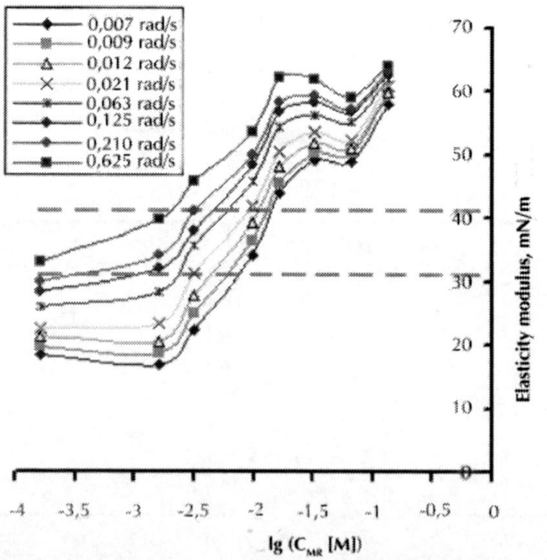

FIGURE 4 Dependence of Lysozyme complex elasticity modulus on MR concentration at frequency 0.007 and 0.625 rad/s. The drop lines indicate the interval of change corresponding parameters for pure LYS.

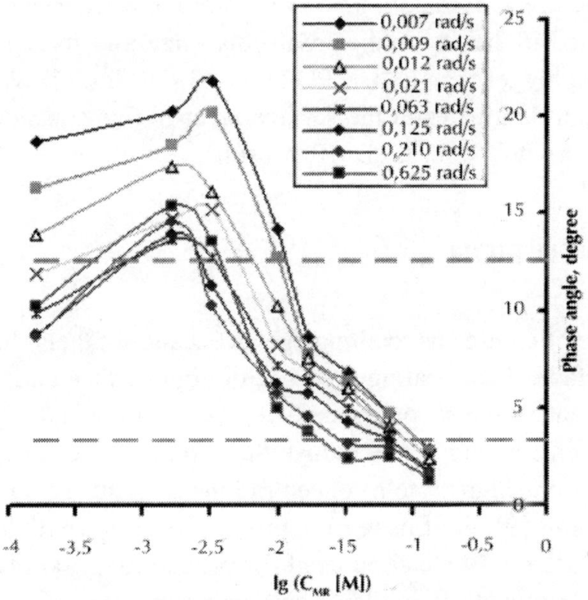

FIGURE 5 Dependence of Lysozyme phase angle on MR concentration at frequency 0.007 and 0.625 rad/s. The drop lines indicate the interval of change corresponding parameters for pure LYS.

There are some informations about monotonous decrease of the equilibrium surface tension, dilatational elasticity, and adsorption of lysozyme for non-ionic surfactant decyl dimethyl phosphine oxide (C_{10}DMPO) as the concentration of surfactant increases in the mixture. However, in the case of mixtures of non-ionic surfactants with more flexible proteins like β-casein, the elasticity of the interfacial layer decreases before passing through a maximum as the concentration of surfactant increases [7]. Possibly, the weaker interfacial network formed by β-casein as compared to globular proteins determines the dilatational response of the mixtures. The same picture was shown for the system β-casein mixed with dodecyl dimethyl phosphine oxide (C_{12}DMPO). For all studied frequencies (0.005–0.1 Hz) the elasticities for adsorption layers have a maximum about 4×10^{-5} mol/l C_{12}DMPO concentration. It was shown the obtained values are very close to those measured for the surfactant alone. Thus, in this concentration region the surfactant dominates the surface layer. In our case we have the close picture with maximum on the complex viscoelastic modulus line (fig. 4). But MR has a smaller molecular size and its concentration of maximum is bigger than C_{12}DMPO (1.7–3.3×10^{-2} mol/l). We can assume the surfactant dominates in the surface layer in this concentration region in agreement with [8] for C_{12}DMPO results.

8.4 CONCLUSION

Proteins are often used as foaming agents because of their ability to unfold at the interface, thus creating layers with high surface elasticity and also steric resistance against coalescence of layers. Although they decrease the interfacial tension and hence reduce the driving force for disproportionation, often quite high protein concentrations are required for the formation of stable foams [9]. For this reason, mixing lysozyme with low-molecular weight surfactants MR can be used for production of antibacterial pharmacological products of emulsion and foam types with enhanced physical stability and period of using while increasing its enzymatic activity and wider spectrum of action.

Methylresorcinol belongs to the class of phenols. It opens more attractive prospects of its possible industrial application. Outstanding representatives of phenolic compounds – antioxidants such as gallic acid, catechin, quercetin can combine both the surface-active properties and act as stabilizers of emulsions. Quercetin improved the dispersion state of the emulsions with the increasing of its concentration. Gallic acid, despite its partitioning in the water phase due to its polarity, delayed the formation of both the hydroperoxides and thiobarbituric acid reactive substances (TBARs) and limited their accumulation. Catechin did not affect the formation of oxidation products whilst quercetin, among the tested antioxidants, caused the lowest formation of both hydroperoxides and TBARs through 33 d of storage [6].

Based on previous studies, we can assume that the MR has antioxidant properties [10]. In this case MR behavior will allows a stabilization mechanism that is based on interactions with protein molecules which lead the formation of a stiff viscoelastic surface film that can also work as barrier towards pro-oxidant species as well as exert some antioxidant properties. This is a subject of further investigation.

KEYWORDS

- **Dynamic light scattering**
- **Lysozyme**
- **Microcalorimetry**
- **Rheological**

REFERENCES

1. Petrovskii, A. S.; Deryabin, D. G.; Loiko, N. G.; Mikhailenko, N. A.; Kobzeva, T. G.; Kanaev, P. A.; Nikolaev, Yu. A.; Krupyanskii, Yu. F.; Kozlova, A. N.; El'-Registan, G. I. Regulation of the functional activity of lysozyme by alkylhydroxybenzenes. *Microbiology*, **2009**, *78(2)*, 146–155.
2. Plashchina, I. G.; Zhuravleva, I. L.; Martirosova, E. I.; Petrovskii, A. S.; Loiko, N. G.; Nikolaev, Yu. A.; El'-Registan, G. I. Effect of Methylresorcinol on the Catalytic Activity and Thermostability of Hen Egg White Lyzozyme. In *Biotechnology, Biodegradation, Water and Foodstuffs*. Nova Science Publishers: N.-Y., 2009; 45–57.
3. Martirosova, E. I. Regulation of hydrolase catalytic activity by alkylhydroxybenzenes: thermodynamics of C-AHB and hen egg white lysozyme interaction. In *Biotechnology and the Ecology of Big Cities.* Nova Science Publishers: N.-Y., 2011; 105–113.
4. Martirosova, E. I.; Plashchina, I. G. Adsorption Behavior of 5-Methylresorcinol and its Mixtures with Lysozyme at Air/Water. In *Biochemistry and Biotechnology: Research and Development.* Nova Science Publishers: N.-Y., 2012.
5. Loglio, G.; Pandolfini, P.; Miller, R.; Makievski, A. V.; Ravera, F.; Ferrari, M.; Liggieri, L. Drop and bubble shape analysis as tool for dilational rheology studies of interfacial layers, In *Novel Methods to Study Interfacial Layers, Studies in Interface Science*. Möbius, D.; Miller, R., Eds., 2001; *11*, 439–484.
6. Di Mattia, C. D.; Sacchetti G.; Mastrocola D.; Sarker D. K.; Pittia P. Surface properties of phenolic compounds and their influence on the dispersion degree and oxidative stability of olive oil O/W emulsions. *Food Hydrocolloid.*, **2010**, *24*, 652–658.
7. Maldonado-Valderrama, J.; Patino, J. M. R. Interfacial rheology of protein–surfactant mixtures. *Curr. Opin. Colloid Inter.,* **2010**, *15*, 271–282.
8. Kotsmar, C. S.; Pradines, V.; Alahverdjieva, V.S.; Aksenenko, E.V.; Fainerman, V.B.; Kovalchuk, V.I.; Krägel, J.; Leser, M.E.; Noskov, B.A.; Miller, R. Thermodynamics,

adsorption kinetics and rheology of mixed protein–surfactant interfacial layers. *Adv. Colloid Inter. Sci.*, **2009**, *150*, 41–54.
9. Alahverdjieva, V. S.; Khristov, Khr.; Exerowab, D.; Miller R. Correlation between adsorption isotherms, thin liquid films and foam properties of protein/surfactant mixtures: Lysozyme/C10DMPO and lysozyme/SDS Colloids and Surfaces A: *Physicochem. Eng. Aspects.* **2008**, *323*, 132–138.
10. Revina, A. A.; Larionov, O. G.; Kotchetova, M. V.; Zimina, G. M.; Zolotarevski, V. I.; El-Registan, G. I. Spectrophotometric and chromatographic investigation of the radiation-induced oxidation products of 3,5-hydroxytoluene aqueous solutions. *Chem. High Energies*, **2004**, *38*, 176–182.

CHAPTER 9

INTRODUCTIONS IN CULTURE *IN VITRO* RARE BULBOUS PLANTS OF THE SOCHI BLACK SEA COAST (SCILLA, MUSCARI, GALANTHUS)

A. O. MATSKIV, A. A. RYBALKO, and A. E. RYBALKO

CONTENTS

9.1 INTRODUCTION

Liliaceae plants have the important meaning for cultivation of the cut off flowers, potting of culture and creation of landscape compositions. Analyzed the literature on problems of micropropagation, receptions of the new forms and virus free plants. The directions of researches on perfection, technology of cultivation of a landing material, improvement from viruses are planned. Many of these plants are medicinal.

In the Sochi Black Sea Coast the different kinds of order Liliales having prospect of use in a pharmaceutical industry grow. They are representatives of family Amaryllidaceae – a snowdrop (Galanthus alpinus Sosn., G. platiphyllus Traub et Moldenke, G. risechense Stern, G. woronowii Losinsk., Leucojum aestivum L.); Hyacinthaceae – species of muscari (Muscari coeruleum Losinsk., M. dolichanthum Woronow et Tron, M. neglectum Guss., M. pallens M. Bieb.), a Scilla bifolia L., S. siberica Haw.); Colchidaceae – (Colchicum speciosum Steven, C. laetum Steven), etc. They have the status vulnerable and require protection. Along with it, they can be used in a national economy as primroses for early (February–March) gardening.

9.2 MATERIAL AND METHOD

Due to the limited quantity of an initial material we carried out methodical development on lily plants. Initial material of Muscari coeruleum Losinsk in the form of a bulb it was received from the Sochi office of the Russian Geographical society. Scilla bifolia L. and Galanthus woronowi Losinsk bulbs are selected in the suburban Sochi woods. Before introduction in culture of in vitro of a bulb previously cleared of pollution under flowing water within an hour. Further with a scalpel deleted the infected sites and sterilized in 25% whiteness solution (during 20 min) and washed out in the 3rd portions of the sterile distilled water. For introduction in culture of in vitro we used MS nutrient medium [6].

9.3 RESULTS

The studied plants – a scilla, galanthus and muskari are vulnerable at the expense of gathering flowers that reduce the reproduction of seeds providing genetic variability. One of preservation factors population is ensuring formation of seeds. We carried out collecting seed boxes and counted up quantity of seeds. It is established by us that in places of collecting flowers there are no daughter plants of these species (Fig. 1) [1, 2]. In a protected environment it is formed seed-boxes and daughter plants (Figs. 2, 3, and 4). We defined the seed productivity of scilla and galanthus. The seed-cases of scilla contain 3.03 ± 1.99 pieces of seed, galanthus – 20.59 ± 11.46.

FIGURE 1 Galanthus woronowi – in a protected environment Seed-cases (1), daughter plants (2).

FIGURE 2 Galanthus woronowi – influence of anthropo-genous factor. No seed-cases (1), daughter plants.

FIGURE 3 Flowering of Scilla bifolia.

FIGURE 4 Seed-cases (1), daughter plants (2) and bulbs of Scilla bifolia.

Thus, we established that one of the factors causing degradation of populations of these cultures is mass collecting flowers. In controllable conditions there is a formation of seeds and germination of affiliated plants. For preservation of these important components of a biodiversity of the Sochi Black Sea Coast creation dwelling place for these and other the small bulb plants is necessary. In our researches when using MS nutrient medium with a various combination of plant growth regulator, the following results are received.

Considering importance of these plants for working out of the program of new generation of preservation of disappearing flora in our region, we have spent work on introduction to culture in vitro for micropropagation of muscari, scilla and galanthus. Studies are before undertaken on micropropagation of lilies [3] and muscari [4]. Preliminary materials on this research are reported on International scientific conference "Pharmaceutical and Medical Biotechnology" (Moscow, on March, 20–22th, 2012) [5]. For realization of researches used the sterile plants of lily and muscari (Figs. 5, and 6), and also scales of bulbs of scilla and galanthus (Figs. 7, and 8) selected the suburban forest Sochi. With that end in view on Murashige and Skoog medium (6) added benzyladenine (1 mg/l) and naphthaleneacetic acid (0.2 mg/l) landed explants of bulb scale. Within 8 weeks on them were formed affiliated bulblets. At their change on the fresh environment again accrued bulblets. At the subsequent division after 8-week cultivation the reproduction factor has averaged lily – 1:10, scilla – 3,6 and muscari – 1:21 piece (Table 1). Such procedure we spend regularly, that gives the chance will receive a considerable quantity of homogeneous landing units. Similar results are received for galanthus. By the same technique we spend micropropagation various industrial cultivars of lilies. The applied standard technique can be used and for other plants of order Liliales.

TABLE 1 Definition of factor of reproduction in culture of in vitro.

No.	Plant	Number of tubes	Total microcutting	Mean
1.	Scilla	13	46	3.5
2.	Muscari	3	64	21.3
3.	Lilium	18	181	10.1

We also used an embriogeniya. Embriogeniya is a new direction in vegetative reproduction of plants [2]. We studied an embriogeniya on an example muscari and scilla. For this purpose eksplant from sterile culture landed on the MS nutritious medium with gibberelic acid (2 mg/l), isopentenil adenine (1 mg/l) and naphthaleneacetic acid (0.2 mg/l). Within 4 weeks of cultivation it is received callus (Fig. 8). Callus will be used for studying of possibility of receiving medicinal components and new forms of plants of these types. Between options of a nutrient medium of an essential difference it is not revealed, therefore this technique is applied to different species.

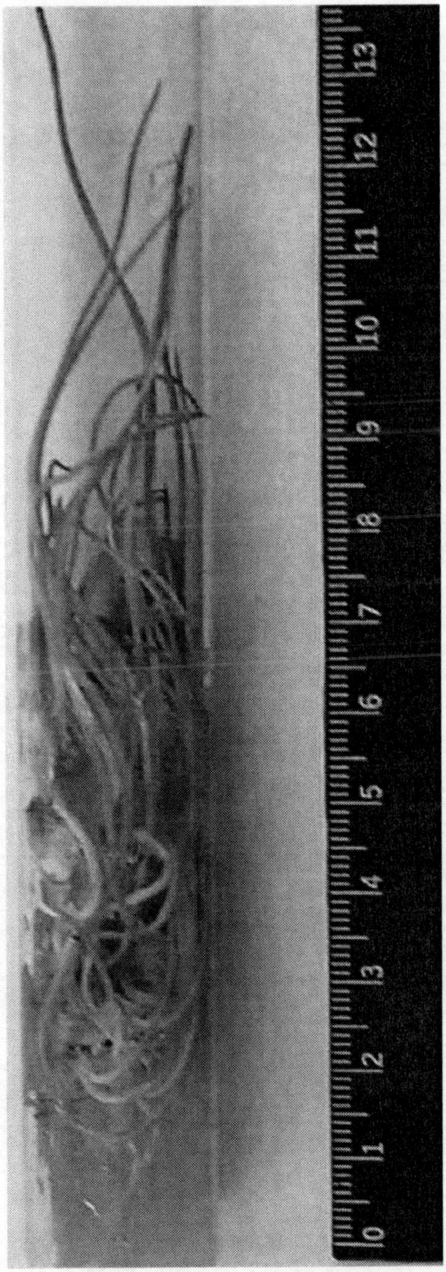

FIGURE 5 Muscari coeruleum in test tubes.

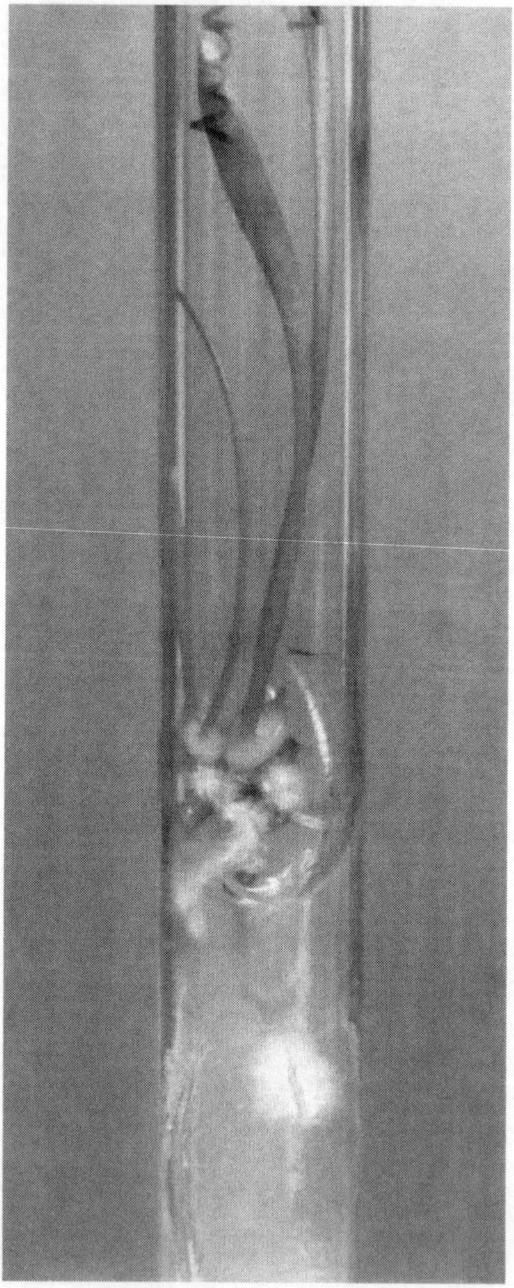

FIGURE 6 Bulblet of Lilium in test tubes.

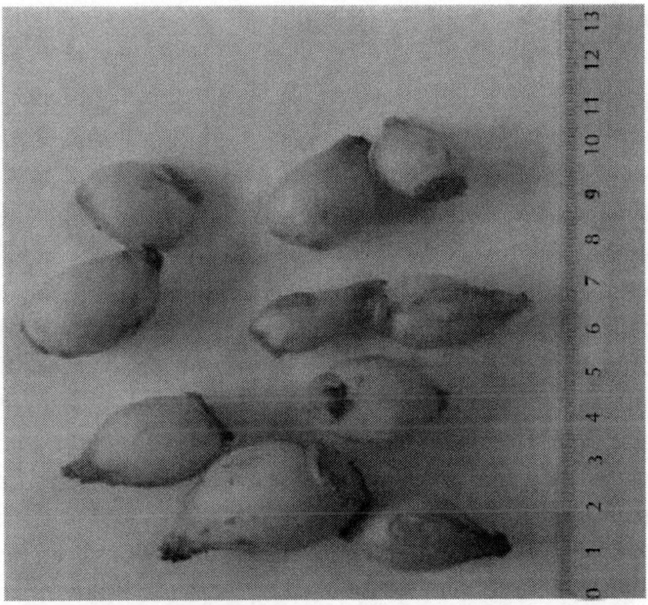

FIGURE 7 Bulb of Scilla bifolia.

FIGURE 8 Bulb of Galanthus woronowi.

9.4 DISCUSSION

The region the Sochi Black Sea Coast is the most capacious enclave of Russia by quantity of species. More than 2,000 species of wild-growing vascular plants here grow. Among them a large number endemic, relicts, rare types demanding protection. Seventy eight from them are included in the Red book of Russia [8]. There are considerable stocks of effective natural medicines [9]. Needs for medicinal raw materials, uncontrollable collecting these valuable resources put threat of disappearance of many versions. It causes of expansion of researches on application of methods of biotechnology for biodiversity preservation. It will provide possibility of use of new types of raw materials in a national economy. Usual methods of reproduction don't correspond to cope with this problem. Large-scale production in culture of in vitro will provide advance on the market at lower prices. Besides, equipment quite simple, also works can be executed by people with average qualification. The micro multiplied plants can be introduced in natural conditions in wild populations, and as to provide material delivery for herbariums [1]. At the same time, legislation improvement is required stopped illegible collecting plants [7].

9.5 CONCLUSIONS

1. Introduced into the culture in vitro geophytes Sochi Black Sea Coast (Scilla bifolia, Muscari coeruleum Losinsk., Galanthus woronowii Losinsk).
2. It was confirmed the dependence of the nature of the concentration of cytokinins morphogenesis.
3. Obtaining sterile culture, which makes it possible to carry out research on the use of these crops in the economy, and specifically in the restoration of natural habitats in the ornamental purposes in the project «wild flora from Russia», in developing the program receive a new generation of drug substances on the basis of domestic raw materials.

KEYWORDS

- **Galanthus**
- **Galanthus woronowi Losinsk**
- **Muscari coeruleum Losinsk**
- **Muskari**
- **Scilla**
- **Scilla bifolia L.**

REFERENCES

1. Afolayan, A. J.; Adebola, P. O. In vitro propagation: A biotechnological tool capable of solving the problem of medicinal plants decimation in South Africa. *African J. Biotech.,* **2004**, *3(12)*, 683–687.
2. Batygina, T. B. Reproduction, propagation and renewal of plants, Embriologiya of floral plants. Terminology and concepts. T. 3: Reproduction systems. SPb.: World and family, 2000; pp. 35–39.
3. Sibileva, V. N.; Titova, S. M.; Rybalko, A. E. Studi of micropropagation of lilies. In *Student scientific researches in the field of tourism and resort business*. Materials of the 2th International research and practice conference on 20–23 May, 2011; pp. 218–219
4. Pavlenko, P. V.; Rybalko, A. E. Influence of cytocinins on micropropagation of muscari. Ibid. pp.175–177.
5. Matskiv, A. O.; Arakeljan, M. A.; Rybalko, A. E. Introductions in culture in vitro rare bulbous plants of the Sochi Black Sea Coast (Scilla, Muscari, Galanthus). Proceeding of the Moscow international scientific and practical conference "Pharmaceutical and Medical Biotechnology" (Moscow, on 20–22 March, 2012) M: Joint-Stock Company "Expo-biohim-technologies," D. I. Mendeleyev University of Chemistry and Technology of Russia, D.I. Mendeleyev University of Chemistry and Technology of Russia, 2012; pp. 455.
6. Murashige, T.; Skoog, F. A revised medium for rapid growth and bioassays with tobacco cultures, *Physiol. Plant*, **1962**, *15(3)*, 473–497.
7. Rybalko, A. A. Features of Teaching of the Ecological Right at Saving Biodiversity of Sochi Black Sea Coast. In *Biochemistry and Biotechnology: Research and Development*. Nova Science Publishers, Inc.: N.Y., **2012**; 173–176.
8. Solodko, A. S.; Kirii, P. V. In *Atlas of the healing flora of Sochi region of the Black Sea Coast*. Moskow, Sochi, 2010; 321 p.
9. Solodko, A. S.; Kirii, P. V. The Red Book of Sochi (rare and vaniching plants). V. 1. *Plants and fungi*. Published by Beskovs. Sochi, 2002; p. 148.

CHAPTER 10

EFFECT OF COBALT CONTENT ON THE CORROSION RESISTANCE IN ACID, ALKALINE AND RINGER SOLUTION OF FE-CO-ZR-MO-W-B METALLIC GLASSES

D. KLIMECKA-TATAR, G. PAWLOWSKA, R. ORLICKI, and G. E. ZAIKOV

CONTENTS

10.1 INTRODUCTION

Continuously increasing interest in metallic glasses based on iron is associated with abnormal physical properties of these materials, and as well as a wide range of potential applications, and relatively low production costs. These materials exhibit high strength, hardness, corrosion resistance, can also be precursors of nanocrystalline soft magnetic material [1]. Exceptionally high vitrification ability have the alloys $Fe_{68-x}Co_xZr_{10}Mo_5W_2B_{15}$ (innovative material received in the Institute of Physics of the Czestochowa University of Technical) [2]. Very high hardness and paramagnetic properties at room temperature, make that $Fe_{68-x}Co_xZr_{10}Mo_5W_2B_{15}$ materials can be used in the production of coatings, including the cutting elements in the manufacture of surgical blades. The use conditions of medical instruments determine the selection of materials used in the various components of tools manufacture, so that their operation satisfies the functionality, security and reliability requirements. Proper selection of materials for the instruments results from criteria that should be provide [2]:

- corrosion resistance under different kind of conditions;
- an appropriate set of mechanical properties;
- reliability and stability of the property within a certain operation time.

Massive amorphous alloys of $Fe_{68-x}Co_xZr_{10}Mo_5W_2B_{15}$ composition are produced by suction arc melted alloys to a water-cooled copper form (form of rods or tubes) or one-way rapid quenching on spinning copper drum (tape). With increase of Co content in the alloy composition the improve of alloy vitrification ability is observed. Another consequence of greater cobalt percentage content is the temperature Curie increase of amorphous materials [2]. However, the disadvantage results from the cobalt content increase is lower corrosion resistance in sulphate environments [2].

The aim of the research was to investigate the corrosion resistance of $Fe_{68-x}Co_xZr_{10}Mo_5W_2B_{15}$ materials in different environments, especially in Ringer solution simulating electrochemical conditions of the human body.

The results presented in this chapter are highly important to the potential application of these materials in medicine.

10.2 CHARACTERISTICS OF THE SAMPLE AND RESEARCH METHODOLOGY

The test material was an amorphous alloy in the form of thin rods (∅ ≈ 1 mm) with the $Fe_{68-x}Co_xZr_{10}Mo_5W_2B_{15}$ (x = 7,9,11) composition (lower indices correspond to atomic percent) obtained by suction casting – SC.

- Sample I – $Fe_{61}Co_7Zr_{10}Mo_5W_2B_{15}$
- Sample II – $Fe_{59}Co_9Zr_{10}Mo_5W_2B_{15}$
- Sample III – $Fe_{57}Co_{11}Zr_{10}Mo_5W_2B_{15}$

To electrochemical research there were used electrodes in the form of rotating disks with working area of approx. 0.02 cm^2. The geometrical surface of electrodes were determined using an optical microscope Zeiss Neophot 32 at magnification about 100×, using the planimeter module. The disks were performed by samples mounting in acrylic glass caps. Electrochemical research were carried out in the vented (Ar saturated) solutions at a temperature of 25°C, the disk rotation speed was 16 rev/sec and at a potential skanningu 10 mV/s, using the offset from the cathode value (–1.0 V) to the anode (+2.0 V). The CHI 680 potentiostat (CH Instruments, Austin, Texas, USA) were used. The values of electrode potentials were measured against a saturated calomel electrode (SCE). Each potentiokinetic polarization curve was performed three times to give generally good repeatability of the runs. In case of discrepancies the additional repetitions were performed.

For samples I, II and III following corrosion tests were performed:

- potentiokinetic polarization curves in 1M HNO_3 and 1M NaOH environments;
- potentiokinetic polarization curves in a physiological fluid (Ringer's solution of the following composition: 8.6 $g \cdot dm^{-3}$ NaCl, 0.3 $g \cdot dm^{-3}$ KCl, 0.243 $g \cdot dm^{-3}$ $CaCl_2$ at 37°C;
- polarization resistance tests in Ringer solution.

10.3 THE RESULTS

10.3.1 CORROSION TESTS IN ACIDIC AND ALKALINE SOLUTIONS

Comparing the potentiokinetic curves shapes of $Fe_{68-x}Co_xZr_{10}Mo_5W_2B_{15}$ material in 1M HNO_3 it can be seen clear differences in the anodic current density (Fig. 1). In this environment the sample II, with the largest content of cobalt, reconstituted with the greatest speed.

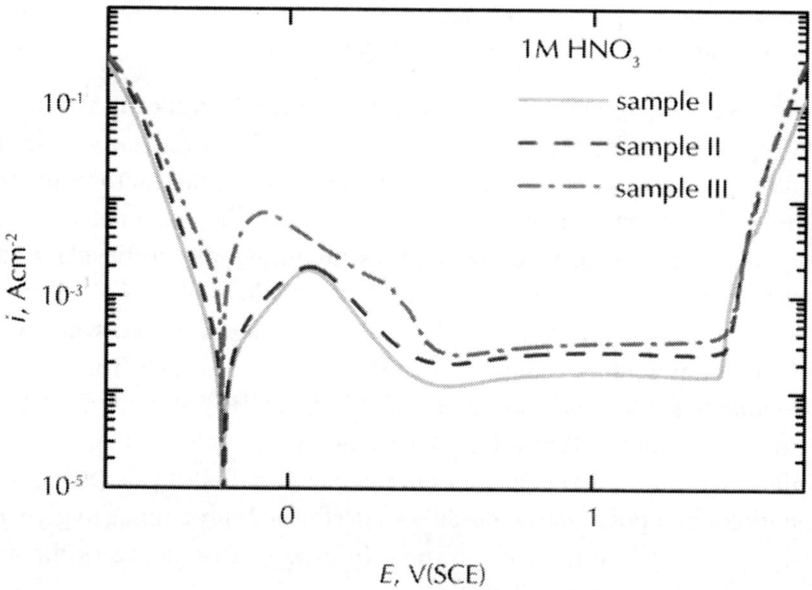

FIGURE 1 The potentiokinetic curves of $Fe_{68-x}Co_xZr_{10}Mo_5W_2B_{15}$ material in 1 M HNO_3 solution.

Also the polarization curve of this sample (III) differs from the others. The current density is significantly higher than the other in terms of both active and passive state. Under the potential about $E > 0.5$ V, the i_{pas} current density of alloy containing 11% Co is two times higher than for the alloy with a low cobalt content (Table 1). The increase in cobalt content of the $Fe_{68-x}Co_xZr_{10}Mo_5W_2B_{15}$ material promotes to active dissolution processes which should be associated with a higher value of exchange current for cobalt than iron (cobalt is thermodynamically noble metal then iron: $E^{o}_{Co^{2+}/Co} = -0{,}28\ V$ whereas $E^{o}_{Fe^{2+}/Fe} = -0{,}44\ V$, thus in the case of equal value of exchange current for Fe and Co, the iron should dissolve first. A similar tendency was observed in sulphate environments [5].

TABLE 1 Corrosion potential and passive current density in 1 M HNO_3 for the $Fe_{68-x}Co_xZr_{10}Mo_5W_2B_{15}$ material.

Sample	E_{corr}, V	i_{pass}, mA/cm²
I	–0.21	0.12
II	–0.21	0.20
III	–0.21	0.26

In an alkaline solution (1M NaOH) test material spontaneously passivates, but there were no differences in polarization curves (Fig. 2). The current density for $E = 0.6$ V, regardless of the amounts of cobalt in the material was equal 0.18 mA·cm^{-2}.

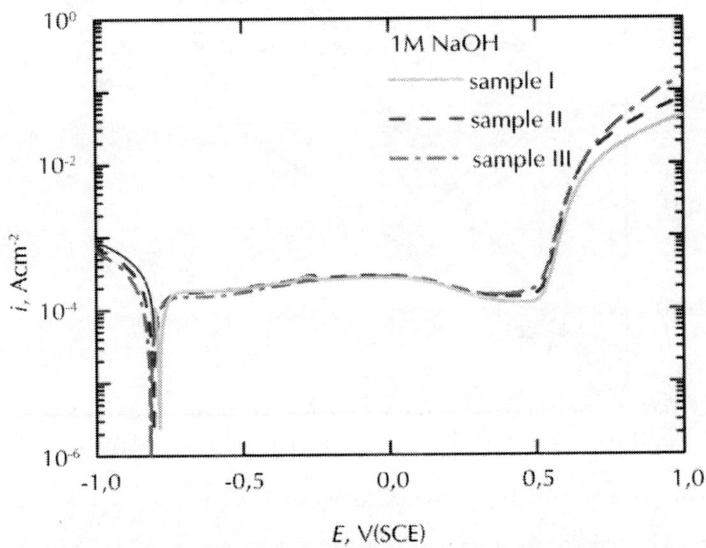

FIGURE 2 The potentiokinetic curves of $Fe_{68-x}Co_xZr_{10}Mo_5W_2B_{15}$ material in 1 M NaOH solution.

10.3.2 CORROSION RESISTANCE IN RINGER PHYSIOLOGICAL SALINE SOLUTION

To evaluate the corrosion resistance of the test amorphous glass in conditions of contact with physiological fluids linear polarization resistance research $\eta = E - E_{kor} = f(i)$ in Ringer solution was conducted. The $\Delta E = f(\Delta i)$ function slope allows the determination of polarization resistance (R_p), which is inversely proportional to the corrosion rate [2]. Polarization resistance of the tested material decreased in the following order:

Sample I (R_p =2738 Ωcm^2)
Sample II (R_p = 1995 Ωcm^2)
Sample III (R_p = 1156 Ωcm^2)

Accordingly to the presented results received in the environment simulating the electrochemical conditions of the human body, the highest corrosion resistance has the alloy with the lowest (7% at.) cobalt content.

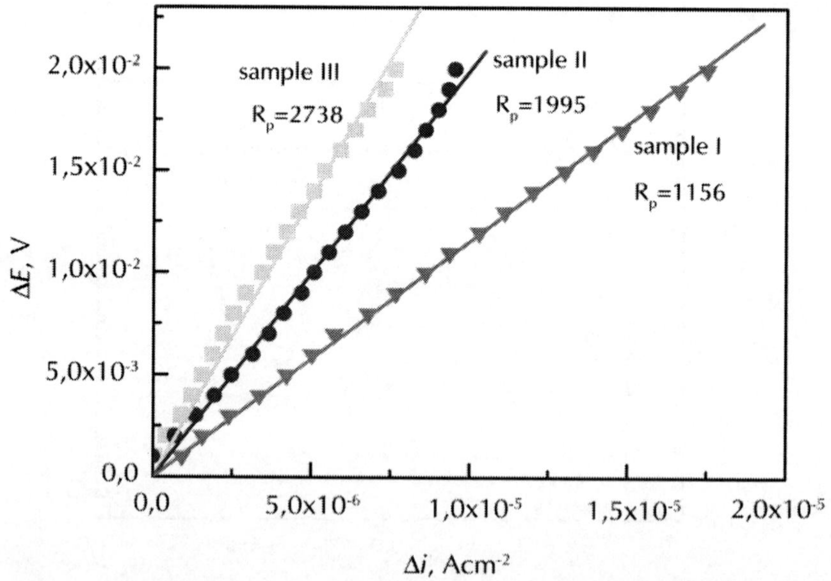

FIGURE 3 The dependence $\eta = E - E_{kor} = f(i)$ of tested material in Ringer solution.

The local damage of the passive layer, often associated with the presence of Cl^- ions in corrosive medium and it can lead to the initiation and the pitting corrosion development [2]. This type of corrosion appears in preference areas of surface (inclusions, defects, dislocation), due to the lack of visible products on the metal surface, in particular it is difficult to identify in the initial stage. The development and propagation of pits (and pitting corrosion) often have a drastic form and create the threat of sudden rupture of the working element or even pulverization of the alloy. In the analyzed material presence of the Mo and W in the alloy and also amorphous structure (due to defects limit) promote effective passivation [3]. In the Ringer solution (containing chloride ions) at a temperature corresponding to the temperature of the human body (37°C) for the potentials higher than 0.3 V, rapid increase in current density associated with the interruption of the passive layer (Fig. 4) was observed. The pits nucleation potentials (E_{pit}) was determined based on the unidirectional polarization runs through the extrapolation of the straight sections of breakdown above potential to intersect the minimum current in the passive range.

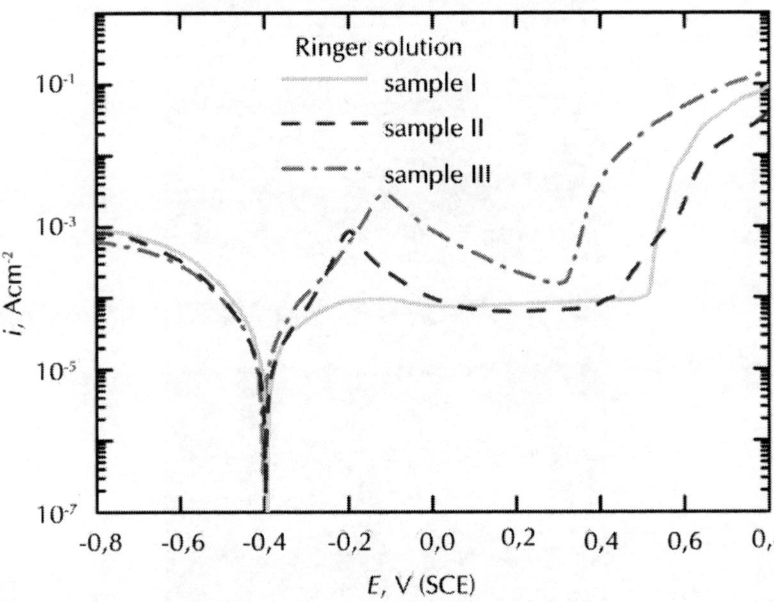

FIGURE 4 The potentiokinetic curves of $Fe_{68-x}Co_xZr_{10}Mo_5W_2B_{15}$ material in Ringer solution.

It can be seen that the passive layer breakdown occurs, easier when there is cobalt in the material (Table 2).

TABLE 2 The pits nucleation potentials of $Fe_{68-x}Co_xZr_{10}Mo_5W_2B_{15}$ material in Ringer solution.

Sample	E_{pit}, V vs. SCE
I	0.52
II	0.42
III	0.32

In Fig. 5 is shown the microscopic images of the surface after exposure in conditions of anode pitting nucleation potential (exposure in Ringer solution). On the material surface there are visible not evenly distributed pits with different sizes. The highest amount of pits is located on the sample III surface. Exposure in the same solution at a lower potential does not lead to the creation of pits.

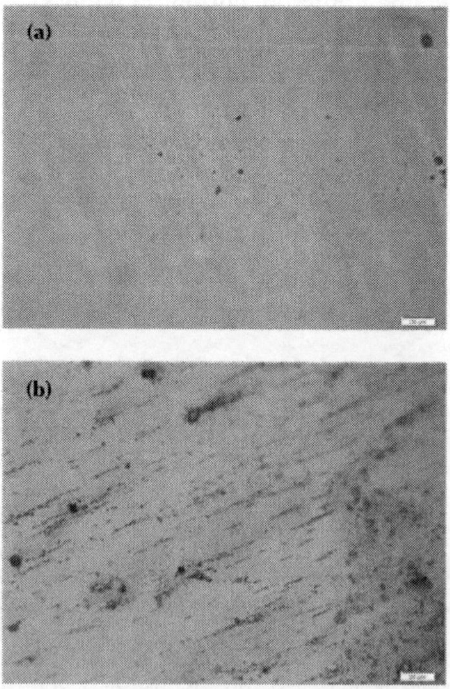

FIGURE 5 *(Continued)*

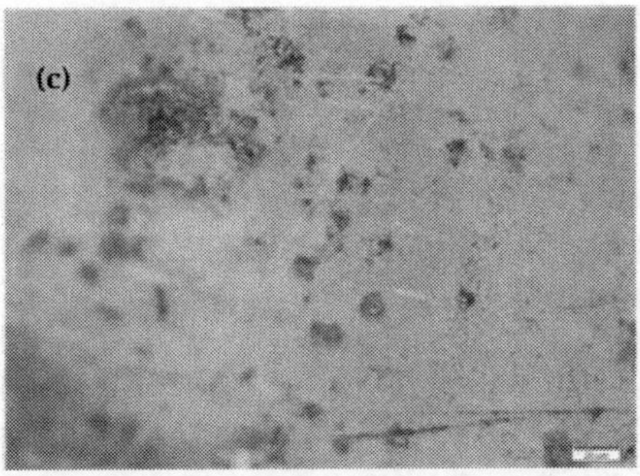

FIGURE 5 The images of $Fe_{68-x}Co_xZr_{10}Mo_5W_2B_{15}$ material surface (a) sample I, (b), sample II, (c) sample III; after 5 min exposure in the Ringer solution ($E = E_{pit}$).

Tested alloys in physiological solutions have the ability to repassivation. The repassivation ability of the material determined from the curve back after exceeding pits nucleation potential. Bidirectional potentiokinetic curves for sample I is shown in Fig. 6 (arrows indicated the direction of the potential change). The repassivation potential for sample I (the most resistant in the environment) is about 0.09 V vs. SCE (Fig. 6). Below this potential passive layer is stable, and the material should not undergo the pitting corrosion.

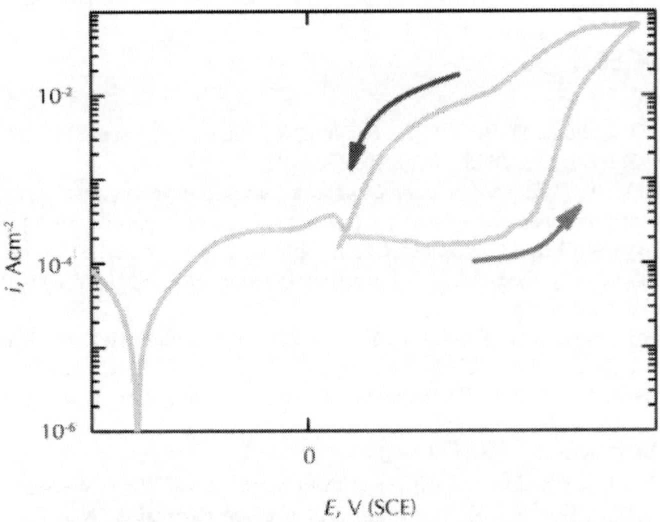

FIGURE 6 Potentiokinetic polarization curves of sample A in Ringer solution.

10.4 SUMMARY

- An increase in cobalt content in the $Fe_{68-x}Co_xZr_{10}Mo_5W_2B_{15}$ material causes the corrosion resistance decrease in 1M HNO_3 and 1 M NaOH solution, and reduces the thermodynamic stability of the passive layer formed on material in Ringer solution.
- Composition $Fe_{59}Co_9Zr_{10}Mo_5W_2B_{15}$ of metalic glasses is the optimal composition, due to the compromise between capacity and improvement of the vitrification and decreasing magnetic properties and corrosion resistance.
- The examined materials Fe-Co-Mo-WB passivate easily what can further enhance their resistance by anodizing the surface and conduct the material in the passive state.

KEYWORDS

- **Bidirectional potentiokinetic curves**
- **Pitting corrosion**
- **Repassivation**
- **Ringer solution**

REFERENCES

1. Chena, Q. J.; Fana, H. B.; Yeb, L.; Ringerc, S.; Suna, J. F.; Shena, J.; McCartneyd, D. G. *Mat. Sci. Eng.*: *A*, **2005**, *402(1–2)*, 188–192.
2. Pawlik, P. In *Rola składu chemicznego i procesu wytwarzania w kształtowaniu właściwości magnetycznych masywnych amorficznych i nanokrystalicznych stopów żelaza*, wyd WIPMiFS Politechniki Częstochowskiej, Monografie 12, Częstochowa, 2011.
3. Paszenda, Z.; Tyrlik-Held, J. In *Instrumentarium chirurgiczne*, wyd. Politechniki Śląskiej. Gliwice, 2003.
4. Pawlik, P.; Pawlik, K.; Davies, H.A.; Wysłocki, J. J.; Leonowicz, M.; Kaszuwara, W. *Inżynieria Materiałowa,* **2007**, *28(3–4)*, 324–329.
5. Pawłowska, G.; Pawlik, P.; Wysłocki, J.; Bala, H. *Rev. Adv. Mater. Sci.,* **2008**, *18*, 41–45.
6. Stern, M. *Corrosion*, **1958**, *14*, 440.
7. Wranglen, G. In *Podstawy korozji i ochrony metali*, wyd. WNT: Warszawa, 1985.
8. Szklarska-Smialowska, Z. In *Pitting and Crevice Corrosion*, NACE International: Houston, 2005.

CHAPTER 11

THE METHODS OF THE STUDY THE PROCESSES OF THE ISSUE TO OPTICAL INFORMATION BIOLOGICAL OBJECT

U. A. PLESHKOVA and A. M. LIKHTER

CONTENTS

11.1 INTRODUCTION

According to the rapid development of the technical cybernetics sphere, based on the biophysical processes implementing for the biological objects behavior management [1, 2, 4, 12], there is a sharp necessity for actual problem solving which is in biocybernetical systems effectiveness increasing biocybernetical system (BCS) [8]. One of the main stages in this problem solving is mathematical model construction, which adequately describes the information transmission process from the signal source to the monitoring object [11].

This task becomes more sophisticated if we take insects for the monitoring object, which are notable for the high level of their behavior uncertainty. As the main information amount, necessary for the insects vital activity, they take with their visual analyzer, so the construction of the model of optical information transmission to the insects with different vision types is the basic during the insects behavior biocybernetical monitoring system construction. Conversely, BCS effectiveness increasing for insects behavior monitoring (Fig. 1) with optical range electromagnetic emission sources can be reached as the result of the implementing the optoelectronic system optimum designing approach and as the result of the calculating of their elements operating parameters on the information quality criteria basis [9], in analytical expressions of which all BCS elements parameters are included: signal/noise function $\tilde{N}/\O$ and informational bandwidth $\ddot{I}$ correspondingly.

$$S/N\,, \tag{1}$$

$$\ddot{I} = \Delta f \cdot \log_2\left(1 + S/N\right), \tag{2}$$

where Δf is a frequency band, sensing by the insect's visual organ.

Taking into account the fact that all noises form natural and artificial emission sources can be additively summed [4], we get the following general noise expression:

$$N = N_S + N_{S.E} + N_{L.E} + N_{S.O} + N_{L.O} \tag{3}$$

where $N_S, N_{S.E}, N_{L.E}, N_{S.O}, N_{L.O}$ *are* noises, caused by the direct sun flash, and by the reflecting sun and artificial emission from the earth surface and from the underage cloudiness.

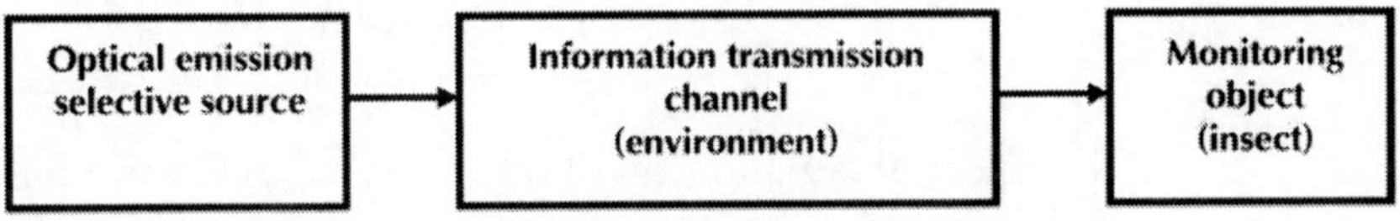

FIGURE 1 Insect behavior biocybernetical monitoring system scheme.

Mathematical modeling of the optical information transmission process into BCS needs to take into account the environmental influence, where artificial and natural emission is spread. The main natural optical clutters, in this case, are celestial bodies, The Earth and its surface, atmosphere.

It is certain that the optical signal sensed by the insects' eye [10] can be written as follows:

$$S = \frac{m}{l^2} \cdot \int_{\lambda_1}^{\lambda_2} r(\lambda) \cdot \tau(\lambda) \cdot \exp\left(-q(\lambda) \cdot l\right) d\lambda, \quad (4)$$

where $l = \sqrt{x^2 + (h_2 - h_1)^2}$, x is the distance between the selective emission source and the monitoring object horizontally, h_1 is the distance between the earth surface and the monitoring object, h_2 is the source height over the earth surface, λ is the wave length, $r(\lambda)$ is the function of the selective light source spectral emissivity, $\tau(\lambda)$ is the function of the insect relative visibility, m is the coefficient registrating the difference between human and insect visibility functions, $q(\lambda) = k(\lambda) + \sigma(\lambda)$, $k(\lambda)$ is the atmospheric transmission spectral coefficient in UV and in the visual spectrum; $\sigma(\lambda) = 0.83NA^3 \cdot \lambda^{-4}$ is a Rayleigh dispersion spectral coefficient [5], N is the molecule number in 1 m^3, A is the molecule cross-section square, m^2.

Taking into account the mathematical model of the information transmission optical channel in the insects behavior monitoring systems [11] expressions for noises $\varnothing_{S.O}, \varnothing_{L.O}$, specified by the Sun emission reflection and artificial selective source from the clouds is the following:

$$N_{S.O}=\frac{m}{\pi}\cdot\left(\frac{R_c}{R_{ç.i}}\right)^2\int_{\lambda_1}^{\lambda_2}\xi(\lambda,T)\cdot k(\lambda)\cdot\tau(\lambda)\cdot Noise(\lambda)\cdot\exp\left(-q(\lambda)\right)\cdot\left(1-\rho(\lambda)\right)d\lambda \quad (5)$$

$$N_{L.O}=\frac{m}{h_2^2\cdot\pi}\int_{\lambda_1}^{\lambda_2}r(\lambda)\cdot\tau(\lambda)\cdot Noise(\lambda)\cdot\exp\left(-q(\lambda)\cdot h_2\right)\cdot\left(1-\rho(\lambda)\right)d\lambda, \quad (6)$$

where $\xi(\lambda,T)$ is a Sun spectral radiation distribution, R_S, $R_{E.O}$ Sun and Orbit radius respectively, $Noise(\lambda)=\left(\mu(\lambda)S_1+\upsilon(\lambda)S_2+\psi(\lambda)S_3\right)$, $\mu(\lambda),\upsilon(\lambda),\psi(\lambda)$ is the spectral characteristics of the soil, water and vegetation reflection respectively, S_1,S_2,S_3 are their weight coefficients, which are fixed with the random numbers generator, making the sequence with the proportional distribution in the given value range (7), $\rho(\lambda)$ is the spectral absorption coefficient of the underedges cloudiness [6, 13].

Random numbers generator is used for imitation of the monitoring systems functioning real conditions [3]:

$$\begin{aligned} S_1 &= rnd(1) \\ S_2 &= rnd(1-S_1), \\ S_3 &= 1-S_1-S_2 \end{aligned} \quad (7)$$

where $rnd(1)$ is a function, allowing to take the equally distributed random number in the given [0,1] range.

As an example for the following calculations we will use the some variations N_1, N_2, N_3 of the S_1, S_2, S_3 weight number sets [Table 1].

As a result of the calculations $N_{S.O}$ and $N_{L.O}$ at (5), (6) we defined that the given noises have a weak influence at the "signal/noise" function value so it is possible to neglect them in the expression (3).

For describing noises N_S, $N_{S.E}$ let's observe the season and the daytime influence on the intensity of the optical range sun emission, reaching the Earth surface. The Earth surface irradiance at the given latitude φ when the weather is fine at the given time *t* equals [7]:

$$E(n,t) = \begin{cases} Q\cos\,\theta(n,t), & \cos\,\theta(n,t) > 0 \\ 0, & \cos\,\theta(n,t) < 0 \end{cases}, \tag{8}$$

where Q is the constant insolation, equal to the sun constant, n is the full earthday number from the beginning of the year, the time t is in the $(0 < t < \tau_0)$ range, where τ_0 is the sun earthday ($\tau_0 = 24$ hours).

The dependence on the sunrays dip θ cosine is the following:

$$\cos\theta(n,t) = \cos\delta(n)\cdot\cos\left[\frac{2\pi}{\tau_0}\left(t + \frac{\tau_0}{2}\right)\right] + \sin\delta(n). \tag{9}$$

Here δ is a declination, which adds the angle between the earth axis and direction to the sun disk center to π/2. Solar declination sine δ as a function of n daytime from the beginning of the year is expressed by the following formula:

$$\sin\delta(n) = \sin\eta\cdot\cos\varepsilon(n), \tag{10}$$

where η is an angle between the earth axis and the vertical to the earth orbit plane ($\eta = 23^0 27'$), $\varepsilon(n)$ is an earth axis azimuth angle, the dependence of which on the season, i.e. on the day number n, is expressed by the formula:

$$\varepsilon(n) \approx \frac{2\pi\tau_0}{\tau_1}(n - 172). \tag{11}$$

In accordance with the taken above assumptions the noise $N_{\tilde{n}}$ from the direct sun illumination can be the following:

$$N_S = m\left(\frac{R_S}{R_{E.O}}\right)^2 \cos\theta(n,t)\cdot\int_{\lambda_1}^{\lambda_2}\xi(\lambda,T)k(\lambda)\tau(\lambda)d\lambda, \tag{12}$$

And the noise at the sun emission reflection from the geological substrate $N_{S.E}$ will be the following:

$$N_{S.E}=\frac{m}{\pi}\left(\frac{R_c}{R_{\c{c}.\hat{i}}}\right)\cos\theta(n,t)\cdot\int_{\lambda_1}^{\lambda_2}\xi(\lambda,T)k(\lambda)\tau(\lambda)\cdot Noise(\lambda)d\lambda \quad (13)$$

Then, substituting Eqs. (3), (12), (13) in Eq. (2), finally we will take:

$$\frac{S}{N}=\frac{m\int_{\lambda_1}^{\lambda_2}r(\lambda)\tau(\lambda)\exp\left(-q(\lambda)\cdot l\right)d\lambda}{l^2\left(N_S+N_{S.E}+N_{L.E}\right)} \quad (14)$$

Let's observe the energy S and informational $S/N,\ddot{I}$ characteristics dependence from the different BCS parameters.

At the Fig. 2 there are the graphics of the "signal/noise" function dependence from the earth surface energy illumination in different seasons and at a different daytime (in this case winter is not observed as it doesn't correspond with the insects vital active phase).

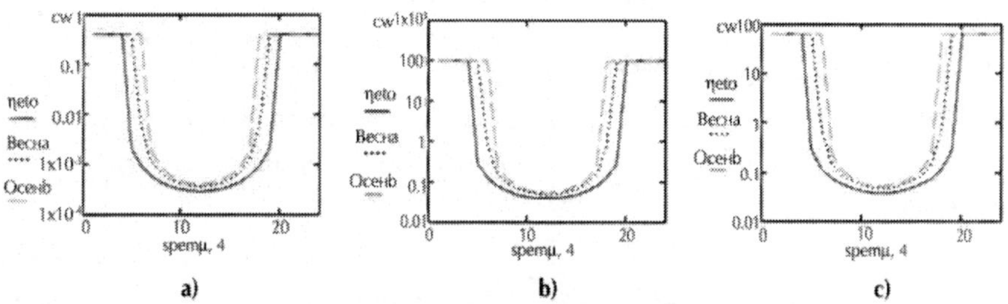

FIGURE 2 The graphics of the "signal/noise" function dependence from the earth surface energy illumination for insects with different vision types: a) monochrome, b) dichrome, c) trichromatic.

From the curve analyses we can notice that the "signal/noise" function changes greatly depending on the season and on the daytime and reaches its maximum at night and then decays quickly and at $t=12\div$ period reaches its minimum.

While modeling the optical information transmission process to the insects it is necessary to define the dependence features of the useful signal $\tilde{N}$ from the altitude h_2 (Fig. 3) and from the distance x (Fig. 4).

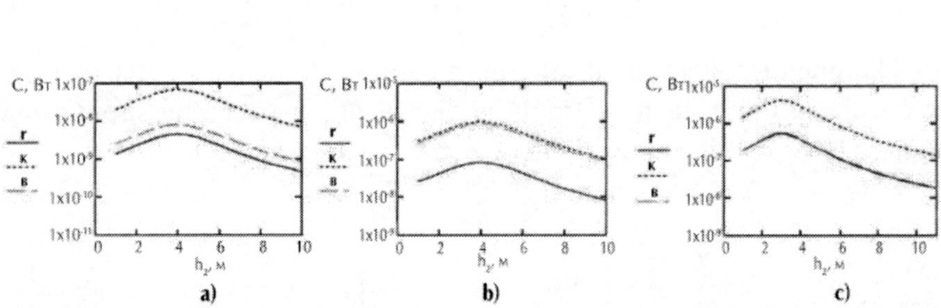

FIGURE 3 The dependence graphics from the altitude h_2, of the useful signal producing by the optical emission selective sources (H–halogen tube, X–xenon lamp, T– tungsten lamp at T=1500 K, x=4м) in the insect's visual organ with the different vision types: a) monochrome; b) dichrome; c) trichromatic.

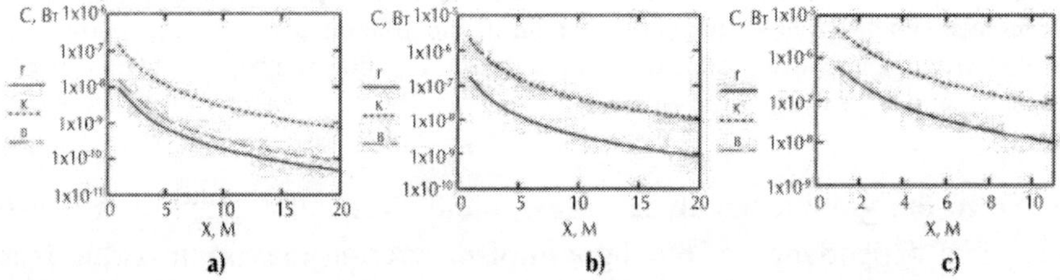

FIGURE 4 The dependence graphics from the distance of the x useful signal, produced by the optical emission selective sources (H– halogen tube, X– xenon lamp, T– tungsten lamp at T=1500 K, h_2=4м) in the insect's visual organ with different vision types: a) monochrome; b) dichrome; c) trichromatic.

As a calculation result there are "signal/noise" function values, which are within 0.003 and 1, which can be seemed as insufficient for effective providing of the insects behavior monitoring process (Fig. 5).

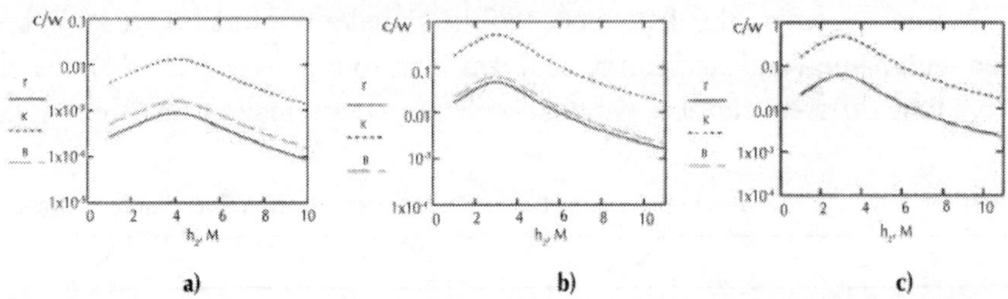

FIGURE 5 The dependence graphics from the altitude h_2 of the "signal/noise" function for the insects with different vision types: a) trichromatic, b) monochrome, c) dichrome and for the electromagnetic emission selective sources (H– halogen tube, X– xenon lamp, T– tungsten lamp at T=1500K, x=4м, N_2).

The graphics analysis at the Fig. 6 shows that there is a dependence of the signal/noise function from the distance x between the monitoring signal source and the monitoring object.

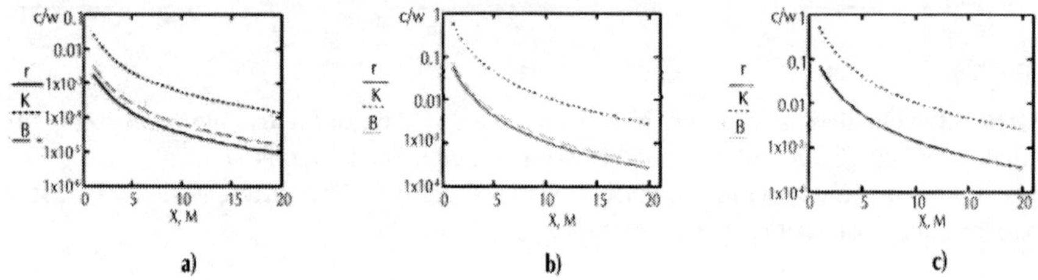

FIGURE 6 The dependence graphics from the altitude x of the signal/noise function for the insects with different vision types: a) trichromatic, b) monochrome, c в) dichrome and for the electromagnetic emission selective sources (H– halogen tube, X– xenon lamp, T– tungsten lamp at T=1500K, x=4м, N_2).

From the graphics analysis (Figs. 5 and 6) it follows that in all cases there is a dependence of the signal/noise function maximum value from the altitude of the electromagnetic emission selective source above the Earth surface what defines its optimal value for the insects with different vision types. For the insects with monochrome vision type there is the signal/noise function maximum value at $h_2 = 4,2ì$ and $x = 1,7ì$, and for the insects with di-, and also with trichromatic vision types it is at $h_2 = 3ì$ and $x = 1ì$. It is also noticed that the signal/noise maximum value is reached while using a xenon lamp for the insects with all vision types, and the signal/noise values for a halogen tube and a tungsten lamp are practically equal for the insects with mono- and dichrome vision types.

While evaluating the dependence of the signal/noise function (Fig. 7) created with natural and artificial optical emission source (*halogen tube*), we observed the different surfaces weight coefficients combinations in (Table 1.)

TABLE 1 Sets of the weight numbers of the natural surfaces reflective characteristics.

	N_1	N_2	N_3
S_1	0.696	0.543	0.211
S_2	0.133	0.199	0.437
S_3	0.171	0.257	0.353

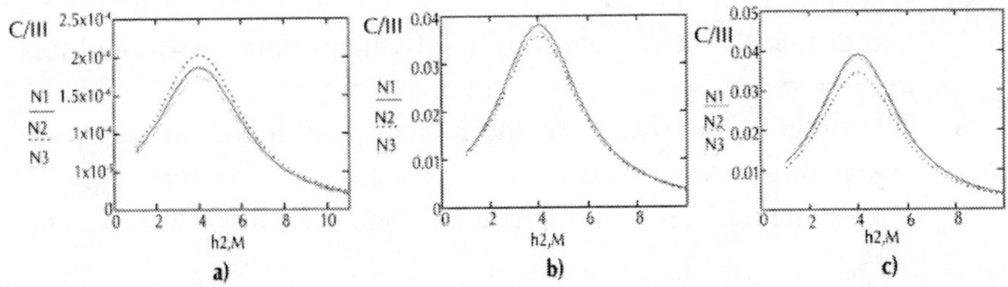

FIGURE 7 The dependence graphics of the signal/noise function from the source altitude above the Earth surface at the random set of the natural surfaces weight coefficients (W–water surface, S–soil, F–flora) for the insects with different vision types: a) monochrome, b) dichrome, c) trichromatic.

From the graphics analysis at the Fig. 7, it follows that for all natural surfaces weight coefficients sets there is the dependence of the signal/noise function maximum value from the electromagnetic emission selective source altitude above the Earth surface. Signal/noise function reaches its maximum at the altitude $h_2 = 4\grave{i}$, than it has a steady decreasing character. Also for all vision types there is a vivid dependence from the random set of the natural surfaces weight coefficients.

11.2 CONCLUSIONS

1. The mathematical model of the process of the optical information transmission in the insects behavior monitoring systems was made;
2. Analytical expressions for energy S and information $S/N, \ddot{I}$ characteristics of the optical information transmission process to the insects were received;
3. The dependence of the energy and information characteristics of the optical information transmission channel to the insects with different vision types from different BCS parameters were investigated;
4. The dependence of the energy and information characteristics from the different BCS parameters of the optical information transmission channel to the insects with different vision types were investigated;

5. It was defined that the natural illumination mode during the year and during the daytime strongly influences onto the optical information transmission process to the insects;
6. The optimal parameters of the related position of the monitoring signal source and receiver of the insects with different and of the electromagnetic emission artificial selective source were calculated;
7. It was stated that the natural surfaces spectral characteristics weight coefficients changes lead to the signal/noise function values change;
8. It was shown that the use of one standard source of the monitoring signal in BCS can be insufficient for providing the effective process of the insects behavior monitoring.

KEYWORDS

- **Biocybernetical system**
- **Earth axis azimuth angle**
- **Halogen tube**
- **Insect's visual organ**
- **Signal/noise function**

REFERENCES

1. Andreevsky, A. S. Certification to the useful model No. 66889 – Insecticide device.
2. Belenov, V. N.; Gazalov, V. S. The analysis of the electrooptical transducer installation influence on the garden plant defense quality, Electrotechnologies and electrical equipment in agricultural production. *Zelenograd*, **2003**, *3*, 30–33.
3. Bobnev, M. P. In *Random Number Generation*. M.: Energy, 1997; 230 pp.
4. Gazalov, V. S. Installations of the electrophysical garden defense from the destructive insects, In *Agriculture Rational Motorization*, M., 1984; pp. 6–9.
5. Van der Hulst, G. In *Light Reradiation by Small Particles.* Foreign Literature Publishing House: M., 1962; 537 pp.
6. Zuev, V. E.; Krekov, G. M. Monography. Atmospheric optics modern problems. V.2. *Gidrometeoizdat*, 1986; 256 pp.

7. Kladieva, A. S.; Dzhalmukhambetov, A. U.; Calculations in the MATLAB environment of the latitude-temporal distribution of the sun energy on the Earth surface. Projecting of the engineering and scientific applications in the MATLAB environment: materials of the IVth Russian scientific conference. Astrakhan: Publishing House «Astrakhan University», 2009; pp.146–155.
8. Lihter, A. M. Optimal projecting of the optic-valved systems. In *Monograph*. Astrakhan: Publishing House «Astrakhan State University», 2004; 241 pp.
9. Lihter, A. M. Modeling of the fishing processes management system. In *Monograph*. Astrakhan: Publishing House «Astrakhan University», 2007; 290 pp.
10. Mazohin–Porshnyakov, G. A. In *Insects Sensors Physiology Manual*. M.: Publishing house of the Moscow University, 1977; 456 pp.
11. Pleshkova, U. A.; Lihter, A. M. The model of the optical information transmission process in insects behavior management systems, *Ecol. Sys. Dev.*. **2010**, *2*, 24–27.
12. Ryibkin, A. P.; Kazakov, V. P. Patent for invention No. 2001100190 – Protective system from the sanguivorous flies.
13. Timofeev, U. M.; Vasyil'ev, A. V. In *Atmospheric Optics Theoretical Basis*. SntP.: Science Publishing House, 2003; 474 pp.

CHAPTER 12

THE EFFECT OF FREE RADICAL OXIDATION ON STRUCTURE AND FUNCTION OF PLASMA FIBRIN-STABILIZING FACTOR

M. A. ROSENFELD, A. V. BYCHKOVA, A. N. SHEGOLIHIN, V. B. LEONOVA, M. I. BIRYUKOVA, E. A. KOSTANOVA, S. D. RAZUMOVSKII, and M. L. KONSTANTINOVA

CONTENTS

12.1 INTRODUCTION

Plasma fibrin-stabilizing factor (pFXIII) is one of the key proteins of blood coagulation. The function of pFXIII is to maintain a hemostasis by the fibrin clot stabilization accompanied by its increasing mechanical strength and the enhanced resistance of the fibrin clot to plasmin. Its reference interval in the human blood plasma equals to 14–28 μg/ml [1]. It is a heterotetramer (FXIII-A_2B_2) with the molecular weight of 320 kDa consisting of the two single-stranded catalytic A subunits (FXIII-A_2) with the molecular weight of subunit of ~83 kDa and the two identical single-stranded inhibitory/carrier B subunits (FXIII-B_2) with the molecular weight of the subunit of ~73 kDa. The subunits are held together by weak non-covalent bonds [2, 3]. The pFXIII activation is a multistage process. The first stage is the thrombin-catalyzed hydrolytic cleavage of an Arg37-Gly38 bond from the amino-terminus of the FXIII-A subunit, this leads to a release of the activation peptide AP. FXIII-A_2B_2 transforms into FXIII-$A_2{'}B_2$ which still does not have an enzymatic activity [4]. The second stage of the activation requires calcium ions. In the presence of the calcium ions the heterosubunits dissociate with the formation of FXIII-$A_2{'}$ and FXIII-B_2. At the last stage FXIII-$A_2{'}$ undergoes conformational changes in the presence of Ca^{2+}, this leads to the Cys314 active site exposure accompanied by the formation of the enzyme FXIII-A_2* (FXIIIa) which belongs to the family of the transglutaminases (*endo*-γ-glutamine: ε-lysine transferases, EC2.3.2.13) [5]. In the presence of FXIIIa and calcium ions, fibrin polymers undergo interchain covalent crosslinking by generation of the ε-(γ-*glu)*lys isopeptide bonds [6]. COOH-terminal sites of γ chains located in the D peripheral regions of the monomeric fibrin molecules are covalently bound to each other to form the intermolecular γ–γ-dimers [7]. Factor XIIIa also catalyses a formation of the isopeptide bonds between α chains of neighbour fibrin molecules. In the sequence of this α *chain* of one *molecule* interacts with α chain of two other molecules what leads to appearance of α-polymers which include more than five α fibrin polypeptide chains [8]. FXIIIa is able to engage fibrin α chains to the covalent linking with other proteins such as α_2-antiplasmin and fibronectin [9]. Today, the catalytic function of FXIII-A_2* has been studied sufficiently well, whereas the role of the noncatalytic FXIII-B subunits is not as clear. These subunits are supposed to be carriers of the catalytic FXIII-A subunits in the bloodstream and protect them against possible proteolytic degradation maintaining a proper level of zymogen in the bloodstream in the same way [10, 11]. Besides, the aforementioned FXIII-B subunits perform a regulatory function controlling the process of the FXIII-A_2B_2 activation by thrombin [12, 13].

pFXIII as well as many other proteins circulating in the blood plasma is known to be a target for reactive oxygen species (ROS) bringing about processes of free radical oxidation of proteins. Free radical oxidation of proteins can be accompanied by a cleavage of polypeptide chains, a modification of amino acid residues and a transformation of proteins to the derived proteins which are highly sensitive to the proteolytic degradation [14]. The modified proteins become accumulated in the body with aging as a result of oxidative stress and a number of diseases [15]. It was demonstrated that fibrinogen, factors V, VIII, X, XIII, and the proteins of fibrinolytic system are sensitive to ROS [16–19]. Fibrinogen is more sensitive to the oxidative modification compared to other main plasma proteins such as albumin, immunoglobulins, transferring and ceruloplasmin [20]. This allows fibrinogen to undergo a reaction of free radical oxidation easier. Induced oxidation of the fibrinogen degradation products formed under the action of plasmin (so called fragments E and D which are chemical and structural homologues of the central and outer regions of fibrinogen) showed that an oxidative capacity of fragment D is essentially higher than an oxidative capacity of fragment E [21]. The comparison of these data and results regarding fibrinogen molecule oxidation [22, 23] allowed us to draw the conclusion that the regions D of fibrinogen are interceptors of free radicals and they perform a protection of the region E against free radical attack providing the key reactions of thrombin binding and fibrinopeptides elimination followed by fibrin self-assembly. Therefore, one may suppose that a similar mechanism is possible for pFXIII to some extent. In other words, besides the functions of FXIII-B_2 mentioned above, the FXIII-B subunits may also inhibit oxidation of FXIII-A_2 containing highly sensitive to oxidation cysteine residue localized in the active center of the catalytic subunit. Therefore, the goal of this study was to determine the influence of free radical oxidation on structural and functional properties of pFXIII at different stages of its activation.

12.2 EXPERIMENTAL METHODS

12.2.1 ISOLATION OF PFXIII.

pFXIII was obtained from human blood plasma by fractional precipitation procedure with ammonium sulphate and subsequent ion-exchange chromatography on DEAE-ToyPearl M650 (Tosoh, Japan) [24]. Transformation of pFXIII to FXIIIa thrombin (Roche, France) in the presence of calcium ions was used as it was described earlier [25]. Activity of FXIIIa was determined by the Lorand method [24] and was equal to 500 units per ml. pFXIII with different way of treatment was oxidized: pFXIII (Sample 1), pFXIII treated with thrombin (Sample 2), pFXIII in the presence of 5 mM Ca^{2+} (Sample 3) and FXIIIa, obtained after the activation of pFXIII with thrombin in the presence of 5 mM Ca^{2+} (Sample 4). The oxidized Samples 1–3 were transformed to FXIIIa and an enzymatic activity of both these samples and the oxidized Sample 4 was compared with the enzymatic activity of non-oxidized Sample 4. Polypeptide chains of the FXIII-A and FXIII-B subunits of oxidized and non-oxidized fibrin-stabilizing factor were shown by polyacrylamide gel electrophoresis to be characterized by molecular weight equal to the native protein molecular weight (Fig. 1). This result points to the absence of both interchain cross linkages and free radical cleavage of the polypeptide chains.

12.2.2 COVALENT LINKING OF FIBRIN CHAINS

Fibrinogen was obtained from citrated human plasma as it was described in previous works [22, 23]. Covalent linking of fibrin chains was catalyzed by different samples of FXIIIa in 0.05 M tris-HCl buffer with pH 7.4 containing 0.15 M NaCl and 5×10^{-3} M $CaCl_2$. A 0.05 ml of thrombin solution (0.25 units NIH) and 0.01 ml of FXIIIa were added to 1 ml of fibrinogen with the concentration of 2 mg/ml. Fibrin crosslinking was terminated after 15 min using 7 M urea and 2% SDS. The existence of covalent crosslinking of the polypeptide chains of the stabilized fibrin formed under the action of factor XIIIa was detected by PAGE (7.5%) of reduced samples in the presence of 1% β-mercaptoethanol. Gels were stained by Coomassie brilliant blue R-250.

12.2.3 OXIDATION OF FIBRIN-STABILIZING FACTOR

Free radical oxidation of fibrin-stabilizing factor was initiated by ozone. The solutions of fibrin-stabilizing factor (2.0 mg/ml) were ozonized in a reactor (3.3 ml) by blowing the ozone–oxygen mixture through the free volume of the reactor [23]. The amount of oxidation agent was equal to 2×10^{-7} M.

12.2.4 SPECTROPHOTOMETRIC MEASUREMENTS

UV spectra of different samples of pFXIII before and after oxidation were recorded on spectrometer SF-2000 (Russia) in 1-cm quartz cells at room temperature.

12.2.5 FTIR MEASUREMENTS

The study of functional groups content was performed on VERTEX-70 Fourier transform IR spectrometer (Bruker, Germany) with a DTGS detector, as it was described in details earlier [21]. The FTIR spectra of the different samples of non-oxidized and oxidized of pFXIII were recorded at room temperature between 4000 and 400/cm at 4/cm nominal resolution accumulating 256–512 scans per spectrum. The samples for measurement were prepared by putting 5–10 μl of aqueous sample solution onto the surface of a silicon plate ($2 \times 2 \times 0.1$ cm^3) followed by evaporation of water at room temperature to constant weight. The evaporation was carried out in a nitrogen atmosphere to prevent the samples against oxidation by ROS pollutants during drying. The FTIR spectra of completely dried samples were recorded in pass through mode using a Micro Focus add-on device (Perkin Elmer, England) equipped with a CaF_2 lens.

12.2.6 RAMAN SPECTROSCOPY MEASUREMENTS

Raman spectra in the range between 3600 and 100/cm at 5–8/cm nominal resolution were registered on Raman scattering microscope Senterra (Bruker, Germany). Laser with wavelength of 785 nm and output power of 25 mW was a light source. Raman spectra were registered using 180° optical geometry. The 50-fold microscope objective and signal accumulation time of 60–120 s was used for excitation of scattering and Raman photon concentration. Tensor-27 and Senterra control device and also FTIR and Raman spectra processing were performed using software package "OPUS" v.6.0 (Bruker).

12.2.7 DYNAMIC LIGHT SCATTERING MEASUREMENTS

Hydrodynamic diameters of different samples of pFXIII before and after oxidation were measured using dynamic light scattering (DLS) on Zetasizer Nano-S (Malvern, England) with a detection angle of 173° as it was described earlier [23]. The results have been processed using the Origin 7.5 computer program.

12.2.8 ESR SPECTROSCOPY MEASUREMENTS

The structural and dynamical changes of different samples of pFXIII in the oxidizing process were measured using electron spin resonance (ESR) spectroscopy of spin labels [21]. Stable nitroxide radical was used as a spin label in this work. The average amount of spin labels on the FXIII macromolecules was 1 label per 4–5 macromolecules. ESR spectra of radicals were registered on X-band spectrometer Bruker EMX 2.7/8 (Germany) at temperature of 25°C, microwave power of 5 mW, modulation frequency 100 kHz and amplitude 1 G. The first derivative of the resonance absorption curve was detected. The samples were placed into the cavity of the spectrometer in a quartz flat cell. Magnesium oxide powder containing Mn^{2+} ions was used as an external standard in ESR experiments. The following parameters were determined from ESR spectra: intensities of the nitroxide triplet extreme lines ($I_{+1,-1}$), linewidth of low-field line of the nitroxide triplet (ΔH_{+1}). These parameters were used for the estimation of rotational correlation time of labels (τ) in interval of 0.1–1.0 ns using equilibrium $\tau = 6.65 \times 10^{-10}\ \Delta H_{+1}\ [(I_{+1}/I_{-1})^{1/2} - 1]$. The error in τ determination is less than 10%. Rotational correlation times of labels as well as a fraction of labels with slow motion ($\tau > 1$ ns) were evaluated using tabulated spectra [26].

12.3 RESULTS

12.3.1 EFFECT OF FREE RADICAL OXIDATION ON ENZYMATIC ACTIVITY OF PFXIII

Ozone-induced oxidation of fibrin-stabilizing factor influences to a considerable extent on its enzymatic activity. It was shown that enzymatic activity of FXIIIa formed from oxidized FXIII-A_2B_2 (Sample 1) was lower compared with non-oxidized FXIIIa (Sample 4) with enzymatic activity of 500 units/ml and was equal to approximately 400 units/ml. At the same time enzymatic activity of FXIIIa obtained from oxidized samples 2 and 3 decreased several times and was equal to ~120 and 60 units/ml, respectively. The decline of enzymatic activity (40 units/ml) took place after oxidation of FXIIIa (Sample 4). These results demonstrate that the reduction of enzymatic activity depends significantly on the stage of the pFXIII conversion into FXIIIa at which oxidation was carried out.

In the presence of FXIIIa the polypeptide fibrin chains undergo covalent crosslinking process (Fig. 1). Under the action of non-oxidized FXIIIa γ polypeptide chains of fibrin are nearly entirely involved into their covalent crosslinking process with formation of γ–γ-dimers with the molecular weight of 95 kDa. α polypeptide chains also undergo enzymatic linking accompanied with α-polymers generation. Since α-polymers incorporate more than five α-polypeptide chains of fibrin [8] their molecular weight is as high as 500 kDa or more. β-chains are not involved in covalent linking process so their content remains constant. γ-chains are known to form γ–γ-dimers for several minutes while the process of α-polymerization is considerably slower and fibrin samples as a rule include some part of intact α polypeptide chains [27]. Oxidation of samples 1–4 leads to a different involvement of both γ and α polypeptide fibrin chains to the process of their covalent linking (Fig. 1). A considerable amount of γ–γ-dimers and the presence of α-polymers in fibrin point to a high degree of preservation of enzymatic activity of FXIIIa formed from oxidized FXIII-A_2B_2 (sample 1) and FXIII-$A_2'B_2$ (sample 2). The samples of fibrin-stabilizing factor oxidized by ozone in the presence of 5 mM Ca^{2+} (sample 3) or activated by thrombin additionally (sample 4) have the least enzymatic activity what can be seen in a total absence of the formation of α-polymers and the presence of only trace quantities of γ–γ-dimers. These results are completely consistent with the data on enzymatic activity of different samples of FXIIIa shown above.

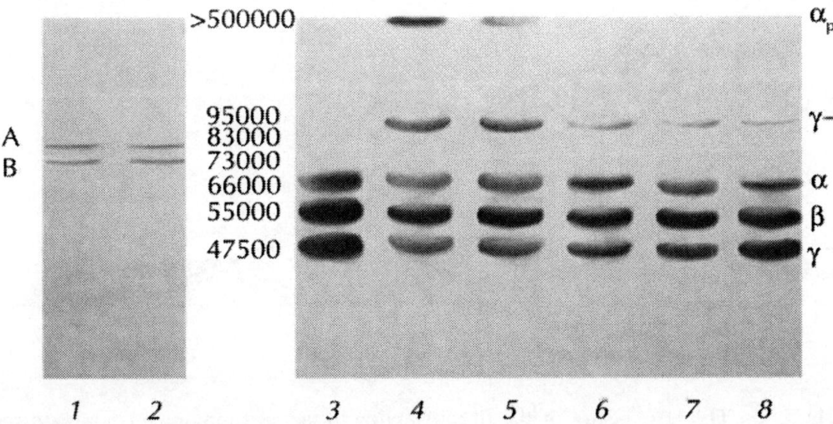

FIGURE 1 PAGE of the unreduced samples for non-oxidized (*1*) and oxidized (*2*) pFXIII and the reduced samples of fibrin: the polypeptide chains of non-stabilized fibrin (*3*); the polypeptide chains of fibrin formed in the presence of the different samples of FXIIIa (*4–8*): non-oxidized FXIIIa, FXIIIa obtained by activation of oxidized pFXIII, FXIIIa obtained by activation of oxidized FXIII-$A_2'B_2$, FXIIIa obtained from pFXIII oxidized in the presence of Ca^{2+}, oxidized FXIIIa, respectively.

12.3.2 TRANSFORMATION OF THE MOLECULAR STRUCTURE OF PFXIII UNDER FREE RADICAL OXIDATION

Changes of the UV spectra of the samples at 280 nm (Fig. 2) point to active interaction between ozone and aromatic amino acid residues (tryptophane, tyrosine and phenylalanine) of the fibrin-stabilizing factor molecules. The absorption spectra maxima shift to short-wave region and reduction of their intensities are caused by formation of the quinoid structures from phenoxil and imidazole cores of amino acid residues under the influence of ozone [23]. Besides, samples 2, 3 and 4 are oxidized to a greater degree than sample 1 what can be seen from the spectra.

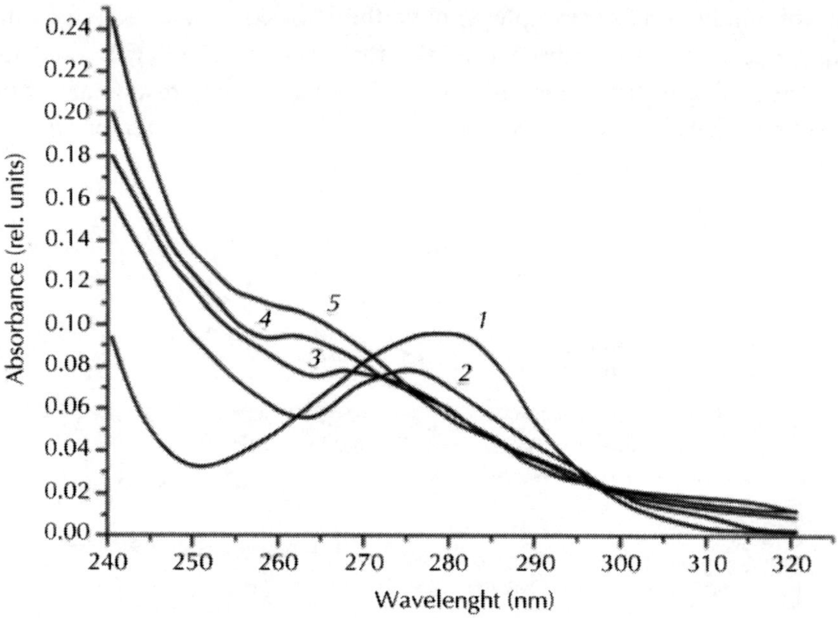

FIGURE 2 The UV spectra of the fibrin-stabilizing factor samples: *1* – non-oxidized pFXIII; *2–5* – ozone-induced oxidized proteins: pFXIII, FXIII-$A_2'B_2$, pFXIII in the presence of Ca^{2+}, FXIIIa, respectively.

Transformation of the molecular structure as a result of the ozone-induced free radical oxidation of the different samples of fibrin-stabilizing factor was studied by techniques of vibrational spectroscopy. On FTIR spectrum of pFXIII there are many well-defined absorption bands in the region of valence vibration of X-H (3000–2400/cm) and in low-frequency region (1300–500/cm). Both of these areas may be a good source of structural information in this case. Particularly the bands of S-H valence vibration of cysteine residues in protein appear in the region of 2520-2600/cm, where there are no other absorption bands. Since pFXIII and FXIIIa have only a small number of SH groups [2] changes in their FTIR spectra are possible to be associated with local changes in the protein structure. FTIR spectra of the non-oxidized and oxidized pFXIII samples demonstrated only slight differences in molecular and supramolecular structures. Therefore differential FTIR spectroscopy was used to define the exact differences in the structure. The results for the most important regions of FTIR spectra of non-activated pFXIII before and after oxidation are presented in the Fig. 3. Positive bands (lying above zero level) in these differential spectra are a measure of decreasing molecular fragment responsible for vibration at the current frequency. Bands lying under zero level point to an appearance of a new element of the molecular structure as a result of oxidation. Spectral analysis displays that ozone-induced oxidation of pFXIII leads to a reduced number of fragments N–H in the peptide backbone (the band 3250–3150/cm with a peak near 3200/cm) and C–H near double bond or in phenolic fragment (3060–3013/cm) (Fig. 3a). At the same time the new molecular fragments C=O of aromatic aldehydes and ketones (negative band in the range of 1720–1660/cm) appeared and a content of carbonyls (two vibration bands of carboxylate ions at 1555–1400/cm) decreased (Fig. 3b). These carbonyls are possibly belonging part of amino acid residues in the protein side chains. A set of bands in the differential spectrum which is responsible for vibrations of derivatives of sulfuric acids illustrates an appearance of the oxidation products of S-containing fragments due to free radical oxidation (Fig. 3c). A set of decreasing and growing bands ν(C–S) which belong to structural fragments of R–S–S–R and R–S–R type, respectively (Fig. 3d) is in good agreement with cleavage of disulfide bridges during oxidation and with oxidation of cysteine fragments S–H.

Differential FTIR spectrum presented in Fig. 4 illustrates the difference between molecular structures of oxidized pFXIII and oxidized FXIIIa. FXIIIa turns out to be more sensitive to ozone-induced oxidation compared with pFXIII. This is proved, for example, by the presence of bands of the additional decrease in a number of molecular fragments C–H, =C–H and S–H, and N–H in a set of amide bands.

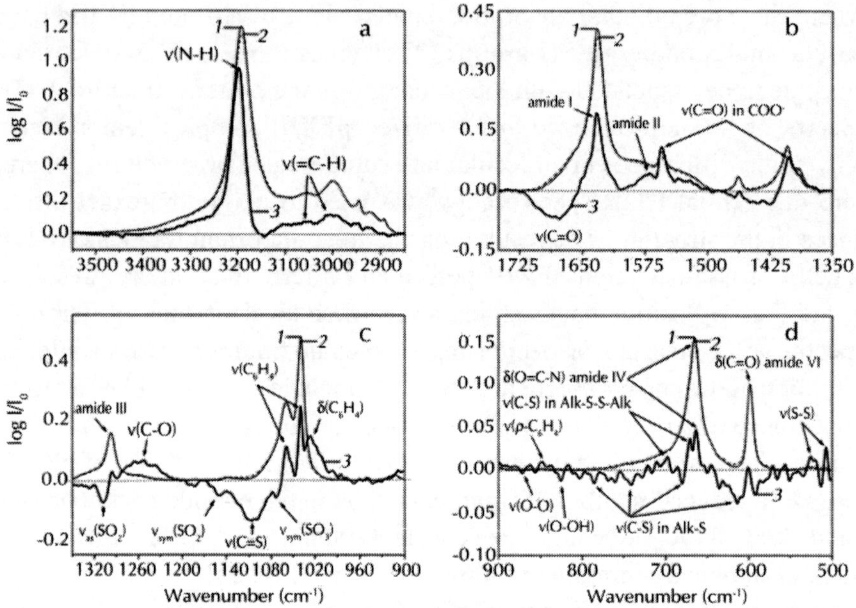

FIGURE 3 The differential FTIR spectra of pFXIII: (a) – in the area of valence vibration of X-H; (b) – in the area of valence vibration of carbonyls and amide I and amide II bands; (c) – in the area of «fingerprints»; (d) – in the area of deformational vibration and valence vibration of heavy atoms. *1* – non-oxidized pFXIII; *2* – oxidized pFXIII; *3* – differential spectrum enlarged along axis Y for demonstrativeness.

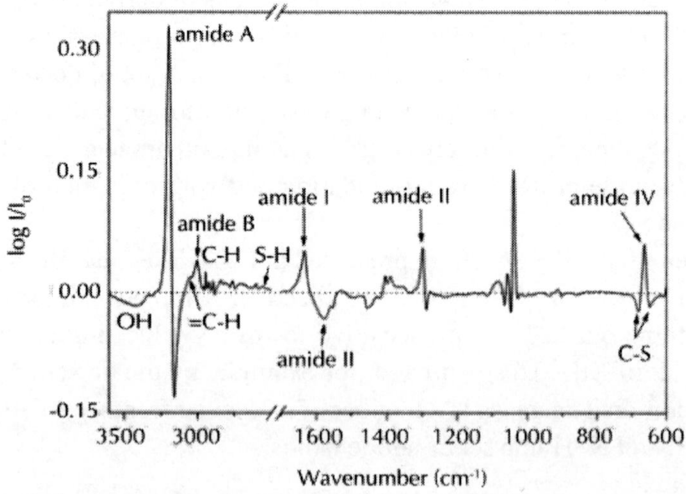

FIGURE 4 The differential FTIR spectrum of ozone-induced oxidized pFXIII relative to oxidized FXIIIa.

Raman spectra are also highly informative in obtaining structural information (Fig. 5). The results revealed an accumulation of aldehyde groups when pFXIII is oxidized. Cysteine groups presented in a different molecular environment in the sample of pFXIII before oxidation (Raman bands at 2587, 2560 and 2545/cm) transform to the other products as a result of oxidation (Fig. 5b) (what corresponds to FTIR analysis results). Non-resonance combinational scattering may be used for studying transformations of the secondary structure of pFXIII under the action of ozone. The spectra have changes in amide I and amide II bands in the region of 1700–1500/cm illustrating essential modifications of the pFXIII secondary structure (Fig. 5c). A transformation of disulphide S–S bonds is observed when pFXIII is oxidized by free radicals (Fig. 5d). An amount of fragments S–S containing in pFXIII had to decrease under the action of ozone. Herewith a set of new bands appears in the oxidized product spectrum. These new bands belong to vibrations ν(C-S) in the different structures of type –C–S–R. The nature of R was interpreted earlier on the basis of FTIR spectroscopy data (Fig. 3c–d).

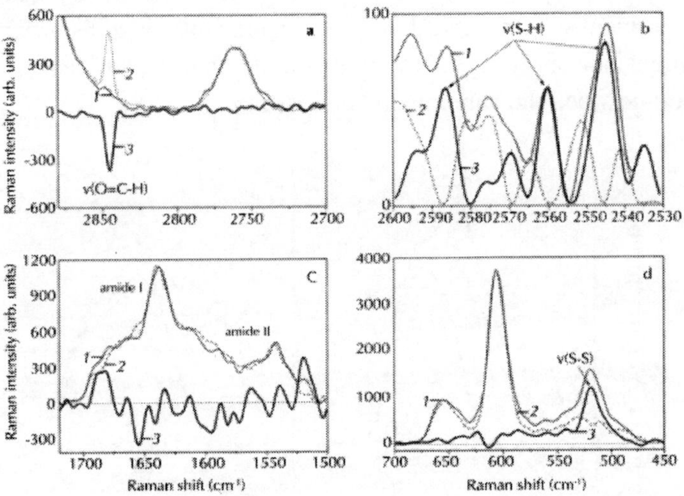

FIGURE 5 The differential Raman spectra of pFXIII: (a) – growth in number of aldehyde groups, band ν(O=C–H) at 2845/cm; (b) – decrease in number of cysteine thiol groups, bands ν(S–H) at 2586, 2560 and 2545/cm; (c) – essential changing of the secondary structure of FXIII, bands of amide I and II in the range of 1700–1500/cm; (d) – decrease in number of molecular disulphide fragments, a band ν(S–S) at 519/cm. *1* – non-oxidized pFXIII; *2* – oxidized pFXIII; *3* – the differential spectrum enlarged along axis Y for demonstrativeness.

12.3.3 CHANGES OF THE PFXIII CONFORMATION DURING OXIDATION

The results of dynamic light scattering demonstrate that all of the non-oxidized and oxidized samples of fibrin-stabilizing factor are characterized by a unimodal size distribution (Fig. 6). Hydrodynamic diameters of non-oxidized molecules of pFXIII and FXIIIa are equal to about 15 and 10 nm, respectively, what corresponds to the data obtained by means of gel-filtration and ultracentrifugation [28]. Activation of FXIII-A_2B_2 by thrombin or calcium ions is accompanied by growth of the macromolecule size what is in agreement with the results obtained earlier [28]. This is associated with structure loosening brought about conformational changes in protein. Oxidation of samples 1–3 is accompanied by increasing in sizes. The maximal size is achieved for the sample 3. A size of the sample 4 does not undergo noticeable changes during oxidation. This allows us to suppose that a generation of the oxygen-containing groups able to form new hydrogen bonds with solvent during the process of free radical oxidation is not of great importance to the phenomenon mentioned above. In other words the size increase for the oxidized samples 1, 2, and predominantly for sample 3 is caused apparently by a disturbance of the three-dimensional packing of the polypeptide chains and weakening of an interaction between the subunits in heterotetramer.

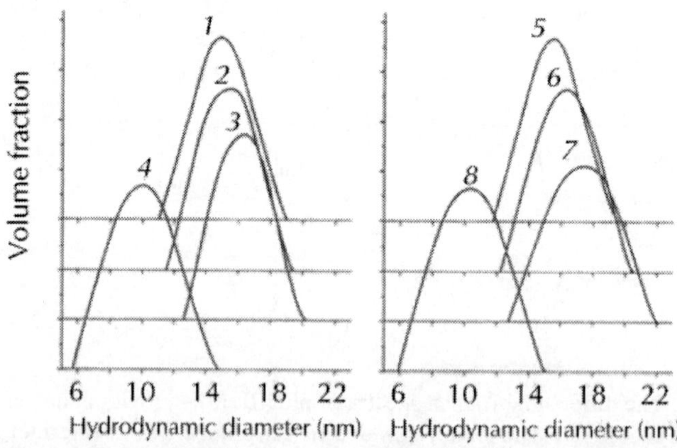

FIGURE 6 The size distribution for the different fibrin-stabilizing factor samples obtained by dynamic light scattering: *1–4* – non-oxidized samples of pFXIII, FXIII-$A_2'B_2$, pFXIII in the presence of Ca^{2+}, FXIIIa, respectively; *5–8* – the same but ozone-induced oxidized samples.

As it was shown by ESR spectroscopy of spin labels the rotational correlation times of the radicals covalently bound to protein macromolecules change as a result of free radical oxidation of these samples. The ESR spectra of radicals covalently bound to amino acid residues of the polypeptide chains of non-oxidized proteins (Fig. 7) are superposition of narrow and wide lines. This indicates the existence of at least two types of areas of label locations; these areas differ in rotational mobility of the labels [28]. Rotational correlation times calculated on the assumption of existence of two main regions of spin label location are equal to $\tau_1 \approx 1 \cdot 10^{-9}$ s and $\tau_2 \approx 2 \cdot 10^{-8}$ s. The portion of labels with a slow rotation is about 70%. Therefore, a considerable number of labels are situated in the internal areas of macromolecules with a high microviscosity. Spectra of non-oxidized samples 1–4 correspond to each other what can be explained by equal microviscosity of the areas where these labels on the samples are positioned. Moreover, it proves that the structure of these areas does not change during the process of enzyme formation. The spectra of spin labels bound to oxidized samples 1–4 essentially differ from the spectra discussed above and from one another (Fig. 7). The portion of labels with slow rotation bound to all of the oxidized samples decreases till 0–50% compared to non-oxidized samples. Labels in the oxidized sample 4 are characterized by the strongest rotation. Rotational correlation time of these labels calculated using equation given above equals to 0.08 ns. The dramatic change of rotational frequency of labels in enzyme as a result of its free radical oxidation is an evidence of the most significant change of the protein conformation (at least in the area of label location) compared with other samples. Strong change of rotational correlation time of labels also occurs when sample 3 is oxidized (0.15 ns). A weakening of interaction between the FXIII-A and FXIII-B subunits of FXIII-A_2B_2 in the presence of calcium ions is likely to lead to further loosening of three-dimensional structure of protein during the process of free radical oxidation. Mobility of labels covalently bound to samples 1 and 2 oxidized under the action of ozone is characterized by rotational correlation times of 0.23 ns and 0.20 ns, respectively. The comparison of ESR spectra also allows one to conclude that there is a larger portion of labels with a slow rotation in the oxidized pFXIII than in the oxidized samples 2–4, which can be given rise to the minimal structural transformations in this sample during oxidation. It should be emphasized that the number of radicals which are covalently bound to samples after oxidation is 2 or 3 times lower than before oxidation. The possible reason for this is the decreasing number of amino groups available to radicals what is in agreement with FTIR data.

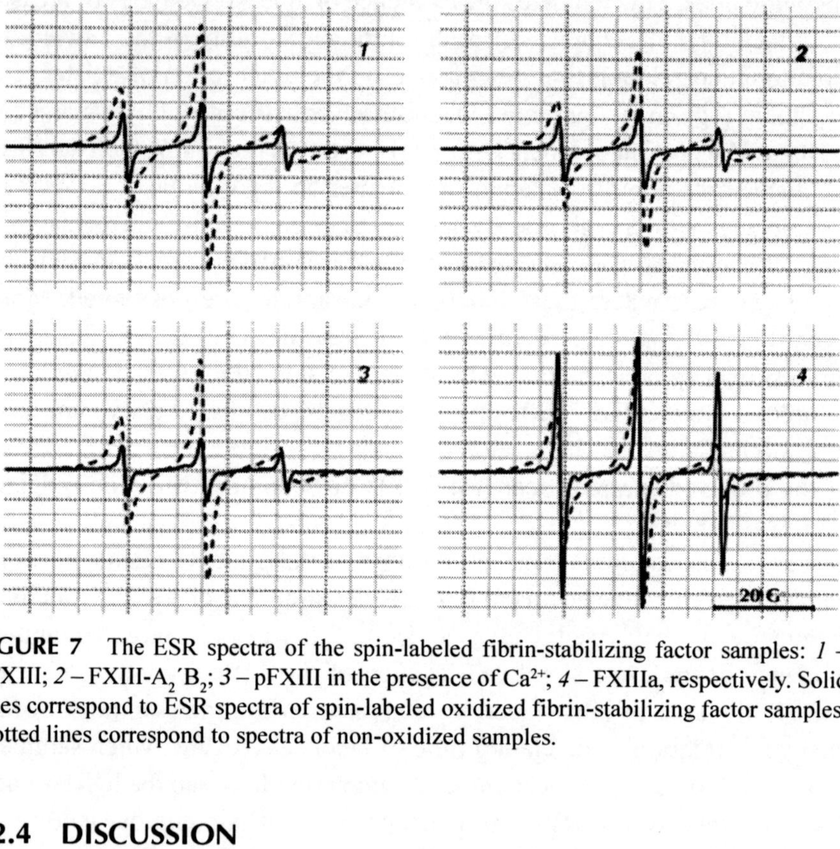

FIGURE 7 The ESR spectra of the spin-labeled fibrin-stabilizing factor samples: *1* – pFXIII; *2* – FXIII-$A_2'B_2$; *3* – pFXIII in the presence of Ca^{2+}; *4* – FXIIIa, respectively. Solid lines correspond to ESR spectra of spin-labeled oxidized fibrin-stabilizing factor samples. Dotted lines correspond to spectra of non-oxidized samples.

12.4 DISCUSSION

The set of results obtained demonstrates that fibrin-stabilizing factor is characterized by different sensitivity to ozone-induced free radical oxidation depending on the stage of its activation. This can be explained by features of the pFXIII structure and a mechanism of its transformation to FXIIIa. As it was shown earlier FXIII-A_2B_2 consists of two pairs of the single-strained FXIII-A and FXIII-B subunits. The FXIII-A subunit consists of 731 amino acid residues including 6 Cys and has no disulphide bonds [13]. At the present time the three-dimensional structure of FXIII-A_2 is understood fairly well [2, 29–31]. The FXIII-A subunit includes N-terminal activation peptide AP-FXIII (Ser1-Arg37) and four domains which are so-called β-sandwich domain (Gly38-Phe184), a catalytic core domain (Asn185-Arg515) and two β-barrel domains which are β-barrel 1 domain (Ser516-Thr628) and β-barrel 2 domain (Ile629-Ser731). The activation peptide and β-barrel 1 domain mask amino acid residues playing a key role in the catalytic process and located inside the catalytic core domain (Trp279, Cys314, His373 and Asp396). Furthermore AP-FXIII of one FXIII-A subunit prevents disclosure of active center situated in another FXIII-A subunit. The catalytic subunit structure is stabilized by several hydrogen bonds and salt bridges [30]. The FXIII-B subunit consists of 641 amino acid residues which include no free thiol groups. The subunit is assembled from ten short consensus repeats, so-called sushi domains, each of them are stabilized by two disulphide bonds [31]. Interaction between homo- and heterosubunits of fibrin-stabilizing factor determines compact and dense structure of the protein.

We have established that pFXIII is the least sensitive to free radical oxidation in comparison with the other samples of fibrin-stabilizing factor. FXIIIa formed from the oxidized pFXIII (Sample 1) retains its enzymatic activity to full extent. This results the accumulation of a considerable amount of γ–γ-dimers and formation of α-polymers from a part of α polypeptide chains during the process of covalent fibrin stabilization. The data obtained from dynamic light scattering demonstrate a minimal growth of hydrodynamic diameter of fibrin-stabilizing factor due to ozone-induced oxidation compared to samples 2 and 3. As it was already noted enlargement of macromolecule dimensions is possibly determined by disturbance of three-dimensional packaging of the polypeptide chains and weakening of the interactions between the subunits in heterotetramer. The size of macromolecule seems to be not affected by protein hydration which occurs due to formation of additional hydrogen bonds with a solvent by oxygen-containing oxidation products. This is confirmed by maintaining the size during oxidation by ozone of non-subunit proteins – fibrinogen and products of its hydrolysis [21, 23]. ESR spectra demonstrate that oxidized pFXIII is characterized by the presence of the largest number of labels with a slow rotation compared to all other oxidized samples. This difference can be explained by minimal structural transformations in zymogen during oxidation. A relative stability of pFXIII to free radical oxidation has two possible reasons. Firstly, maximal compactness of the structure of FXIII-A_2B_2 compared to activated forms of pFXIII. In the native conformation Cys314 is closed by AP-FXIII and Tyr560 side chain located on a β-turn of β-barrel 1 domain [2]. The salt bridge located between Asp343 of catalytic domain and Arg11 of the opposite FXIII-A subunit keep AP-FXIII in position in which active center is fully closed. Besides, oxygen atom of Tyr560 forms a strong hydrogen bond with sulphur atom of catalytic cysteine; this protects the latter from substrate attack [32]. The other key amino acid residues His373, Asp396 and Trp279 localized within the core domain also appear to be inaccessible for a substrate. Screening of the catalytic domain by the activation peptide and β-barrel 1 domain can protect the first one from being involved to free radical oxidation process. Amino acid residues are known to differ on sensitivity to oxidation and particularly ozone-induced oxidation. The sensitivity decreases in a series: Cys, Met, Trp, Tyr, His, Phe [33]. This indicates extra necessity to protect the catalytic domain containing amino acid residues Trp279, Cys314 and His373, which are highly sensitive to oxidation. Transformation of the molecular structure of pFXIII during ozone-induced free radical oxidation revealed by FTIR and Raman spectroscopy (Figs. 4, and 6) does not apparently involve its key amino acid residues in oxidation. Secondary, due to protective role of the FXIII-B subunits. According to the data obtained earlier [28], the subunit FXIII-B is a filamentous, flexible strand. It makes possible for the subunits B to be folded and partially wrapped around the globular subunits A. Such a three-dimensional structure of FXIII-A_2B_2 provides both maximal compactness to the protein and increased strength of bonds between FXIII-A and FXIII-B subunits. Since the FXIII-B subunits close the FXIII-A subunits and the catalytic core domain spatially they not only prevent the catalytic FXIII-A subunits from the proteolytic inactivation [10, 11] and serve as inhibitors of the FXIII-A subunit activation in blood plasma [13], but also can be interceptors of free radicals protecting the catalytic subunits A from oxidation. The results of FTIR and Raman spectroscopy give evidence of this mechanism indicating chemical transformation of disulphide groups during free radical oxidation of the

FXIII-B subunits. In this regard it is interesting to compare the chemical modification of the FXIII-B subunits with a modification of fibrinogen and fibrinogen degradation products – D and E fragments under the action of ozone. These proteins have no free thiol groups in their structure but contain only disulphide groups like the subunits FXIII-B. However free radical oxidation of these proteins is not accompanied by transformation of disulphide groups [21, 23]. This also proves accessibility of S-S bonds of the subunits FXIII-B in FXIII-A_2B_2 to oxidation through their flexible strand-like structure.

Our study showed that pFXIII stability to oxidation decreased essentially after its activation. Activation of FXIII-A_2B_2 being a multi-stage process is initiated by thrombin which hydrolyzes the Arg37-Gly38 peptide bond in every FXIII-A subunit. As a result AP-FXIII is released with formation of the factor FXIII-$A_2'B_2$. After AP-FXIII cleavage, β-barrel 1 domain shifts with Tyr560 being removed from Cys314. However Cys314 is still non-active. Proteolytic cleavage largely weakens the interaction between FXIII-A_2' and FXIII-B_2 [5, 34]. As it was shown by dynamic light scattering, FXIII-$A_2'B_2$ formation is accompanied by growth of hydrodynamic diameter of the macromolecule what is a direct evidence of FXIII-$A_2'B_2$ structure loosening after AP-FXIII removal. The data from dynamic light scattering and ESR spectroscopy point to a further loss of a native compaction of the structure during FXIII-$A_2'B_2$ oxidation (Sample 2) which makes the protein molecule to be more sensitive to oxidation compared with non-activated FXIII-A_2B_2. However the key amino acid residues of the catalytic domain may be inaccessible for the substrate at the first stage of activation. This allows FXIII-$A_2'B_2$ to keep a certain part (about 25%) of its enzymatic activity when oxidized.

Heterosubunits of FXIII-$A_2'B_2$ dissociate with formation of FXIII-A_2' and FXIII-B_2 in the presence of calcium ions. At the last stage of the process also proceeding in the presence of Ca^{2+} FXIII-A_2' undergoes the most significant conformational transformations that lead to exposure of cysteine active center with enzyme FXIII-A_2* formation [4]. Removing of FXIII-B_2 from the complex is of essential importance for Ca^{2+}-dependent transformation of FXIII-A_2' to FXIII-A_2* [35]. Transformation of non-active FXIII-A_2 to FXIII-A_2* is accompanied by significant changes in the subunits structure [2]. In consequence of such structural transformations including adjustment in the region of the core domain both initially closed catalytic Cys314 and other cysteine amino acid residues Cys238, Cys327 and Cys409 located in the catalytic domain become spatially accessible. Significant changes in the enzyme structure are confirmed by ESR spectroscopy results. An abrupt change of rotational frequency of label on enzyme during its free radical oxidation points to the most significant modification of protein conformation, when compared to other samples at least in the label microenvironment. According to the above discussion the decreasing of enzymatic activity of oxidized FXIII-A_2* seems to be absolutely evident. The results of differential FTIR spectroscopy demonstrate that FXIIIa (Sample 4) proved to be more sensitive to ozone oxidation action than pFXIII. Particularly this is illustrated by decrease in a content of molecular fragments C–H, =C–H, S–H, and also N–H in a number of amide bands. The amount of thiol groups decreases mainly as a result of oxidation with formation of sulphones and sulphates. This corresponds to the obtained data revealing that the lost of enzymatic activity of FXIII-A_2* takes place due to NO-induced free radical oxidation of amino acid residues Cys in reaction of S-nitrosylation [17]. It cannot be totally excluded that oxidation of the aromatic amino acid residues (Fig. 2) also contributes to the loss of enzymatic activity. A decrease of enzymatic activity of oxidized FXIII-A_2* can be brought about the spatial disjoining of FXIII-B_2 and FXIII-A_2* as well. It makes screening of the catalytic subunit from free radicals impossible.

The results regarding oxidation of FXIII-A_2B_2 in the presence of calcium ions (Sample 3) are the most interesting. As it was shown in Fig. 1 oxidation of this protein is accompanied by the most significant loss of its enzymatic activity compared to samples 1 and 2. Its activity practically corresponds to activity of oxidized sample 4. It may seem that retaining of both AP-FXIII in FXIII-A_2B_2 structure and initial spatial position of Tyr560 would help keep pFXIII tolerance to free radical oxidation. However, the experimental data do not support this. Addition of calcium ions to FXIII-$A_2'B_2$ is known to be the most important and absolutely necessary for exposure of the active center of plasma fibrin-stabilizing factor and expression of its enzymatic activity. Oxygen atom of carbonyl group of amino acid residue Ala457 located in the catalytic core domain is the main binding site for Ca^{2+}. This region in complex with less affine carboxylic groups of amino acid residues Asp438, Glu485 and Glu490 forms a cavity for binding of calcium ions. The other amino acid residues do not directly interact with Ca^{2+}, but bind ions by means of water molecules [2, 32]. The dynamic light scattering data demonstrated that the most noticeable loosening of the spatial structure of both non-oxidized and oxidized FXIII-A_2B_2 takes place in the presence of calcium ions. Besides, ESR spectroscopy data indicate the most significant change (excepting sample 4) of rotational correlation time of spin label occurs on oxidized pFXIII in the presence of calcium ions, when compared with samples 1 and 2. All these facts lead to a conclusion that the loss of the enzymatic activity of FXIIIa formed from oxidized sample 3 are mainly associated with weakening of interactions between FXIII-A_2 and FXIII-B_2 subunits. The structure of FXIII-A_2B_2 in the presence of calcium ions becomes more open and vulnerable for free radical attack. Hereupon the FXIII-B_2 subunits essentially lose their ability to serve as interception of free radicals and protect the catalytic subunits against oxidative damage. We realize that antioxidant role of FXIII-B_2 is still a working hypothesis and needs further data to be refined.

12.5 CONCLUSIONS

For the first time it has been shown that free radical oxidation of pFXIII causes damages to both its chemical structure and three-dimensional organization as well as functional activity depending largely on the stage of pFXIII conversion into FXIIIa at which oxidation was carried out. The most sensitivity to oxidation was revealed to have either pFXIII in the presence of calcium ions or FXIIIa given rise both to the maximum availability of the key amino acid residues in active site of the FXIII-A subunit and possible weakening the antioxidant role of the FXIII-B subunits. The example of pFXIII oxidation in conjunction with the data obtained earlier regarding the peculiarities of fibrinogen oxidation [21–23] provides a basis to suggest the ability of proteins circulating in the blood plasma to antioxidant self-defense. This mechanism may be among the leading factors that determine protein properties to maintain the native structure and function during ROS formation.

KEYWORDS

- **Aromatic amino acid**
- **Blood coagulation**
- **Fibrinogen oxidation**
- **Hemostasis**
- **Plasma fibrin-stabilizing factor**
- **Unimodal size distribution**

REFERENCES

1. Katona, E.; Haramura, G.; Karpati, L.; Fachet, J.; Muszbek, L. A Simple, quick One step ELISA Assay for the Determination of Complex Plasma Factor XIII (A_2B_2): *Thromb. Haemost.*, 2000, *83*, 268.
2. Komaromi, I.; Bagoly, Z.; Muszbek, L. Factor XIII: Novel Structural and Functional Aspects, *J. Thromb. Haemost.*, 2011, *9*, 9.
3. Bagoly, Z.; Koncz, Z.; Hársfalvi, J.; Muszbek, L. Factor XIII, Clot Structure, Thrombosis. *Thromb. Res.*, 2012, *129*, 382.
4. Takagi, T.; Doolittle, R. F. Amino Acid Sequence Studies on Factor XIII and the Peptide Released During Its Activation by Thrombin. *Biochemistry*, 1974, *13*, 750.
5. Lorand, L. Factor XIII: Structure, Activation, and Interactions with Fibrinogen and Fibrin. *Ann. N. Y. Acad. Sci.*, 2001, *936*, 291.
6. Pizano, J. J.; Finlayson, J. S.; Peyton, M. P. Cross-Link in Fibrin Polymerized by Factor 13: epsilon-(gamma-glutamyl)lysine. *Science*, 1968, *60*, 1892.
7. Chen, R.; Doolittle, R. F. γ-γ Cross-Linking Sites in Human and Bovine Fibrin. *Biochemistry*, 1971, *10*, 4486.
8. Mckee, A.; Mattock, P.; Hill, R. L. Subunit Structure of Human Fibrinogen, Soluble Fibrin and Cross-Linked Insoluble Fibrin. *Proc. Natl. Acad. USA*, 1970, *66*, 738.
9. Tamaki, T.; Aoki, N. Cross-Linking of α_2-Plasmin Inhibitor and Fibronectin to Fibrin by Fibrin-Stabilizing Factor. Biochim. *Biophys. Acta.*, 1981, *661*, 280.
10. Mary, A.; Achyuthan, K. E.; Greenberg, C. S. β-Chains Prevent the Proteolytic Inactivation of the a-Chains of Plasma Factor XIII. Biochim. *Biophys. Acta.*, 1988, *966*, 328.
11. Seelig, G. F.; Folk, J. E. Noncatalytic Subunits of Human Blood Plasma Coagulation Factor XIII, Preparation and Partial Characterization of Modified Forms. *J. Bio.l Chem.*, 1980, *255*, 8881.
12. Ariens, R.; Lai, T-S.; Weisel, J. W.; Greenberg, C. S.; Grant, P. J. Role of Factor XIII in Fibrin Clot Formation and Effects of Genetic Polymorphisms. *Blood*, 2002, *100*, 743.
13. McDonagh, J.; Ikematsu, S.; Skrzynia, C. Detection and Regulation of the Subunit Proteins of Plasma Factor XIII. *Thromb. Haemost.*, 1981, *46*, 241.

14. Stadtman, E. R.; Levine, R. L. Free-Radical Protein Oxidation and Its Relationship with the Functional State of the Body. *Amino. Acids*., 2003, *2*, 207.
15. Stadtman, E. R. Protein Oxidation and Aging. *Free Radic. Res*., 2006, *40*, 1250.
16. Stief, T. W.; Kurz, J.; Doss, M. O.; Fareed, J. Singlet Oxygen Inactivates Fibrinogen, Factor V, Factor VIII, Factor X, and Platelet Aggregation of Human Blood. *Thromb. Res*., 2000, *97*, 473.
17. Catani, M. V.; Bernassola, F.; Rossi, A.; Melino, G. Inhibition of Clotting Factor XIII Activity by Nitric Oxide. *Biochem. Biophys. Res. Commun*., 1998, *249*, 275.
18. Feng, Y-H.; Hart, G. In vitro Oxidative Damage to Tissue-Type Plasminogen Activator: a Selective Modification of the Biological Functions. *Cardiovasc. Res*., 1995, *30*, 255.
19. Rosenfeld, M. A.; Bychkova, A. V.; Shegolihin, A. N.; Leonova, V. B.; Kostanova, E. A.; Biryukova, M. I.; Razumovskii, S. D.; Konstantinova, M. L. Free Radical Oxidation of Plasma Fibrin-Stabilizing Factor. *Doklady Biochem. Biophys*., 2012, *446*, 213.
20. Shacter, E.; Williams, J. A.; Lim, M. Differential Susceptibility of Plasma Proteins to Oxidative Modification: Examination by Western Blot Immunoassay. *Free Radic. Biol. Med*., 1994, *17*, 429.
21. Rosenfeld, M. A.; Leonova, V. B.; Shegolihin, A. N.; Razumovskii, S. D.; Konstantinova, M. L.; Bychkova, A. V.; Kovarskii, A. L. Oxidized Modification of Fragments D and E from Fibrinogen Induced by Ozone. *Biochemistry* (Moscow), 2010, *75*, 1285.
22. Rosenfeld, M. A.; Leonova, V. B.; Konstantinova, M. L.; Razumovskii, S. D. Mechanism of Enzymatic Crosslinking of Fibrinogen Molecules. Biology Bulletin (Moscow), 2008, *35*, 578.
23. Rosenfeld, M. A.; Leonova, V. B.; Konstantinova, M. L.; Razumovskii, S. D. Self-Assembly of Fibrin Monomers and Fibrinogen Aggregation during Ozone Oxidation. *Biochemistry* (Moscow), 2009, *74*, 41.
24. Lorand, L.; Gredo, R. B.; Janus, T. J. Factor XIII (Fibrin-Stabilizing Factor). *Method. Enzymol*., 1981, *809*, 333.
25. Rosenfeld, M. A.; Vasileva, M. V. Mechanism of Aggregation of Fibrinogen Molecules: the Influence of Fibrin-Stabilizing Factor. Biomed. *Science*, 1991, *2*, 155.
26. Bychkova, A. V.; Sorokina, O. N.; Kovarski, A. L.; Shapiro, A. B.; Rozenfel'd, M. A. Interaction of Fibrinogen with Magnetite Nanoparticles. *Biophysics* (Moscow), 2010, *55*, 544.
27. McDonagh, R. P.; McDonagh, J.; Blomback, M.; Blomback, B. Crosslinking of Human Fibrin: Evidence for Intermolecular Crosslinking Involving α-Chains. *FEBS Lett.,* 1971, *14*, 33.
28. Carrell, N. A.; Erickson, H. P.; McDonagh, J. Electron Microscopy and Hydrodynamic Properties of Factor XIII Subunits. *J. Biol. Chem*., 1989, *264*, 551.
29. Yee, V. C.; Pedersen, L. C.; Le Trong, I.; Bishop, P. D.; Stenkamp, R. E.; Teller, D. C. Three-Dimensional Structure of a Transglutaminase: Human Blood Coagulation Factor XIII. *Proc. Natl. Acad. Sci. USA*., 1994, *91*, 7296.
30. Ahvazi, B.; Kim, H. C.; Kee, S. H.; Nemes, Z.; Steinert, P. M. Three-Dimensional Structure of the Human Transglutaminase 3 Enzyme: Binding of Calcium Ions Changes Structure for Activation. *EMBO J*., 2002, *21*, 2055.
31. Souri, M.; Kaetsu, H.; Ichinose, A. Sushi Domains in the B Subunit of Factor XIII Responsible for Oligomer Assembly. *Biochemistry*, 2008, *47*, 8656.

32. Iismaa, S. E.; Holman, S.; Wouters, M. A.; Lorand, L.; Graham, R .M.; Husain, A. Evolutionary Specialization of a Tryptophan Indole Group for Transition-State Stabilization by Eukaryotic. *Natl. Acad. Sci. USA*, 2003, *100*, 12636.
33. Berlett, B. S.; Levine, R. L.; Stadtman, E. R. Comparison of the Effects of Ozone on the Modification of Amino Acid Residues in Glutamine Synthetase and Bovine Serum Albumin. *J. Biol. Chem.*, 1996, *271*, 4177.
34. Radek, J. T.; Jeong, J. M.; Wilson, J.; Lorand, L. Association of the A Subunits of Recombinant Placental Factor XIII with the Native Carrier B Subunits from Human Plasma. *Biochemistry*, 1993, *32*, 3527.
35. Hornyak, T. J.; Shafer, J. A. Role of Calcium Ion in the Generation of Factor XIII Activity. *Biochemistry*, 1991, *30*, 6175.

CHAPTER 13

FUNCTIONING SIMILARITY OF PHYSICOCHEMICAL REGULATORY SYSTEM OF THE LIPID PEROXIDATION ON THE MEMBRANE AND ORGAN LEVELS

L. N. SHISHKINA, M. A. KLIMOVICH, and M. V. KOZLOV

CONTENTS

13.1 INTRODUCTION

The surface-active properties of phospholipids (PL) are widely used for the formation of liposomes, which are a model of biomembranes and also a technique for the study of cells and exposure to them [1, 2]. Earlier it has been established the substantial role of lipids which contain in biosorbents for the regulation of properties of lipids in medium [3]. It was also found that scale and character of interrelations between the different fractions in lipids of the mice tissue had a dependence on physicochemical properties of lipids [4, 5], and the physicochemical properties and composition of natural lipids have an influence on those of liposomes formed from them [6]. Besides, earlier it was shown the existence of physicochemical regulatory system of the lipid peroxidation (LPO) both a membrane and organ levels [7, 8]. It allows us to suggest the existence of a uniform mechanism of the regulatory in the LPO system both on the membrane and the organ levels.

The aim of this work is to study interrelations between the physicochemical properties and the composition of lipids of liposomes formed from the different natural lipids and of the organ lipids of mice.

13.2 EXPERIMANTAL METHODS

Formation of liposomes was carried out by an ultrasound dispersant UZDN-2T in 0.1 M C_2H_2OH aqueous solution of lipids from the liver and brain of mature outbreed mice (female, 12–13 wk aged) and soy bean lecithin (native – experiments No. 1 and No. 3 – and oxidized – experiment No. 2). That procedure was in detail presented in [6, 9]. The total number of animals in experiments with liposomes was 60. Experiments were repeated two tines: in September (experiment No. 1) and May (experiment No. 2). Besides, the 80 white outbreed mice (females) were choice as the experimental animals to investigate interrelations between the different parameters of the LPO regulatory system in the murine organs with the different antioxidant status. These experiments were repeated three times (in November–December, May–June and

September–October) to modify the antioxidant status in organs of mice. In these cases all studied parameters were individually determined for each mouse. The murine organs were placed in ice cooled weighing bottles immediately after decapitation.

Isolation of lipids was done by the Blay and Dyer method in Kates modification [10]. Qualitative and quantitative composition of PL was analyzed by thin layer chromatography [11], as it was described in [12]. It was used type G silica gel (Sigma, USA), glass plates 9×12 cm^2 and mixture of solvents chloroform/methanol/glacial acetic acid/water (25/15/4/2) as a mobile phase. All solvents were of specially pure or chemically pure grade. The ratio of the sums of the more easily oxidizable to the more poorly oxidizable fractions of PL was calculated by the formula [12]: ΣEOPL/ΣPOPL = (PI + PS + PE + PG + CL + PA)/(LFPL + SM + PC), where PI is phosphatidylinositol, PS is phosphatidylserine, PE is phosphatidylethanolamine, PG is phosphatidylglycerine, CL is cardiolipin, PA is phosphatidic acid, LPC is lysoforms of PL, SM is sphingomyelin, PC is phosphatidyl choline.

The sterol content in lipids was determined spectrophotometrically at 625 nm wavelength by method described in [13]. Serva Company (FRG) cholesterol was used for plopping the calibration curve. The spectrophotometrical measurements were carried out on a KFK-3 device (Russia). The content of diene conjugates (DC) and ketodienes (KD) was determined by UV-photometry from the ratio of the optical density of the lipid solution in hexane (0.05–0.3 mg/ml) at 230 ± 2 nm and 270 ± 2 nm to 205 ± 2 nm wavelengths respectively, using a spectrophotometer "Shimadzu UV 3101 PC" (Japan). The peroxide content in lipids was determined by the current iodometric titration method. The antiperoxide activity (APA) of lipids, i.e., the ability of lipids to decompose peroxides, was assessed by the ratio of the difference in the concentrations of peroxides in the oxidized methyl oleate and in the lipid solution in this methyl oleate to the amount of the added lipids [3].

The experimental data were processed with a commonly used variational statistic method [14] and by KINS program given in [15]. The experimental data are presented in form of arithmetic means with the indication of the mean square errors of the arithmetic means ($M \pm m$).

13.3 RESULTS AND DISCUSSION

Earlier it was found that the PL composition and physicochemical properties of lipids from the liver of the laboratory mice significantly differ depending on the season [4–6, 16]. Lecithin (L) is a mixture of the natural lipids among which PC is the main fraction of PL (86–90%), however, the different samples of L also have not identical composition of lipids [9]. In this connection it is necessary to note the presence of the substantial differences between the initial lipid composition and lipids of the formed liposomes [6, 9]. As seen from data presented in Table 1, there are the reliable changes both the ratio of the sums of the more easily oxidizable to the more poorly oxidizable fractions of PL which characterize the ability of lipids to the oxidation and the PC to PE ratio (PC/PE) which is one of the parameters characterizing the membrane structural state between the initial and formed liposome lipids [6, 7].

TABLE 1 The ratio of sums of the more easily oxidizable to the more poorly oxidizable fractions of PL (∑EOPL/∑POPL) and ratio of phosphatidyl choline to phosphatidylethanolamine (PC/PE) in the initial lipids of organs and formed from them liposomes in the different experiments.

Experimental condition		∑EOPL/∑POPL	PC/PE
A1	liver	0.653 ± 0.014	2.307 ± 0.036
	liposomes	0.757 ± 0.042	2.828 ± 0.092
A2	liver	0.736 ± 0.023	1.949 ± 0.039
	liposomes	0.635 ± 0.016	2.015 ± 0.039
B1	brain	1.246 ± 0.0195	1.096 ± 0.01
	liposomes	0.647 ± 0.042	1.82 ± 0.117
B2	brain	1.348 ± 0.025	0.912 ± 0.012
	liposomes	0.817 ± 0.009	1.526 ± 0.015

The scale of the all changes substantially depends on the values for the initial lipids. So, the formation of liposomes from the murine organ lipids brought out the enhancement of the membrane rigidity (the increase of the PC/PE ratio), the more pronounced under the formation of liposomes from the brain lipids (Table 1), while the diminution of the ability of lipids to the oxidation is obtained in case the ΣEOPL/ΣPOPL ratio is more than 0.7 in lipids of the murine organs (Table 1).

These changes lead to the reliable diminutions of the APA of lipids under the formation of liposomes from the liver and brain lipids (Fig. 1) and the appearance of peroxides in the lipids of liposomes formed from L [6].

One of the initial stage of the LPO is the accumulation of DC in the fatty acid chains of PL, especially in the residues of linoleic and linolenic acids which reacts with reactive oxygen species to form a carbonyl bond, due to the enhance the KD content. In this connection it is necessary to note the existence of the common direct correlation between the content of the DC and KD both in lipids of organs of mice and soy bean lecithin (Fig. 2,a) and in lipids formed from them liposomes (Fig. 2,b). However, the coefficient of the linear regression for this interrelation in case of the lipids from liposomes is twofold less than one for the initial lipids owing to the more wide variability in the DC content in the liposomes lipids.

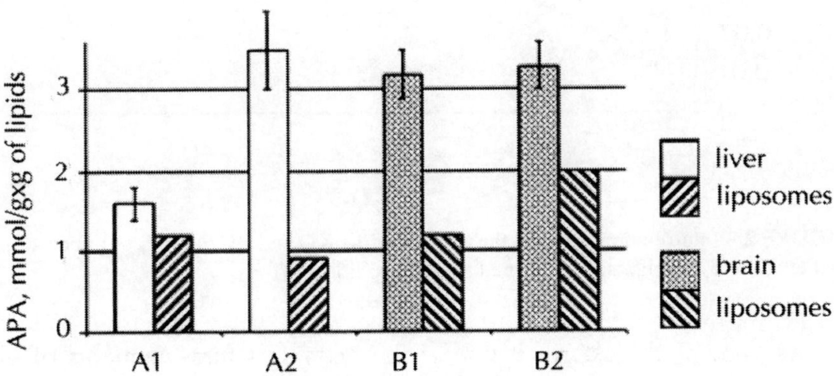

FIGURE 1 Antiperoxide activity of lipids from the murine organs and formed from them liposomes in the different experiments.

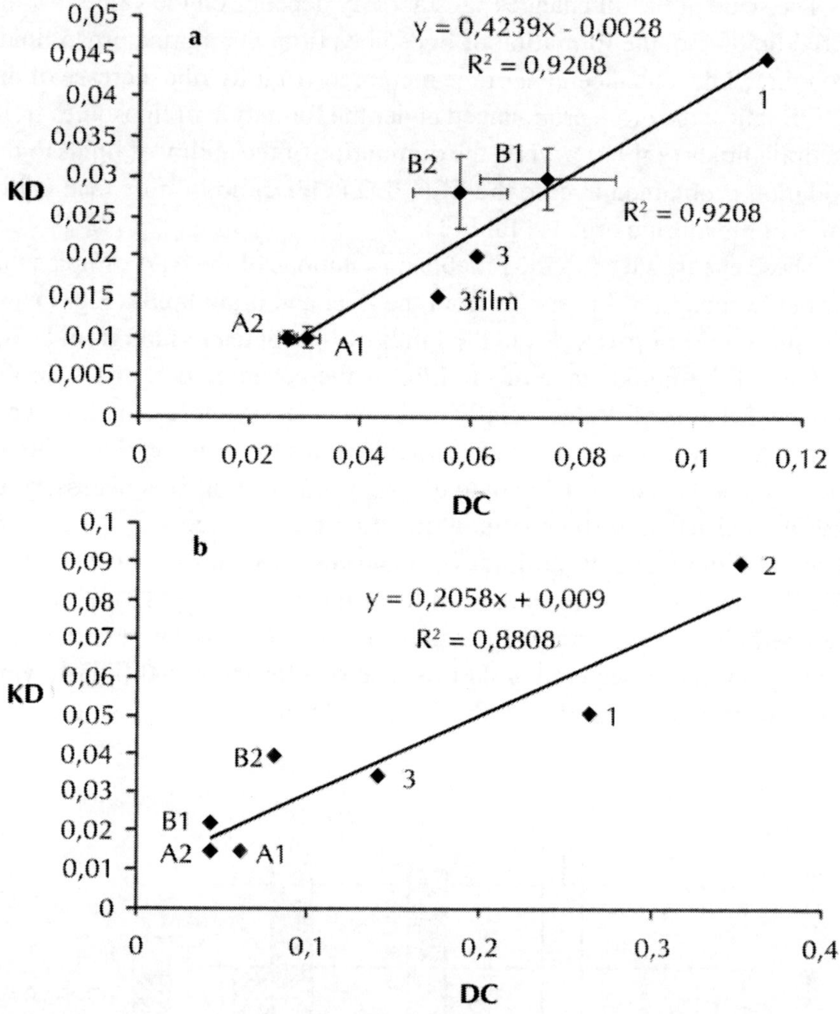

FIGURE 2 Interrelation between the content of the KD and DC in lipids of mice's organs and lecithin (a) and liposomes formed from these lipids (b).

As known, cholesterol is one of the main structural elements of biomembranes. It is obtained that the DC content contributes in the inclusion of cholesterol in the lipid bilayer. It is seen from data, which is presented in the Fig. 3. However, the coefficient of the linear regression of the direct correlation between the cholesterol share in the total lipid composition and the DC content is also higher for the lipids isolated from organs of mice in comparison with one for lipids formed from them liposomes. Besides, there is a shift of this correlation line for the liposome lipids owing to increasing the DC content.

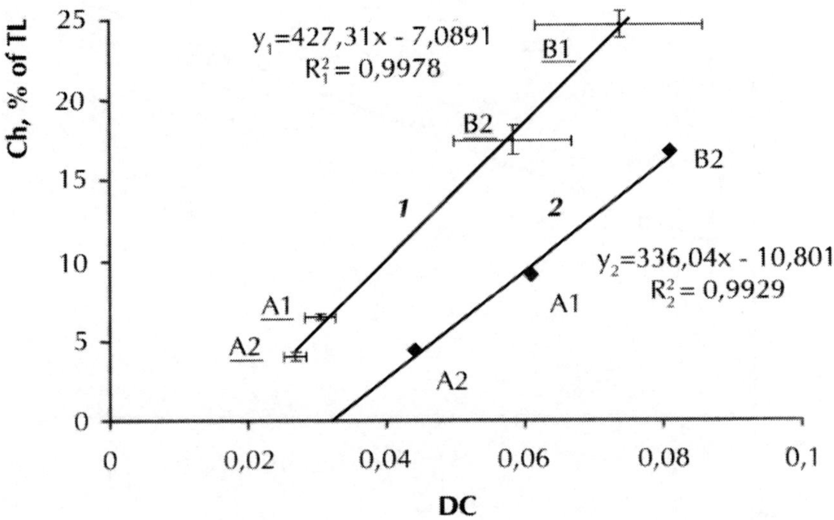

FIGURE 3 Relation between the cholesterol share in the total lipid composition and the DC content in lipids of the murine organs (underlined) and formed from them liposomes (diamonds).

As already emphasized, there is an appearance of peroxides in lipids of liposomes formed from lecithin. Besides, there is the direct correlation between the peroxide concentration and the KD content in lipids of these liposomes (R^2 = 0.9994) while for liposomes formed from the liver and brain lipids of mice there is the direct correlation between the APA and KD content in their lipids (R^2 = 0.9287).

The enhancement of the SM share in PL under the simultaneous increase the relative content of lysoforms is usually considered as the adaptation reaction of the membrane system on the different actions [17]. In this connection it is interestingly to note following result. The common direct correlation between the LPC content and SM share in PL is revealed for lipids of the liver and brain mice (experiment 2) and the native lecithin, and also formed from these lipids liposomes in experiments which were performed in May–June (Fig. 4).

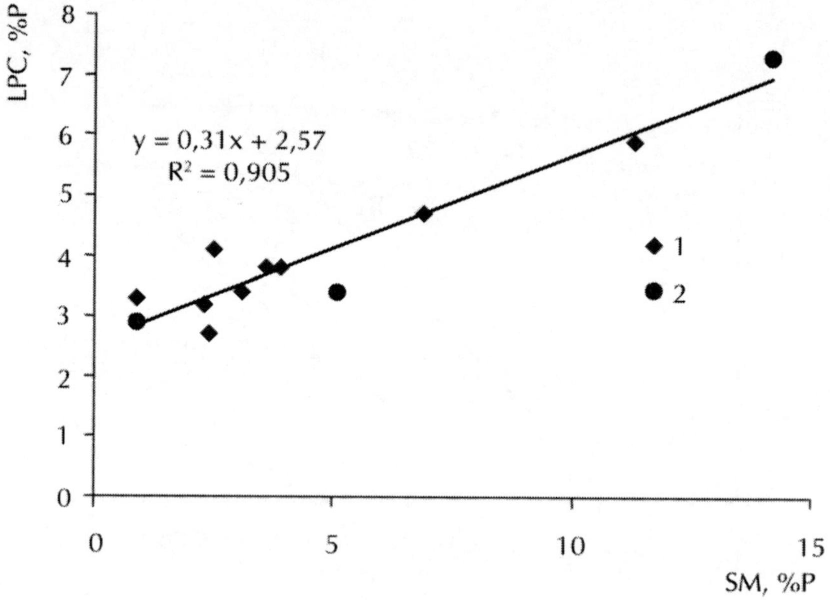

FIGURE 4 Interrelation between the relative content of lysoforms and SM share in phospholipids from organs of mice and the native lecithin (1) and in PL of liposomes formed from them (2).

13.4 CONCLUSION

As above noted, the physicochemical regulatory system of the LPO functions both a membrane and organ levels [7, 8]. In these experiments the steady and similar interrelations between the content of the diene conjugates and ketodienes, of the diene conjugates and sterols in lipids, the lysoforms and sphingomielin shares in phospholipids, of the peroxides content in lipids and/or the lipid antiperoxide activity and the amount of ketodienes both in the lipids of the mice organs and the liposomes formed of the natural lipids were revealed. It allows us to conclude that the functioning of the physicochemical regulatory system of the lipid peroxidation on the biomembrane and organ levels has the uniform regulatory mechanism.

KEYWORDS

- **Diene conjugates**
- **Ketodienes**
- **Phospholipids**
- **Qualitative composition**
- **Quantitative composition**
- **Sterols**

REFERENCES

1. Margolis, L. B.; Bergel'son, L. D. In *Liposomes and Their Interaction with Cells.* Nauka: Moscow, 1986; 240 p. (*in Russian*).
2. Gregoriadis, G.; Allison, A. C., Eds.; In *Liposomes in Biological Systems.* Wiley and Sons: N.Y., 1980; 422 p.
3. Men'shov, V. A.; Shishkina, L. N.; Kishkovskii, Z. N. Effect of Biosorbents on the composition, content and antioxidative properties of lipids in medium. *Appl. Biochem. Microbiol,* **1994**, *30(3)*, 359–369.
4. Kozlov, M. V.; Urnisheva, V. V.; Shishkina, L. N. Interconnection of parameters of Regulation System of Lipid Peroxidation and Morphophysiological parameters of Mouse Liver, *J. Evol. Biochem. Phys.,* **2008**, *44(4)*, 470–475.
5. Khrustova, N. V.; Kozlov, M. V.; Shishkina, L. N. Effect of Physicochemical Properties of Murine Liver Lipids on the Interrelation Between the Parameters of Their Composition, *Biophysics*, **2011**, *56(4)*, 656–659.
6. Klimovich, M. A.; Shishkina, L. N.; Paramonov, D. V.; Trofimov, V. I. Interrelations between the Physicochemical Properties and the Composition of Natural Lipids and the Liposomes Formed from Them, *Oxid. Commun.,* **2010**, *33(4)*, 965–973.
7. Burlakova, Ye. B.; Pal'mina, N. P.; Mal'tseva, Ye. L. In *Membrane Lipid Oxidation*; Carmen Vigo-Perfley, Ed.; CRC Press: Boston, **1991**, *3*, 209–237.
8. Shishkina, L. N.; Kushnireva, E. V.; Smotryaeva, M. A. A New Approach to Assessment of Biological Consequences of Exposure to Low-Dose Radiation, *Radiat. Biol. Radioecol.* **2004**, *44(3),* 289–295 (*in Russian*).
9. Klimovich, M. A.; Paramonov, D. V.; Kozlov, M. V.; Trofimov, V. I.; Shishkina, L. N. Interrelation between the Peroxidation Parameters of Natural Lipids and the Lipo-

somes Characteristics Formed from Them. In *Modern Problems in Biochemical Physics*. New Horizons. Nova Science Publishers: N.Y., 2012; *Chapter 32*, 255–262.
10. M. Kates. The Technique of Lipidology. Moscow: Mir, 1975. 214 p. (*Russian version*).
11. Findlay, J. B. C.; Evans, W. H., Eds.; In *Biological Membrane. A Practice Approach.* Mir: Moscow, 1990; 424 p. (*Russian version*).
12. Shishkina, L. N.; Kushnireva, Ye. V.; Smotryaeva, M. A. The combined Effect of Surfactant and Acute Irradiation at Low Dose on Lipid Peroxidation Process in Tissues and DNA Content in Blood Plasma of Mice, *Oxidat. Commun.*, **2001**, *24(2)*, 276–286.
13. Sperry, W. W.; Webb, M. A Revision of the Schoenheimer-Sperry Method for Cholesterol Determination, *J. Biol. Chem.*, **1950**, *187(1)*, 97–106.
14. Lakin, G. F. Biometry. 3rd edition; Vysshaya shkola: Moscow, 1990; 293 p. (*in Russian*).
15. Brin, E. F.; Travin, S. O. Model of mechanisms of the chemical reactions, *Chem. Phys. Rep.*, **1991**, *10(6)*, 830–837 (*in Russian*).
16. Khrustova, N. V.; Shishkina, L. N. Effect of Lipid Physicochemical Characteristics on Interrelations Between Lipid Composition and the Mouse Liver Index, *J. Evol. Biochem. Phys.*, **2011**, *47(1)*, 37–42.
17. Shishkina, L. N.; Shevchenko, O. G. Lipids of Blood Erythrocytes and their Functional Activity, *Adv. Recent Biol.*, **2010**, *130(6)*, 587–602. (*in Russian*).

CHAPTER 14

INVESTIGATION OF TENSILE STRENGTH OF THREE NEW SOFT SILICONE ELASTOMERS AND COMPARISON WITH RESULTS OF PREVIOUSLY TESTED MATERIAL

W. WIĘCKIEWICZ, R. ORLICKI, and G. E. ZAIKOV

CONTENTS

14.1 INTRODUCTION

For many years various soft materials have been used to line prostheses in order to eliminate irritating effect of prosthesis on the prosthetic base. Different materials have been used for this reason: soft acryl, acryl-silicone materials, and for some time, more and more often, silicone elastomers [1–5]. These latter materials are basis for investigation of physical mechanical and physical chemical properties [6]. One of the physical mechanical investigations in case of elastomers is examination of the tensile strength. It should be explained that the tensile strength of a silicone elastomer exerts a decisive effect on the intramolecular forces, which prevent movement of molecules in relation to each other, and the force of chemical bonds [7]. The tensile strength, calculated on the basis of interatomic and intermolecular forces, is not comparable with the tensile strength calculated in practice. The differences can be explained by heterogeneities – microdefects in the material, i.e., structural discontinuities. They cause concentration of applied tensions primarily on the borders. The originating defect may initiate spacing, cracking and rupture of the sample.

The nature of soft silicone elastomers should also be explained. They constitute a group of materials containing organic silicone compounds referred to as polyalkyl siloxanes with structure cross-linked into networks. Their skeleton includes silicone atoms joined in a chain or a network. The organic rodents usually represent $–CH_3$, or $–C_6H_5$, while the degree of network cross-linking is expressed by the radical-silicone atoms ratio and equals 1.2. Important properties of silicones include: thermal resistance, very good dielectric properties, good resistance to oxidation, effect of atmospheric factors, resistance to corrosive effect of gases, resistance to acids, hydrophobicity, adhesiveness, organic elasticity and low hardness. Lipophilicity is a negative feature [8, 9].

Silicones may be used for medical purposes and then they are referred to as silicone elastomers. Elastic materials may be divided into acrylic, silicone materials, alternative soft polymers and materials for biological renewal of tissues. Soft materials on the basis of acrylic contain a plastificator. They consist of an ethylene-polymethacrylate powder with an addition of dibenzoil. So-called monomer, i.e., *N*-buthyl methyl-ethyl methacrylate, or *N*-buthyl methacrylate with an addition of ethyl octane is another component. It should be emphasized that soft acrylic is characterized by a good bonding to an acrylic plate, however it hardens rapidly. Another disadvantage of this material is that its elasticity decreases rapidly, while its fragility and porosity increase. The above properties are significantly affected by oral environment [10–12]. Evaluating silicone-lining materials, it should be stressed that their bonding to the prosthesis plate is less durable despite the use of brand-name glues, which still lack perfect quality [13, 14]. Soft alternative polymers are materials of a polyurethane type, and their disadvantage includes high absorption of water. This hydrophobicity may lead to three-fold increase of the volume of basic materials. Soft metarials for renewal of tissues are used for temporary relining of partial removable, complete and post-surgical prostheses [15–17]. They are produced on the basis of acrylic resins and do not contain any methacrylate monomers, but a plastic mixture of esthers and ethyl alcohol. It should be stressed that these materials are placed on low-mucosal part of a properly prepared prosthetic plate, inserted into the oral cavity and shaped functionally for about 72 hr, simultaneously relieving the prosthetic base.

The aim of the study was to investigate the tensile strength of three silicone elastomers: A-Soft line 30, Elite Super Soft and Bosworth Dentusil and compare the findings with previously evaluated material, Ufi Gel SC.

14.2 MATERIALS AND METHODS

For the evaluation of tensile strength of new materials, 36 samples were prepared, 12 for every material. Each of the sample consisted of two parts: an acrylic cylinder with the length of 20 mm and diameter of 4 mm, and a silicone cylinder with the length of 20 mm and diameter of 4 mm, joined together with glue. The investigation of tensile strength was performed on a testing machine Houndfield H5KS model THE S/N D83281.

The previously tested material Ufi Gel SC was also prepared in 12 samples consisting of two parts joined with glue: acrylic and silicone, each of them with the length of 20 mm and diameter of 4 mm. The investigation was carried out on a testing machine FPZ 10/1 manufactured by Heckert Rauenstein. The prepared samples were inserted and fixed into the machine evaluating the tensile strength and tested.

14.3 RESULTS

Analysis of the tensile strength tests of the A-Soft Line 30 material revealed findings ranging from 6.15N to 14.35N. Calculated mean value was 9.63N (Table 1). IN case of Bosworth Dentusil the tensile strength ranged from 1.00N to 12.35N, and the mean value was 7.50N (Table 2). Evaluation of Elite Super Soft revealed the tensile strength ranging from 5.15N to 9.65N, while the mean value was 7.29N (Table 3).

TABLE 1 Investigation of tensile strength of A-Soft Line 30.

Number of test	The test result [N]
1.	7.15
2.	6.85
3.	14.35
4.	11.50
5.	5.50
6.	11.00
7.	8.15
8.	12.50
9	10.35
10.	6.15
11.	11.00
12.	11.00
	Average: 9.63

TABLE 2 Investigation of tensile strength of Bosworth Dentusil.

Number of test	The test result [N]
1.	12.00
2.	6.50

TABLE 2 *(Continued)*

Number of test	The test result [N]
3.	1.00
4.	7.65
5.	12.00
6.	12.35
7.	4.35
8.	6.35
9	3.65
10.	8.50
11.	6.00
12.	9.65
	Average: 7.50

TABLE 3 Investigation of tensile strength of Elite Super Soft.

Number of test	The test result [N]
1.	6.00
2.	9.35
3.	7.85
4.	5.15
5.	9.65
6.	6.15
7.	8.85
8.	8.50
9	6.85
10.	7.65
11.	6.15
12.	5.35
	Average: 7.29

The findings of tensile strength tests of Ufi Gel SC ranged from 16.72N to 22.06N, while the mean value was 18.61N (Table 4).

TABLE 4 Investigation of tensile strength of Ufi Gel SC.

Number of test	The test result [N]
1.	17.90
2.	22.06
3.	18.00
4.	16.72
5.	18.37
6.	18.60
7.	20.72
8.	18.92
9	16.96
10.	17.90
11.	18.76
12.	18.37
	Average: 18.61

14.4 CONCLUSIONS

Ufi Gel SC is characterized by the best tensile strength, followed by A-Soft Line 30, Bosworth Dentusil and Elite Super Soft.

Each of the estimated material can be used in prosthodontic treatment in patients, depending on the necessity of relieving the prosthetic base.

KEYWORDS

- **Acryl-silicone**
- **Bosworth Dentusil**
- **Polyalkyl siloxanes**
- **Prosthetic base**
- **Soft acryl**

REFERENCES

1. McCord, J. F.; Donaldson, A. C.; Lamont, T. J. A Contemporary update on 'soft' linings. *Dent. Update.*, **2011**, *38(2)*, 102–104.
2. Tanimoto, Y.; Saeki, H.; Kimoto, S.; Nishiwaki, T.; Nishiyama, N. Evaluation of adhesive properties of three resilient denture liners by the modified peel test method. *Acta. Biomater.*, **2009**, *5(2)*, 764–769.
3. Kasperski, J.; Chladek, W.; Karasiński, A. Laboratoryjna ocena zmian właściwości materiału elastyfikowanego Softerex. *Prot. Stom.*, **2000**, *50(5)*, 293–295.
4. Piotrowski, P.; Wasilewicz, P. Badania sztywności i wytrzymałości płyt akrylowych podścielonych Molloplastem B. *Prot. Stom.*, **2000**, *50(1)*, 52–55.
5. Piotrowski, P. Poprawa retencji protez całkowitych dolnych przez podścielenie elastomerem silikonowym Molloplastem B. *Stom. Współcz.*, **1999**, *6(4)*, 44–47.
6. Tasopoulos, T.; Jagger, R. G.; Jagger, D. C.; Griffiths, A. E. Energy absorption and hardness of chair-side denture soft lining materials. *Eur. J. Prosthodont Rest. Dent.*, **2010**, *18(4)*, 189–194.
7. Mutluay, M. M.; Ruyter, I. E. Evaluation of bond strength of soft relining materials to denture base polymers. *Dent. Mater.*, **2007**, *23(11)*, 1373–1381.
8. Mutluay, M. M. A prospective study on the clinical performance of polysiloxane soft liners: one-year results. *Dent. Mater. J.*, **2008**, *27(3)*, 440–447.
9. Marciniak, J. *Biomateriały*. Gliwice, Polska: Wydawnictwo Politechniki Śląskiej; 2002.
10. Mante, F. K.; Mante, M. O.; Petropolous, V. C. In vitro changes in hardness of sealed resilient lining materials in various fluids. *J. Prosthodont.*, **2008**, *17(5)*, 384–391.
11. Minami, H.; Suzuki, S.; Minesaki, Y.; Kurashige, H.; Tanaka, T. In vitro evaluation of the effect of thermal and mechanical fatigues on the bonding of an autopolymerizing soft denture liner to denture base materials using different primers. *J. Prosthodont.*, **2008**, *217(5)*, 392–400.
12. Santawisuk, W.; Kanchanavasita, W.; Sirisinha, C.; Harnirattisai, C. Dynamic viscoelastic properties of experimental silicone soft lining materials. *Dent. Mater. J.*, **2010**, *29(4)*, 454–460.

13. Lassila, L. V.; Mutluay, M. M.; Tezvergil-Mutluay, A.; Vallittu, P. K. Bond strength of soft liners to fiber-reinforced denture-base resin. *J. Prosthodont.*, **2010**, *19(8)*, 620–624.
14. Meşe, A.; Güzel, K. G.; Uysal, E. Effect of storage duration on tensile bond strength of acrylic or silicone-based soft denture liners to a processed denture base polymer. *Acta. Odontol., Scand.*, **2005**, *63(1)*, 31–35.
15. Skorek, A. Wybrane aspekty zintegrowanego postepowania chirurgiczno-protetycznego u chorych leczonych z powodu złośliwych nowotworów szczęki. *Czas Stom.*, **2001**, *54(12)*, 800–804.
16. Więckiewicz, W.; Bogucki, A. Z. Płytka obturująca – proteza stosowana w rehabilitacji pacjentów po zabiegach chirurgicznych. *Prot Stom.*, **2002**, *52(4)*, 234–237.
17. Więckiewicz, W. Tworzywa miękkie stosowane do podścielania protez u pacjentów po zabiegach operacyjnych. *Prot Stom.*, **1996**, *45(4)*, 233–235.

CHAPTER 15

ANALYSIS OF THE INFLUENCE OF THE EARTH'S ELECTROMAGNETIC RADIATION ON WATER BY ELECTROCHEMICAL METHOD

A. A. ARTAMONOV, V. M. MISIN, and V. V. TSETLIN

CONTENTS

15.1 INTRODUCTION

Change of pure water conductivity has been studied in special-purpose electrochemical cell during 365 days. Lack of thermal factors action on recorded changes of water conductivity has been proven. Data on water conductivity change has been compared to data on geomagnetic radiation of geosphere. Effect of environmental electromagnetic interference on pure water conductivity value is shown. Common nature of dynamics of currents' behavior in electrochemical cell and electromagnetic interference has been revealed. The aim and scope of this analysis is searching new concept of mechanism of mild exposure of external factors on physical-chemical properties of water.

It is being systematically researched biological effect of electromagnetic interference on living organisms [1–3]. It is often mentioned in these papers negative action of various kinds of non-ionizing radiation even if very low doses of radiation including electromagnetic interference (EMI). For this reason, WHO recommends to be committed to preventive policy of maximum possible decreasing of exposure time of organisms by EMI of technogenic origin [4].

Close relationship between change of electrical conductivity of water and solar activity that is correlation of solar activity with various processes in biosphere has been established in [5]. Method of electrochemical analysis used by the authors has revealed high sensitivity of water to faint external impacts. Unfortunately, it is not shown in the paper how influence of the Sun on water may occur.

The said facts might be caused by the following reason. Most of biochemical reactions in organisms of humans, animals and plants are connected to behaviour of electrical processes in liquid water media [6]. Since all nerve signals are transmitted via electric impulses the electrical currents playing significant role in functioning of organisms are always circulating in the organisms. For this reason there are common basic laws of existence of organisms' high sensitivity to faint external effects as is described in Ref. [7].

Nature of initial event of interference action onto biological objects has not still been explained satisfactorily. Lack of such a theory allows no explanation for experiments being performed under conditions of objects' being shielded from EMI, in constant magnetic field and others. There exists no concept of mechanism of mild exposure of cosmic and geophysical factors, either.

Response of water might be probable initial event of an organism's response to mild exposure; this is probably connected to high sensitivity of oxidation-reduction processes in water media to action of external factors. Such features as electronic work function, zero charge potential, electrode potential, etc. are connected to concept of electrochemical processes. For this reason, the structure of near-electrode layer will depend on nature of electrodes' material and specific nature of its interaction with solvent [8].

Effect of EMI on properties of water can be an example of mild exposure. This fact has inspired us to research electrochemical processes that occur in pure water and to find possible dependence of these on EMI.

There exists common law for all organisms of response to supermild exposures. Basic laws of high sensitivity to mild external exposures are described in Ref. [7].

On the other hand, the conclusion is made based on literature data analysis performed by the authors that there is no conventional acceptable description of structure of liquids and, particularly, water, nowadays. Such a non-conflicting conventional description would allow to estimate response of water to physical exposure of low intensity including natural variations of electromagnetic and radiation background. Lack of such a theory allows no explanation for experiments being performed under conditions of objects' being shielded from EMI, in constant magnetic field and others. There exists no concept of mechanism of mild exposure of cosmic and geophysical factors, either.

Close relationship is established in paper [5] between change of conductivity of water and solar activity. On their opinion such relation might be reason for correlation of solar activity and various processes in biosphere. However, the authors have not discovered in their paper how the Sun can influence via water. Method of electrochemical analysis used by the authors has shown high sensitivity to mild external factors. Electrochemical contact is the area in which main electrochemical processes take place. Many of electrochemical characteristics such as electronic work function, zero charge potential, electrode potential and others are connected to this concept. Effect of material's nature and specific character of its interaction with solvent is built into parameters of the structure of near-electrode layer [8].

15.2 EXPERIMENTAL METHOD

Round-the-clock measuring of electric currents in interelectrode spacing filled with high purity water and at constant stabilized voltage supplied to electrodes forms the basis of method of water conductive properties' monitoring. The experiments have been performed at the plant described in Ref. [9].

Measuring has been performed in electrochemical cells made in the shape of glasses of Pirex glass. Electrodes have been inserted through holes in cells' covers. Electrodes have been made in the shape of platinum tape or in the shape of thin sheets of food-grade stainless steel with dimension of 35×20 mm^2. Interelectrode spacing changed within 5 mm to 30 mm. A 18 ml of high purity water was poured into the cell. Constant stabilized voltage within 2.4 V was supplied to electrodes. Features of the installation design allowed round-the-clock monitoring operation throughout 365 days under conditions preset of stable power supply. Isoline of voltage supplied to cells allowed concluding that stability of supplied voltage was equal to 1.3 mV. Electrochemical cells have been put into temperature controlled cabinet where temperature can be maintained within the given range with accuracy of ±0.1°C. Water has been purified in two stages. At first, distillate has been obtained at distillation plant in accordance with the State Standard (GOST) R 6709-72. Then distillate has been purified in water deionization plant to remove ionic impurities (basic ions and acid ions). Grade of water purification has been inspected in conductometric cell using the instrument HI 983322-1 "HANNA". Conductivity of deionized water with the required degree of purity amounted to σ= 0.1–0.2 uS/cm.

More detailed data on the plant operation features and features of water purification are given in paper [9].

15.3 RESULTS AND DISCUSSION

Results of round-the-clock continuous monitoring of current during 365 days (the signal has been taken from electrochemical cell with platinum electrodes) are given at Figs. 1–4. The cell has been placed into temperature-controlled cabinet with constant temperature of 25±0.1°C in shadowed place. As is seen from the figures, the intensity and behaviour of day-to-day variations of current being recorded change daily. Most intensive day-to-day variations have been observed in spring (Fig. 2). Sharp surges and falls that might last for several days are evident in overall dynamics of currents.

Winter

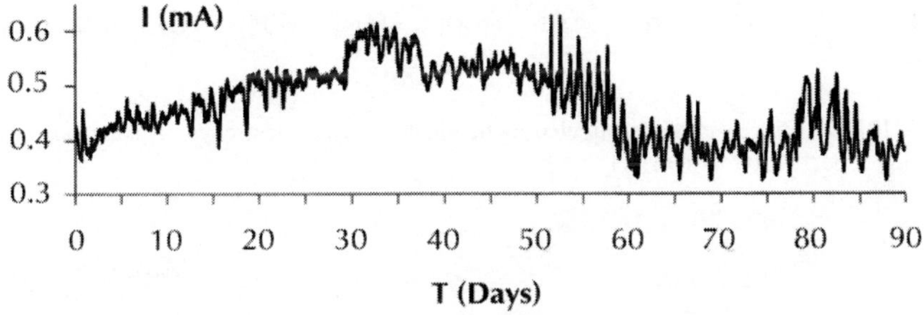

FIGURE 1 Current recorded in electrochemical cell with electrodes of platinum for the period December 1, 2008–February 29, 2009.

Spring

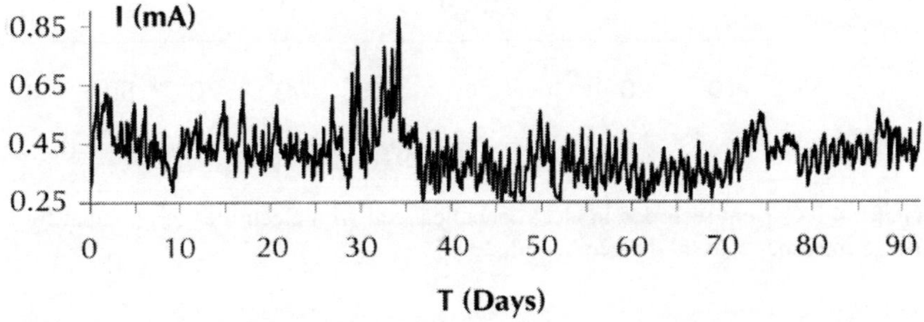

FIGURE 2 Current recorded in electrochemical cell with electrodes of platinum for the period March 1, 2009–May 31, 2009.

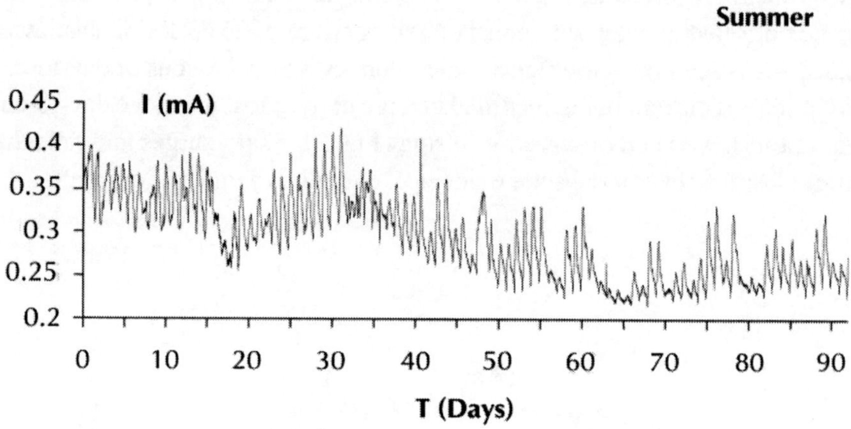

FIGURE 3 Current recorded in electrochemical cell with electrodes of platinum for the period June 1, 2009–August 31, 2009.

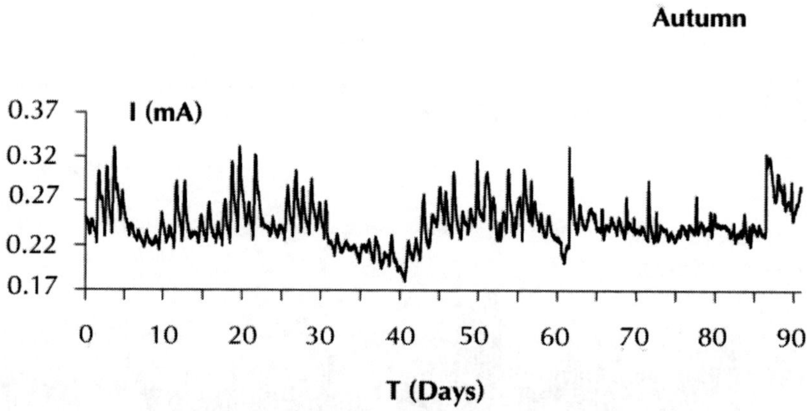

FIGURE 4 Current recorded in electrochemical cell with electrodes of platinum for the period September 1, 2009–November 30, 2009.

In many cases we haven't succeed to obtain distinct averaged curve because of time-varying position of global extreme points at the current curve. But, in some months it has been possible to obtain the resulting curve for the whole month representing well the general dynamics of currents. For example, Fig. 5 shows the averaged currents curve in relative units for April 2008 as per data recorded in the cell with steel electrodes.

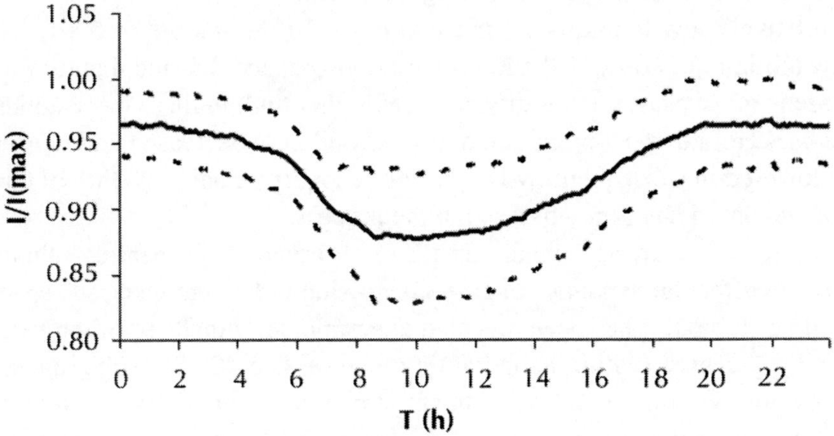

FIGURE 5 Averaged daily variations of current. Dash-dot line shows standard deviation.

Effect of air humidity on variations amplitude has been observed when long-term measurements of daily variations of currents. Slight variations of currents recorded from electrochemical cells filled with water have been observed at high humidity. This fact can be explained by absorbing ability of moisture present in the air in suspension state. Figure 6 illustrates action of high humidity on variation of currents in electrochemical cell with steel

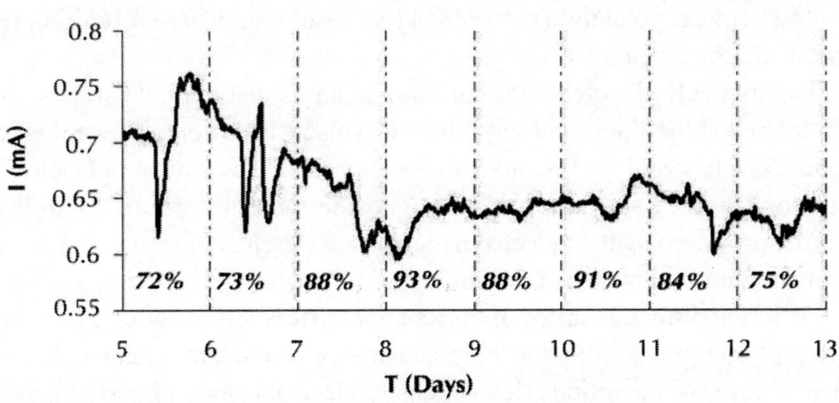

FIGURE 6 Current recorded in electrochemical cell with electrodes of steel for the period March 6, 2009–March 13, 2009. Daily average humidity is shown with digitals.

electrodes. It is clearly seen at the figure that daily variations are observed at relatively low humidity in first two days. To the middle of the fourth day, minimum is observed and no variations observed at the diagram for subsequent two days. Humidity has been rather high in these days against the background of 10-point cloudiness. Cloudiness decreased and humidity lowered in subsequent two days, and recovery of daily cycling of currents recorded has been observed in the process.

In order to correctly explain results of observations probability of temperature effect on dynamics of currents flowing in the interelectrode space shall be studied. It has been revealed experimentally that rise of temperature by 1-degree (within temperature range of 4 to 40°C) results in step-up by not over than 1%. Environment temperature fluctuations within 24 hours in closed room can reach maximum of 5°C in summer therefore 5% increase of currents driven by temperature can be expected. That is why recording of currents in temperature-controlled cell completely excludes theory of rhythms' origin as a result of environment temperature variations. Another experimental fact completely denying rhythm-forming factor of environment temperature daily variations points to physical and chemical origin of currents' daily variations. It has been established experimentally for different cells with different chemical composition of electrodes that currents are of opposite direction within 24 hours (Fig. 7). The said fact is explained by the point that electrode of platinum has positive electrochemical potential ($\varphi_{Pt} = +1,2\ \hat{A}$) and iron electrode has negative electrochemical potential ($\varphi_{Fe} = -0,44\ \hat{A}$) that results in different limiting reactions on the electrodes.

The first cell (denoted with red line) contains platinum electrodes; the second cell (blue line); and the third one (black line) contains steel electrodes. The first and the second cells are placed side by side in light-blocking wooden box, and the third one is inside the temperature controlled cabinet protected against electromagnetic interference of technogenic origin with aluminium sheet 1.5 mm thick.

Different daily dynamics of currents on different electrodes and under different conditions points to high sensitivity of contact potential difference to outward conditions (low-intensity electromagnetic interference of the Earth) and also to material touching or being dissolved in water.

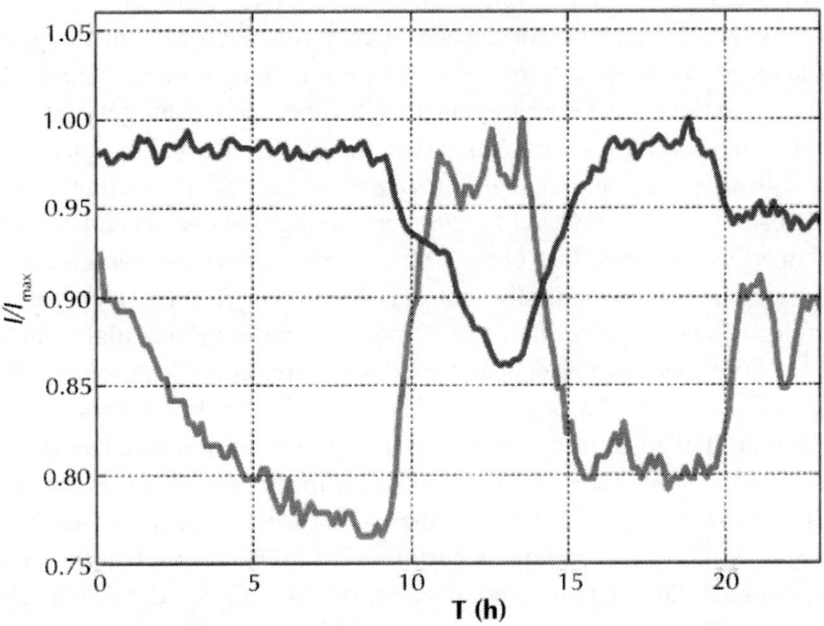

FIGURE 7 Daily variations of currents in electrochemical cells containing different electrodes: red line – cell with electrodes of platinum, blue line – cell with electrodes of steel.

For explanation of further experimental results the following operational hypothesis is proposed. Electronic work function of a material changes under the influence of various environmental factors: light, temperature, various vibrations, and all modes of radiation. Electromagnetic interference both of technogenic origin and interference being generated by mantles of the Earth shall be outlined. This interference has direct effect on electronic work function. It becomes particularly apparent when there is contact of two or more substances different by chemical composition. Semiconductor devices that have very developed contacting surface can serve the example. Solutions for which molecule of a dissolved substance forms the contacting surface of solvent around itself can be classified as most developed contacting surfaces. Value of contact potential difference between vacuum and substance is numerically equal to electronic work function to vacuum. Contact potential difference forming the variations of physical and chemical properties resulting in various consequences unexplainable by existing conceptions appears under conditions of contact of two heterogeneous substances. Special role is given in this hypothesis to electromagnetic interference of the Earth being formed at tidal rubbing of mantles of the Earth against each other caused by gravitational interaction of the Sun and the Moon with the Earth and also through global intra-Earth tectonic and seismic processes. Influence of external super-mild cosmic and geophysical factors upon living organisms can be easily explained by the point that change of electrons' activity in molecules of water medium causes naturally changes of oxidizing (electronic) potential of molecules of water medium and change of membrane potential value in cells of organisms as consequence. For living organisms, control of cell's potential value is the basis of regulation of all core vital processes.

Diurnal dynamics of currents in water in various geographical points of the Earth: in Moscow, in the Crimea (at the territory Crimean Astrophysical Observatory and at the Northern Caucasus in Pyatigorsk has been obtained within the experiments performed (Fig. 8). Start of all the measurements of diurnal dynamics of currents has been synchronized with

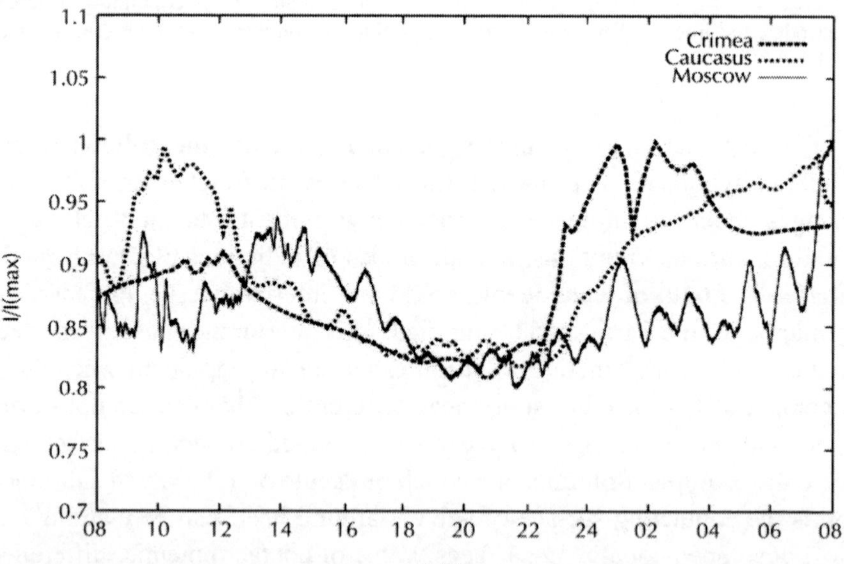

FIGURE 8 Variation of electric currents in Moscow, in the Crimea and at the Caucasus measured simultaneously during 24 hours in the interval December 10, 8 a.m.–December 11, 8 a.m. Moscow time.

start of measurements in laboratory rooms in Moscow in identical electrochemical cells. To enhance reliability, the measuring system contained two identical cells that have been put into heat-insulating jacket of plastic foam and have been equipped with system of temperature control within the range 25–30°C. Figure 8 shows similarity of processes progressing in water situated in different geographical zones and also highlights globality of external influencing factor. In virtue of difference of geographical environment we don't have complete coincidence of curves of currents but common dynamics of currents' variation is distinct.

We have compared the results on EMI of geospheres recorded by executives of FSUE Kavkazgeolsiyemka in Kislovodsk survey point with the results on currents in electrochemical cell obtained in Pyatigorsk city (measurements have been performed using portable plant described above). Variations of EMI of geospheres have been observed with "MGR-01 M" multi-channel geophysical recording device intended for measuring of temporal and space variations of parameters of natural pulse electromagnetic field of the Earth's geospheres. To demonstrate the results, two diagrams are given at Figs. 9 and 10. Common dynamics is seen at comparing both figures; however, two peaks equal by value (at 12 o'clock

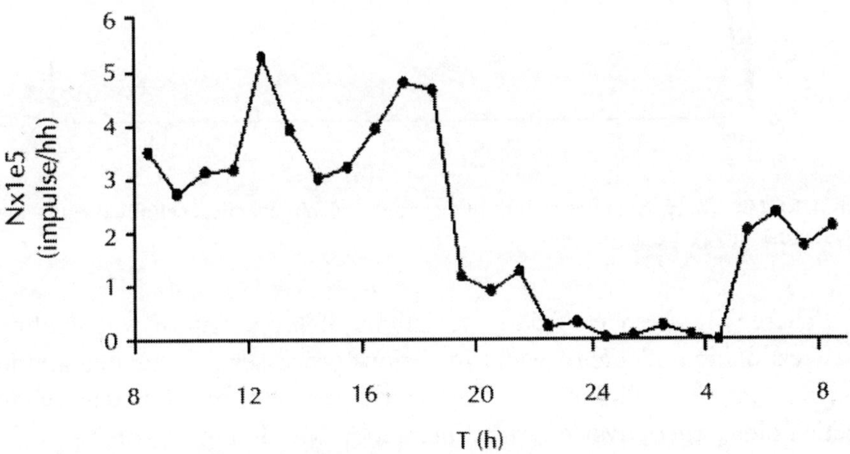

FIGURE 9 Daily variation of number of pulses N of magnetic component at frequency 1–3 kHz. Kislovodsk city is the survey point.

and at 17 o'clock) can be seen at the diagram with EMI variations (Fig. 9), and peaks differ substantially by amplitude at the diagram of currents' measurement (Fig. 10). Intense variation of currents is apparent in the period from 20 o'clock till 04 o'clock (Fig. 10) and such intense variation is not present at EMI diagram (Fig. 9). The divergence observed might be explained by the point that EMI measurements have been performed within narrow band (1–3 kHz) and it seems likely that considerable part of spectrum has not been measured by experimental plant.

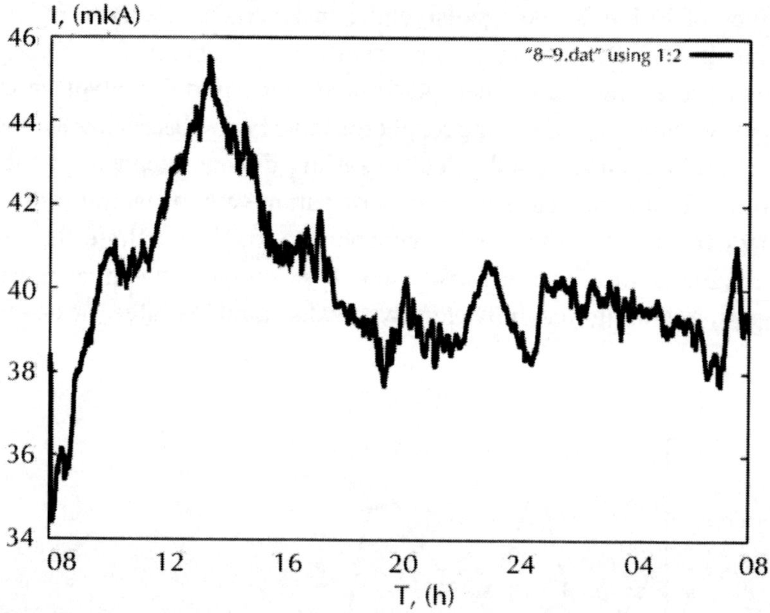

FIGURE 10 Daily dynamics of current recorded in electrochemical cell filled with water. Pyatigorsk city is the survey point.

Thereby, existence of daily variations of properties of water allows new estimation of role of water in various processes of animated and inorganic nature. With that water can be the conductor of weak electromagnetic fields of geophysical origin interaction with living organisms.

15.4 CONCLUSIONS

1. Daily variation of water conductivity in electrochemical cells has been demonstrated.
2. Opposite direction of currents' variation within 24 hours for cells with different chemical composition of electrodes of platinum and stainless steel has been revealed.
3. Effect of air humidity on amplitude of currents' daily variations has been observed.
4. Common nature of behaviour (or variation) dynamics of currents in electrochemical cell and electromagnetic interference is shown from comparison of results on electromagnetic interference of geospheres with results of water conductivity measurement.

KEYWORDS

- **Conductometric cell**
- **Isoline of voltage**
- **Kislovodsk survey point**
- **Pirex glass**
- **Technogenic origin**

REFERENCES

1. Mileva, K.; Georgieva, B.; Radicheva, N. About the biological effects of high and extremely high frequency electromagnetic fields. *Acta. Physiol. Pharmacol. Bulg.*, **2003**, *27(2)*, 89–100.
2. Goodman, E. M.; Greenebaum, B.; Marron, M. T. Effects of electromagnetic fields on molecules and cells. *Int. Rev. Cytol.* **1995**, *158*, 279–338.
3. Naarala, J.; Hoyto, A.; Markkanen, A.; Cellular effects of electromagnetic fields. *Altern. Lab. Anim.* **2004**, *32(4)*, 355–360.
4. Rudiger, M. International Commission on Non-Ionizing Radiation Protection. Guidelines for limiting exposure to time-varying electric, magnetic, and electromagnetic fields (up to 300 GHz). *Health Phys.*, **1998**, *74(4)*, 494–522.

5. Ageev, I. M.; Shishkin, G. G. The correlation of solar activity with the electrical conductivity of water. *Biophysics.* **2001**, *46(5)*, 829–832.
6. Grigoriev, Yu.; Grigoriev, A. Building a dialogue about the risks of electromagnetic fields. Radiation protection program department of the human environment. Geneva, Switzerland. *World Health Organization.* 2004; 79 p.
7. Kislovsky, L. D. In *Biological Effects of EMF*. Krasnogorskaya, N. V., Ed.; Nauka: Moscow, 1984; *2*, 17.
8. Salem, R. R. The theory of the double layer. Fizmatlit: Moscow, 2003; p. 44.
9. Artamonov, A. A. Investigation of variation the redox properties of water during the day. In *Monograph*. LAP LAMBERT Academic Publishing Gmb & Co.: Saarbrucken, Germany, 2012; 100 p.

CHAPTER 16

FORMATION OF GLYCOSIDE BONDS MECHANISM OF REACTIONS

J. A. DJAMANBAEV, A. D'AMORE, and G. E. ZAIKOV

CONTENTS

16.1 INTRODUCTION

The reactional ability of the glycoside center with the purpose of elaboration of the effective methods of introduction of the urea rests and its derivatives in the first position of the carbohydrate ring have been presented in this work. It was shown that the protoning of *N*-glycoside in the conditions of acid catalysis leads to the growth of the effective, positive charge on C_1 and it increases its reactional ability with reference to attacking nucleophiles, that was confirmed by the quantum-chemical calculations.

The reactional ability of the first carbon atom of monosaccharides and glycosides has been examined in works [1, 2]. It is well known that reactional ability of glycoside center C_1 in the processes of nucleophilic connection and replacements is defined mainly by the size of positive charge on C_1. The calculations of electronic charges carried out by the method of inductive parameters Del-Re [3] and modern quantum-chemical methods of calculation of electronic structure [HyperChemPro7, MNDO] show that among all carbon atoms of a carbohydrate molecule the greatest effective positive charge really possesses C_1. Effective charges on atoms and lengths of bonds of the urea's carbohydrate derivatives received by method MNDO are presented in Fig. 1.

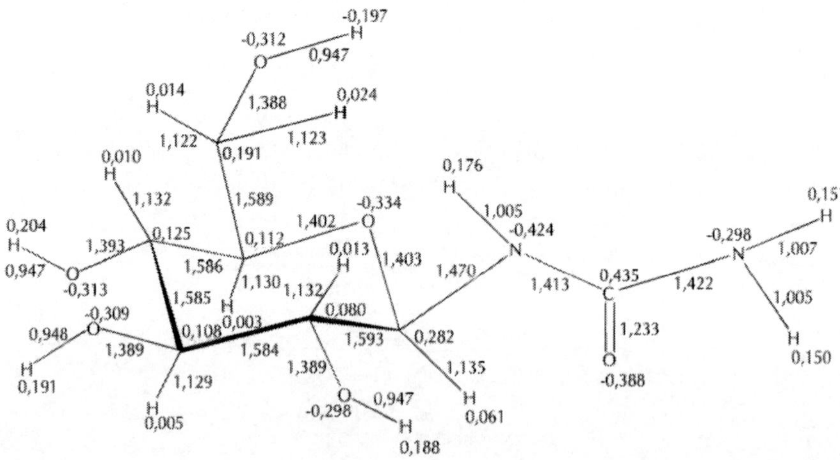

(a)

FIGURE 1 *(Continued)*

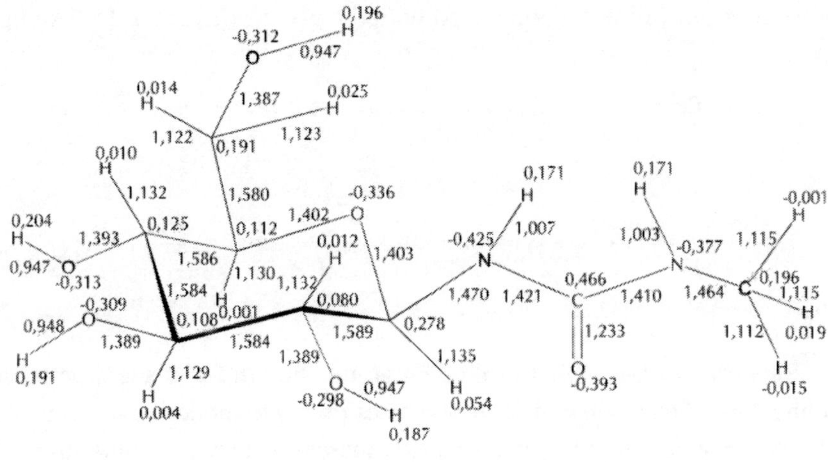

(b)

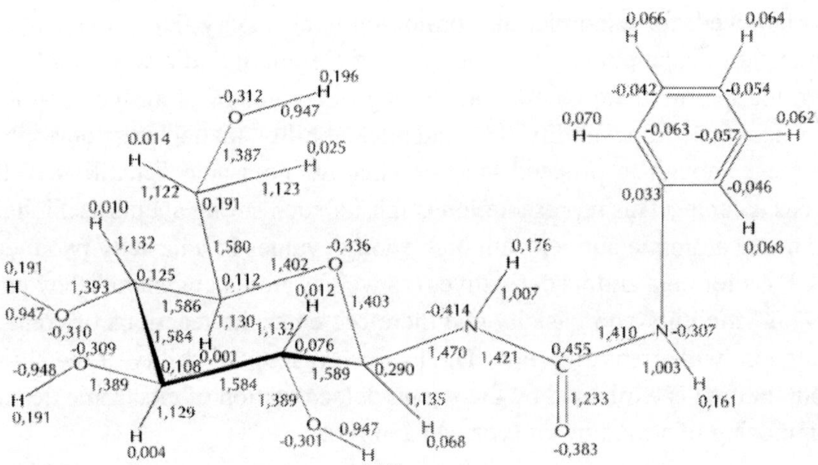

(c)

FIGURE 1 Charges on atoms and lengths of bonds in molecules of derivatives glycosylurea received by optimization of the geometrical parameters by quantum-chemical method MNDO: a) glycosylurea; b) glycosylmetylurea; c) glycosylphenylurea.

Nevertheless, in the reactions of formation of glycosyl bonds takes always place the acid catalysis, which is connected with effects of protoning of heteroatoms, directly connected with the glycoside center, for example:

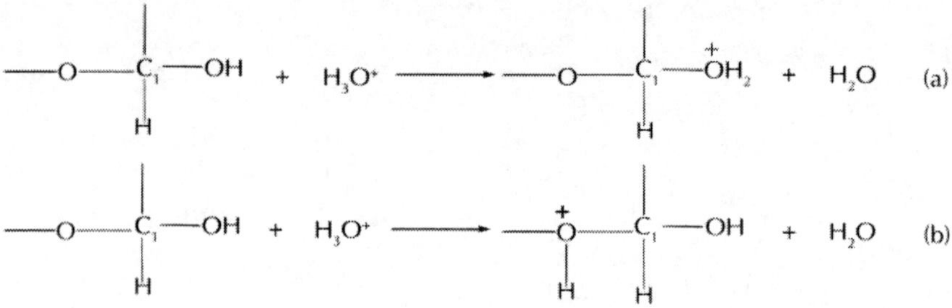

Regardless of, on what of oxygen atoms there takes place a protoning, connection of a proton considerably raises owing to inductive effect the size of a positive charge on C_1 and due to it increases its reactional ability with the reference to nucleophilic agents. This point of view on the nature of activating action of acids is the most widespread. There are also another representations connected with studying of reactions of carbonyl group [4–6]. It is believed, for example, that protoning is necessary not at the stage of connection nucleophile, but at the stage of elimination of a water molecule from the intermediate compound, having the structure of aminoalcohol. It means that reactional ability of C_1 and nucleophility of attacking nucleophile are great enough to proceed the first stage without acid activation of the glycosid center. This representation is fair for such strong agents as aliphatic and many aromatic amines with high enough values of a basicity parameter (pK_a). As for urea and its derivatives (as well as amids in general) they have low nucleophility and basicity and therefore enter the reactions of connection on C_1 with great difficulty. The lowered reactional ability of the amino group in urea is explained by the strong delocalization of electronic density or imposing of structures of type (A) and (B):

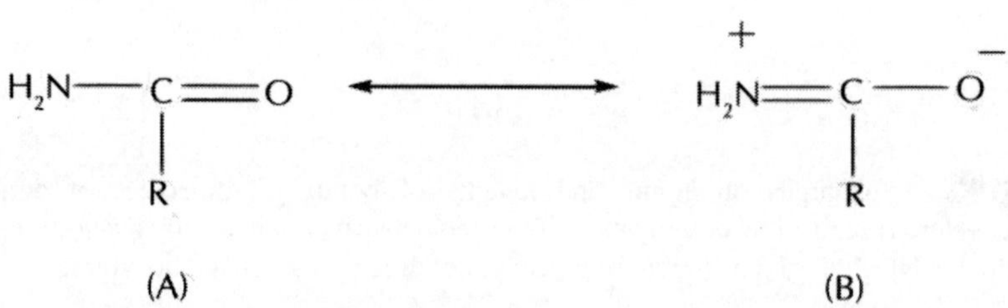

The structure (B) is actually interfaced acid and as the results of potensiometric titrations show, amides of carbonyl acids are more weak acids, than bases [4]. As basicity and nucleophility are the values, mutually connected with each other, the decrease of basicity leads to the decrease of nucleophilic reactional ability of amino group of urea. The effective atomic charges and lengths of bonds in molecules of derivatives of the urea received by optimization a of geometrical parameters by semi-empirical, quantum-chemical method MNDO are shown in Fig. 2.

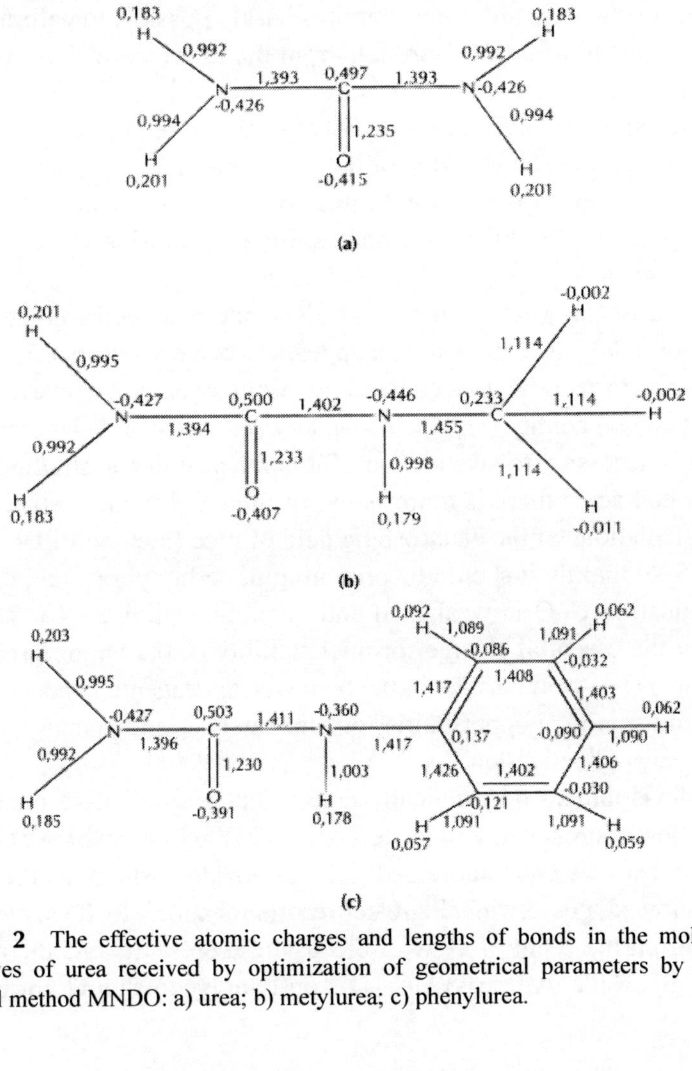

FIGURE 2 The effective atomic charges and lengths of bonds in the molecules of derivatives of urea received by optimization of geometrical parameters by quantum-chemical method MNDO: a) urea; b) metylurea; c) phenylurea.

What approaches are possible for activation of urea and its derivatives and simplification of reactions of formation *N*-glycosylureas?

The variant at which the molecule of urea is deformed is theoretically possible and gets such space arrangement at which *n*-orbital of nitrogen atom are not blocked with *n*-orbitals of C=O group. In this case the basicity and nucleophility of amino group should increase sharply. It is known, for example, the cyclic lactam in which uncoupled electrons of nitrogen atom are not blocked with – electrons of carbonyl group [4]. As the result of it is the increase in basicity parameter at five orders in comparison with usual (not cyclic) lactams. However, it is hardly possible to deform a molecule of urea and to deduce -orbitals from the interface with *n*-orbitals in conditions of coordination catalysis.

On account of stated one can believe, that attempts to activate the carbamide fragments with using of homogeneous acid or coordination catalysis is the work of the future. In this connection searching of ways for increasing the nucleophilic reactional ability of glycoside center of sugars is actual.

Because of the weak protoning of glycoside and cyclic oxygen atoms usual acids catalysis of proton type appears to be very poorly effective – it is necessary to use the concentrated solutions of acids. However, in the field of Gamete acidity (H_0) the secondary reactions of sugar start to develop. It is necessary to take into consideration also that in solutions of the concentrated acids there is a protoning of urea and its derivatives. Thus, as the calculations of the electrostatic field of urea (method PPDP/2) have shown [5–6] mainly it should be protoning of carbonyl oxygen, therefore communication C–O is weakened and communication C–N is strengthened, but the essential change for nucleophility of the amino group does not occur. These peculiarities in the behavior of reagents in sour environments limit synthetic opportunities of acid catalysis and demand to search for new ways of activation.

For development of practically convenient and effective methods of introduction of urea rests and its derivatives in the first position of the carbohydrate ring we have addressed to N-glycosides, which, as the kinetic studies showed, possess much greater reactional ability to the nucleophilic agents. From the point of view of electronic representations the replacement of glycoside hydrocsyl HC_1–OH on N-aglycon should not increase the electrophilic reactional ability of C_1. Indeed, as it was shown by the carried out calculations of -electronic charges by the method of inductive parameters Del-Re, during the formation of N-glycosid bond it occurs an appreciable increase in the effective positive charge on C_1, therefore the electrophilic reactional ability of the glycoside center should increase too.

It should be considered, however, one important circumstance – high protono-acseptornic ability of nitrogen atom in comparison with the atom of oxygen. Protoning of this center in N-glycosides in conditions of acid catalysis will lead to the growth of an effective positive charge on C_1 and to the increase of its reactional ability in relation to attacking nucleophils. These reasons specified a real opportunity for the development of effective methods of synthesis of glycosides with little basicity on the basis of N-glycosides in which N-aglycon has high enough basicity [7–9].

Let's explain above-mentioned on an example of the simple scheme for acid catalysis reactions of nucleophilic replacement at C_1 [1, 2]. The presented mechanism includes two basic stages of reaction – protoning of initial glycosid on the atom of nitrogen of N-glycosid bond (a) and as the result of it interaction of the protoning form of glycosid with attacking nucleophyl (b) the replacement of initial N-aglycon on a new aglycon takes place (the Scheme 1).

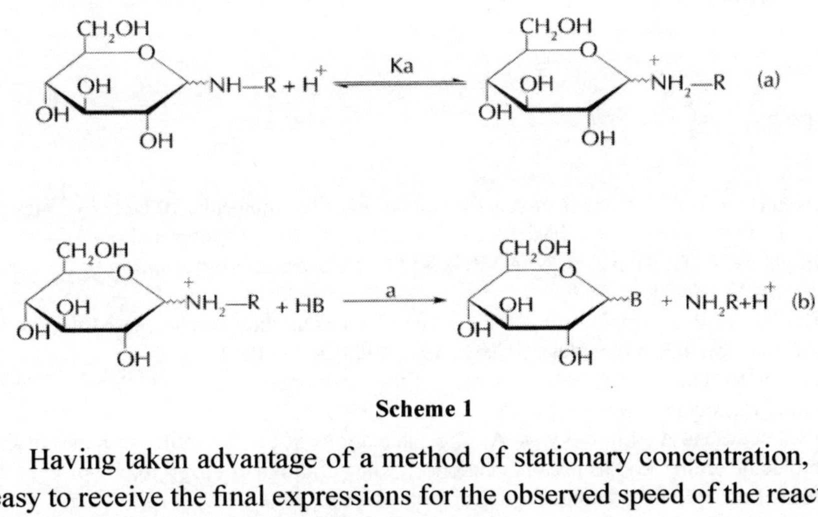

Scheme 1

Having taken advantage of a method of stationary concentration, it is easy to receive the final expressions for the observed speed of the reaction (1) and for the observed action constant (2).

$$V_{observ.} = a\,[-HC_1NH_2^+R] + [HB] = k_{observ.} = [-HC_1NHR]\,[HB] \quad (1)$$

$$k_{observ.} = a\,[H^+] / k_a \quad (2)$$

According to last expression, the less the constant of dissociation K_a of the protoning form of glycoside (i.e. the more basicity initial the N-aglycon, expressed in the size pK_a), the lower acidity of solution the reaction (b) of replacements initial N-aglycon on a new aglycon will proceed.

Therefore, knowing the size of a parameter of basicity pK_a of initial N-glycoside it is possible to estimate approximately that area of values pH in which the reaction of replacement will proceed with good speeds for practice.

KEYWORDS

- **Gamete acidity**
- **Glycosides**
- **Monosaccharides**
- **N-aglycon**
- **N-glycoside**

REFERENCES

1. Afanasjev, V. A. The mechanisms for activation of the molecules of carbohydrates in the conditions of the acid-basicity catalysis. Frunze.: Ilim, 1974; p. 137.
2. Bochkov, A. F.; Afanasjev, V. A.; Zaikov, G. E. *Formation and Splitting of Glycosid Bonds*. M.: Sciences, 1978.
3. Del Re, G. R. A simple MO LCAO method for calculating the charge distribution in saturated organic molecules, *J. Chem. Soc.*, 1958; 4031–4040.
4. Arnett, M. The quantitative comparison of the weak organic bases, *The New Problems of the Physical Chemistry*. M.: Peace, 1967; 195–341.
5. Panteleev, Yu. A.; Lipovsky, A. A. The calculations of the electronic structure of the molecule of urea and its protonic kations. I. The definition of the protoning place. The role of the electrostatic interaction, *J. Struct. Chem.*, **1976**, *17(1)*, 4–8.

6. Panteleev, Yu. A.; Lipovsky, A. A. The calculations of the electronic structure of the molecule of urea and its protonic kations. II. The change in the structure of the molecule of urea by protoning, *J. Struct. Chem.*, **1976**, *17(1)*, 9–14.
7. Afanasjev, V. A.; Djamanbaev, J. A. The synthesis of glycosylurea on the basis of N-glycosid m-nitroanilin as a glycosilic agent, *The News of Academy of Sciences of Kyrgyz SSR*. **1973**, *2*, 64–65.
8. Afanasjev, V. A.; Djamanbaev, J. A. The effective method of obtaining N-glycosids of urea, *Chem. Nature Comp.*, **1974**, *2*,176–178.
9. Djamanbaev, J. A.; Abdurashitova, J. A.; Zaikov, G. E. Sugar carbamides. Organic and physical chemistry using chemical kinetics. In *Prospects and Development*, Medvedevskikh, Yu. G.; Valente, A.; Howell, R. A.; Zaikov, G. E. Nova Science Publishers: New York, 2008; pp. 291.

CHAPTER 17

NEW ADDITIVES FOR BURNING OF THE FUELS

E. GIGINEISHVILI, L. ASATIANI, N. LEKISHVILI, A. D'AMORE, and G. E. ZAIKOV

CONTENTS

17.1 INTRODUCTION

It is known from literature that some derivatives of ferrocene and ferrocene-containing carbinoles as catalysts of special purpose for burning of solid fuel [1] have found a wide application. Possessing such properties as thermal stability, low toxicity, good solubility in organic solvents producing nearly smokeless burning of organic compounds, they are valuable additives to all fuels (gas, liquid, solid).

One of widely used organometallic catalysts of burning SRF is *n*-butylferrocene. In this direction numerous researches were carried out revealed plenty of other derivatives of ferrocene, superior in their catalytic efficiency to *n*-butylferrocene. Among them are saturated and ferrocenecontaining unsaturated carbinoles [2–7]. For example, cyclopentadienyl[2-(1-oxyethyl) cyclopentadienyl], displays not only catalytic properties interchangeably with n-butylferrocene but also it manages to stay strongly as a part of charge [4].

As it shown in Ref. [3] the series of ferrocene-containing derivatives with –CH(OH)R groups are considered and are recommended as softeners and catalysts of fuel-binding propellant components.

Based on the above, for creation of new effective burn catalysts and softeners of SRF, before us there was a task to synthesize and investigate unknown earlier derivatives of ferrocene containing mono- and divinylacetylene alcohols.

17.2 EXPERIMENTAL

The currents of all reaction were tested by Method of Thin-Layer Chromatography-TLC. NMR spectra were run on UNITY 400(400 MHZ for ^{1}H). Elemental analyses were performed in the micro-analytic laboratory of Tbilisi State University (Georgia). Infrared spectra were recorded on "Specord 75 IR" at Tbilisi State University. All solvent were vigorously dried with an appropriate drying agent and distilled before use.

17.2.1 1-FERROCENYL-1-PHENYL-1-OXY-2-PENTYN-4-EN-FK-1G (I)

To the solution of lithium vinylacetylene (prepared from 0.5g (0.07 mol) Li and vinyl-acetylene in 100 ml dry THF under Helium (He) at room temperature with stirring 10 g (0.031 mol) of benzoylferrocene in 100 ml dry THF was added. The reaction finished right after the adding the benzoylferrocene. The reaction mixture was poured into cold water and extracted with ether. The ether extract was washed by water and dried over anhydrous sodium sulphate; then the ether was evaporated. The result is sticky substance, which after adding the hexane gives yellow crystals. The crystals was washed by hexane and drayed; mp. 94–95^{0}C; yield 10.3 g (95.4%), For $C_{21}H_{18}Fe$. Calculated: 73.68% C: 5.26% H; 16.37% Fe; Found: 73.30% C, 5.49% H, 16.38% Fe.

17.2.2 1,1'-BIS(-1-PHENYL-1-OXY-2-PENTYN-4-ENYL) FERROCENE-FK-2G (II)

(II)-method of synthesis of this compound analogical the method of synthesis of compound I, using) 0.4 g (0.05 mol) Lithium, vinylacetylene and 5 g (0.0126 mol) dibenzoylferrocene. The result is yellow crystals with m.p. 112–114°C; yield 6 g (95%); For $C_{32}H_{26}Fe$; Calculated: 77.10% C; 5.22% H; 11.26% Fe; Found: 77.47% C, 5.22% H; 11.29% Fe.

17.2.3 1-FERROCENYL-1-PHENYL-1-METOXY-2-PENTYN-4-EN (III)

A 0.5 g (0.001 mol) catalyst $HgSo_4$ was added to the solution of 2 g (0.0058 mol) FK-1G (I) in 50 ml methanol. The reaction took place right after adding the catalyst and finished after 30 min at room temperature, and then the solution was filtered by Buchner filter; after separating of the solvent yellow crystals were precipitated. They were washed by pentane and dried. m.p. 80-81°C; yield 1,9g (91%); For $C_{22}H_{20}Fe$; Calculated: 74.15% C; 5.51% H; 15.73% Fe; Found: 74.17% C, 5.60% H; 15.68% Fe.

17.2.4 1,1'-BIS (1-PHENYL-1-METOXY-2-PENTYN-4-ENYL FERROCENE (IV)

Method of synthesis of this compound analogical the method of synthesis of compound IV, using 1 g (0.002 mol) FK-2G (II) and 0.5 g (0.001 mol) $HgSO_4$. The reaction mixture was filtered by gel filter distilling of solvent the very sticky substance is received, which was cleaning by column chromatography (adsorbent – silicagel (100–250), solvent – hexane, eluent – ether/hexane.1:5); ether distilling solvent the yellow crystals obtained with m.p. 87–89°C; yield 0.9 g (85%); For $C_{34}H_{30}Fe$; Calculated: 74.57% C; 5.70% H; 10.64% Fe; Found: 77.25% C, 5.80% H; 10.11% Fe.

17.2.5 1-FERROCENIL-1-PHENYL-1,4-PENTADIEN-3-ONE (V)

A 0.6 g (0.006 mol) of $ClCH_2COOH$ was added to the solution of the solution of 2 g (0.0058 mol) FK-1G (I) in 30 ml dry benzene. The reaction mixture was stirred at room temperature for 30 min, than was filtered by layer Al_2O_3. The filtrate consisted of two compounds, which were separated by column chromatography (adsorbent-silicagel (100–250), solvent-hexane, eluant-ether/hexane 1:4); were formed two following products – V and VI – di (1-ferrocenyl-1-phenyl-2-yn-4-pentyn)ether:

V is sticky substance of dark red color with R_f– 0.78 (hexane/ether1:3) yield 1.2 g (60%); For $C_{21}H_{18}OFe$ Calculated: 73.68% C; 5.26% H; 16.73% Fe; Found: 73.15% C, 5.70% H; 16.25% Fe;

VI is yellow stickle substance with R_f –0.85 (ether/hexane 1:1; yield-insignificant amount. For $C_{32}H_{25}OFe_2$; Calculated: 75.66% C; 5.10% H; 16.81% Fe; Found: 75.76% C, 5.25% H; 16.31% Fe.

17.2.6 1,1'-BIS(1-PHENYL-1,4-PENTADIEN-3-ON)FERROCENE (VII) AND 1,3-BIS-PHENYL-1,3-BIS(VINYL-ETHINYL)-2-OXA [3]-FERROCENOPHANE (VIII)

Method of synthesis of this compounds analogical the method of synthesis for compounds V and VI, using 1 g (0.002) alcohol II (FK-2G) and

0.4 g (0.004 mol) $ClCH_2COOH$. The result dark red oil (VII) with R_f – 0.423 (hexane/ether 1:1); yield 0.43 g (43%) and yellow-brawn oil (VIII) with R_f – 0.75 (ether/hexane 1:1); yield 0.15 g (15%);

For $C_{32}H_{26}OFe$ (VII); Calculated: 77.10% C; 5.22% H; 11.26% Fe; Found: 77.43% C, 5.14% H; 10.83% Fe;

For $C_{32}H_{26}OFe$ (VIII); Calculated: 80.00% C; 5.00% H; 11.66% Fe; Found: 80.20% C, 5.34% H; 11.25% Fe.

17.3 GENERAL RESULTS

For development the synthetic chemistry of new ferrocene burn catalysts of SRF, reaction of mono- and dibenzoylferrocene with metal derivatives of vinylacetylene [8–9] was investigated. It is known that the most convenient method for receiving tertiary vinylacetylene alcohols of aliphatic and aromatic series is interaction of magnesium-brom-inevinylacetylene with appropriate ketones [10–12]. However, this method in our case did not turn out effective, since the magneziumbrominevinylacetylene does react neither with benzoyl-, nor with dibenzoyl ferro-cene. Variation of solvent (diethyl ether, tetrahydrofuran, benzene, and toluene), temperature and ratio of reactants did not render influence to reaction, which is obviously connected to spatial shielding of carbonylic carbon atom.

We replaced the magneziumbrominevinylacetylene with the lithiumvinylacetylene because that actively attacks to spatial carbonyl-containing substances in comparison with organo-magnezium reactants [2].

Lithiumvinylacetylene was prepared by simplified method - receipt of butyllithium was excluded from initial stag of the process -, and in such a manner two stage process of receiving lithium vinylacetylene was brought to one-stage; In particular, lithium vinylacetylene was prepared by direct interaction of vinylacetylene with small-sized cut lithium in the dry tetrahydrophurane.

As one would expect, lithiumvinylacetylene reacts well with both – benzoylferrocene and dibenzoylferrocene, and the corresponding mono- and divinylacetylene alcohols FK-1G (1-ferrocenyl-1-phenyl-1-oxy-2-pentin-4-en) and FK-2G [1,1'-bis-(1-phenyl-1-oxy-2-pentin-4-enyl) ferrocene] were obtained with yields 90–95%. The reaction proceeds according to the following scheme:

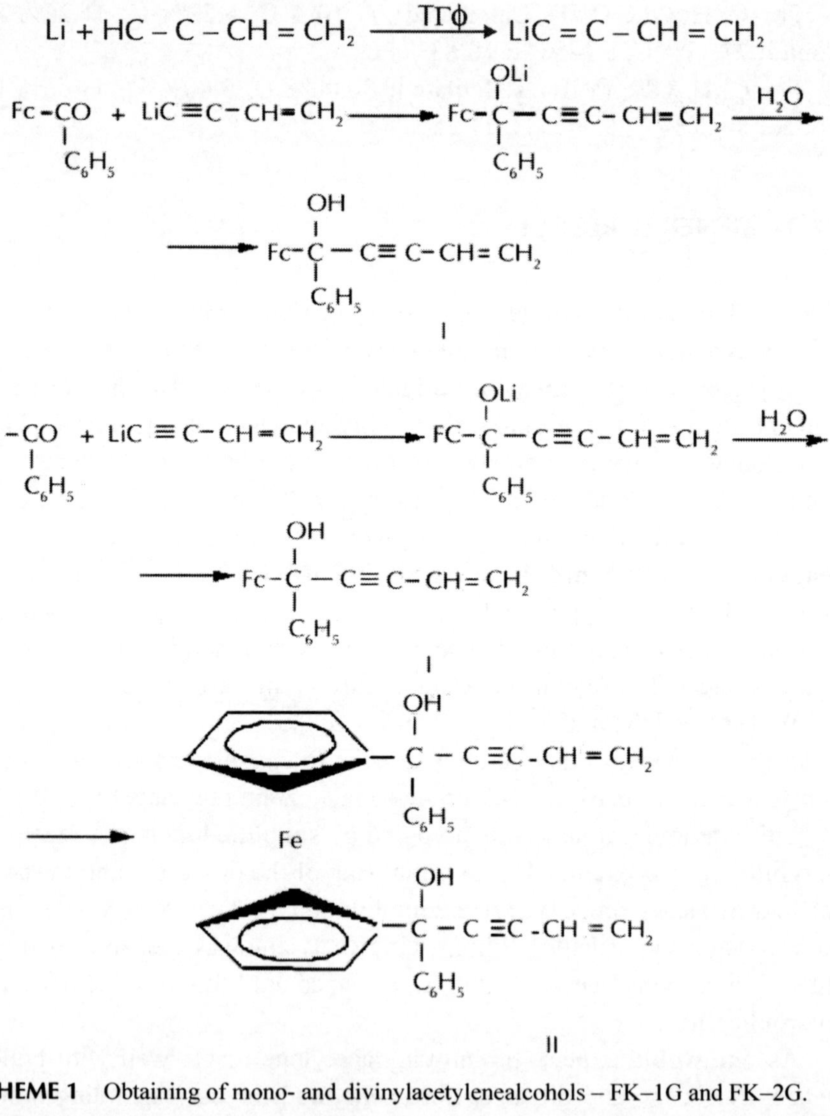

SCHEME 1 Obtaining of mono- and divinylacetylenealcohols – FK–1G and FK–2G.

The aforementioned method was developed by this chapter authors in a convenient not only as preparative method, but also from technological point of view, since the installation of receipt of ferrocene containing vinylacetylenealcohol is simple. The process itself, as it was already noted above, was a two-stage process but was shrunk down to one continuous method without waste, is environmentally safe and does not require difficult-to-obtain expensive initial substances [8].

Composition and structure of received alcohols – FK-1G and FK-2G were established according to data of element analysis, NMP and IR spectra. IR spectra there were found the following maximums of the absorptions: 1610/cm (C=C), 3110/cm, (C–H of the ferrocene), 3600–3200/cm (O–H), and weak peak in the area 2225/cm are (C≡C) (Fig. 1).

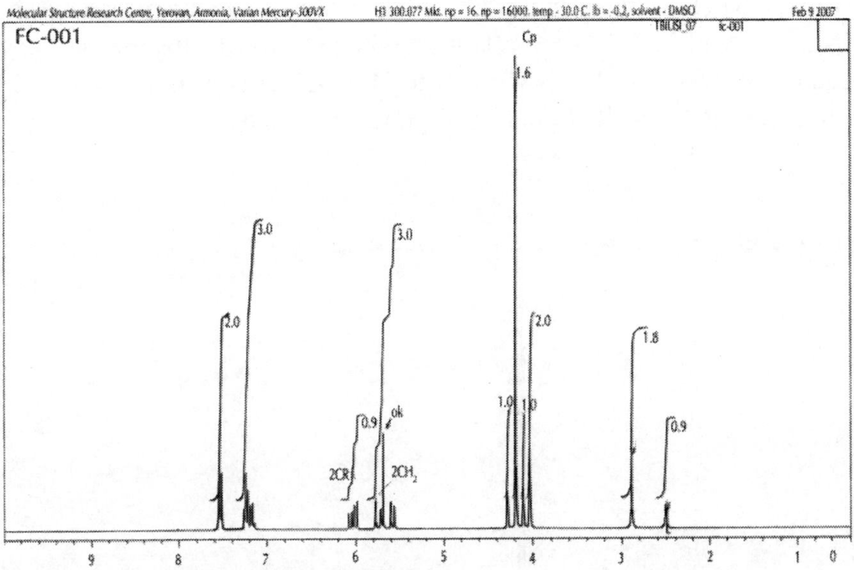

FIGURE 1 Spectrum ^{1}H NMR of FK-1G.

In ^{1}H NMR spectrum of connection FK-1G (Fig. 1) singlet resonance signals with chemical shift (c.s.) 4.2 ppm are found out related to protons of un-replaced cyclopentadiene, singlet protons resonance signals at 3rd and 4th carbon and 3rd and 5th carbon in the cycle, with c.s. 4.0 ppm, 4.1 and 4.3 ppm, respectively. In the spectrum as well resonant signal with c.s. is found out. 5.68 ppm related to the protons of tertiary hydroxy group, as well as signal with c.s. 5.58 ppm (with hem-constant – 2.38 Hertz and *cis*-constant – 11.11 Hertz) related to protons of vinylgroup. Signal with c.s. 5.74 ppm (trance I with constant of 17.46 Hertz) and resonance signal with c.s. 6.03 ppm non-equivalent N-protons (B) and N (C) in vinyl group (Outline 2), respectively.

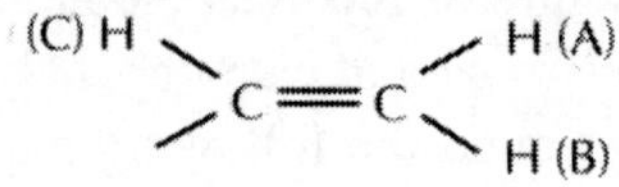

SCHEME 2 Formula of vinyl group with direction of three of non-equivalent protons.

In the ^{1}H NMR spectrum of FK-1G there was found the resonance singlet signal with c.s. 7.1–7.6, relating to the protons of substituted phenyl group. The spectrum ^{13}C NMR of FK-1Γ (Fig. 2) is in total conformity with spectrum ^{1}H NMR of the corresponding compounds.

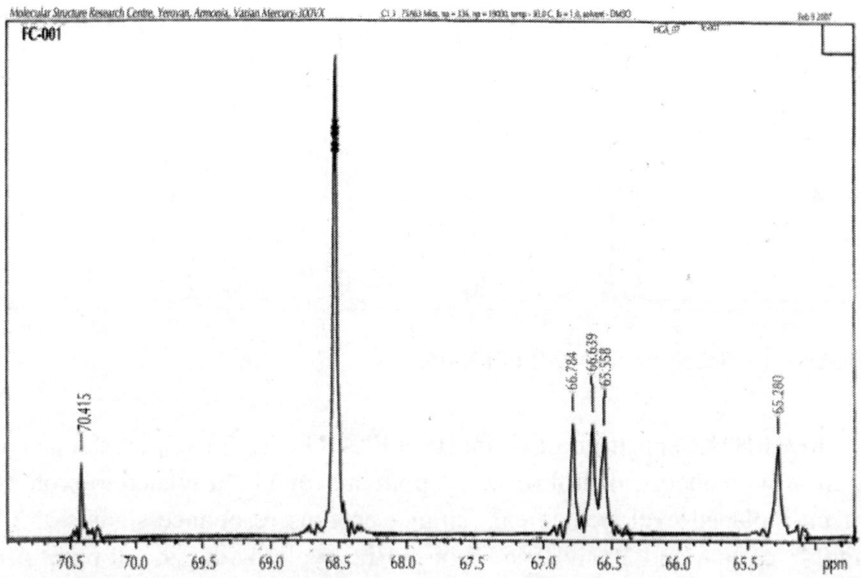

FIGURE 2 Spectrum ^{13}C NMR of FK-1G.

In the spectrum ^{13}C NMR of FK-1G resonance signals with c.s. 125.26–126.9 ppm of nuclei of carbon atom of phenyl group were found; In the spectrum one can found resonance signals of nuclei of carbon for: C–O – 82.129 ppm, C≡C – 96.53 ppm and 94.18 ppm, also the resonance signal of the carbon nuclei uninvolved carbon atom in cyclopentadiene – 68.52 ppm as well as involved the nuclei of carbon atoms C1 (70.42 ppm C2 (65.28 ppm), C3 (66.56 ppm).

Reaction ability of obtained vinylacetylene alcohols, (FK-1G and FK-2G), in particular the reaction of anionotropic rearrangement in various conditions was investigated.

It was established that as opposed to easy isomerization tertiary vinylacetylene carbinols with formation of divinylketones (in presence of $HgSO_4$) [10–12] reaction of anion-tropic re-arrangement (in solution of mathanole) of the structures of FK-1G and FK-2G proceeds anomalously. Particularly takes place reaction of the methylation of the hydroxy group. The reaction anion-tropic rearrangement did not proceed. As a result with quantitative exit the appropriate simple ethers are obtained:

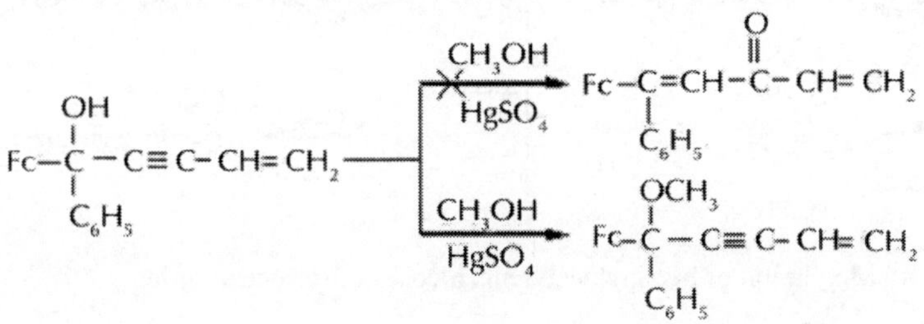

III

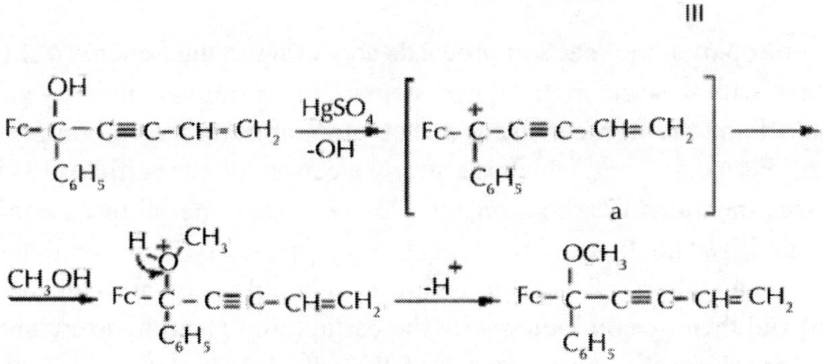

III

SCHEME 3 Metilation of monovinylacetylenic alcohol of ferrocene series.

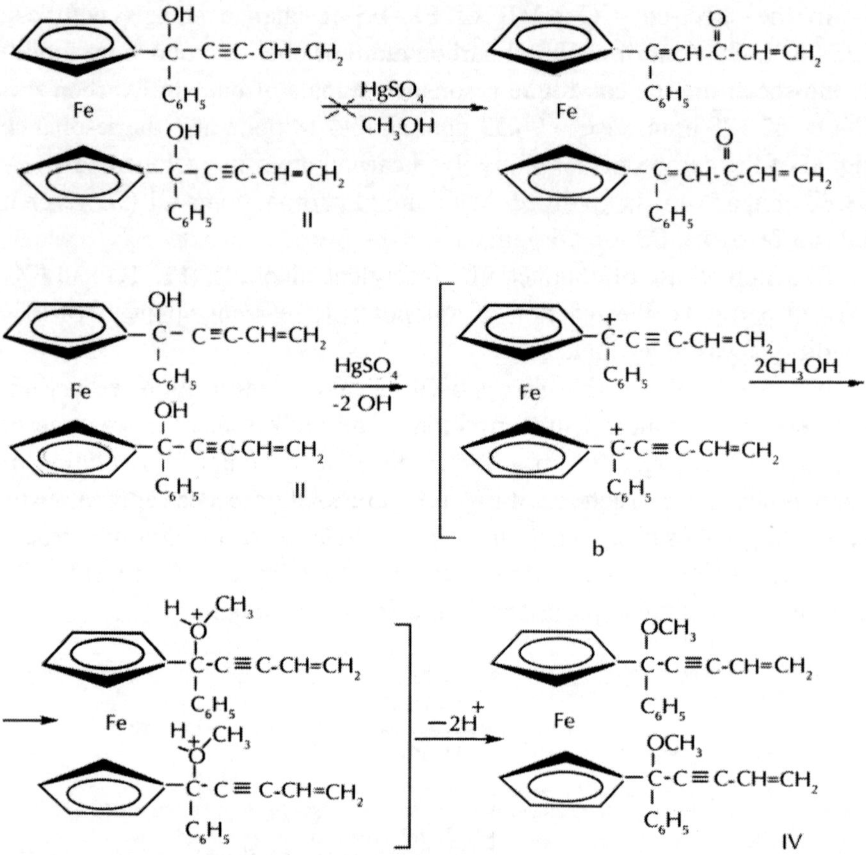

SCHEME 4 Metylation of bisvinylacetylenic alcohol of ferrocene series.

In our opinion the reaction proceeds according to the Scheme 4. In accordance with this schematic at first stage of the reaction a hydroxy group is torn off and the ferrocene ring carbon atom appears the carbonium ion (**a** or **b**). Ferrocene ring, which is a strong electronodonor particle [13–14], stabilizes mentioned carbonium ion [15–18] well. Therefore carbonium ion is stabilized not by turning of the allenyc type ion (which usually takes places, if neighbors of cationic centre are the alkyl- or aryl-substituent group) and then by combining with the earlier eliminated hydroxy group, but by attacking the oxygen atom of the solvent's hydroxy group. The result is a production of intermediate type of oxonium ions, deprotonation of which gives ethers (III and IV).

It must be noted that reaction with participation FK-2G proceeds hardly longer in comparison with FK-1G and finished during 0.5 hr. Reason of such delayed – reaction way is apparently somewhat a weak stabilization carboneum ion "b" (Scheme 4) as compared to carboneum ion "a" (Scheme 3), as electronodonor influence of ferrocene ring in the case "b" should cover two cation centers; therefore efficiency of such influence with respect to each centre is lost.

In order, to direct these reactions to the desirable rearrangement course, they were carried out in solvents, which did not have a nucleophylic centre, to prevent the attack on to the intermediate carbonium ion obtained during the reaction. As the solvent, we chose dry benzene. It turned out that in benzene in the presence of $HgSO_4$, the reaction does not proceed, while in the case of $ClCH_2COOH$ as a catalyst, the reaction follows two courses – two new substances – the product of rearrangement of FK-1G were isolated – unsaturated ketone V (as a main product – 60% yield) and the product of intermolecular dehydratation – VI (with insignificant amount). A part of reactionary mass is gummed:

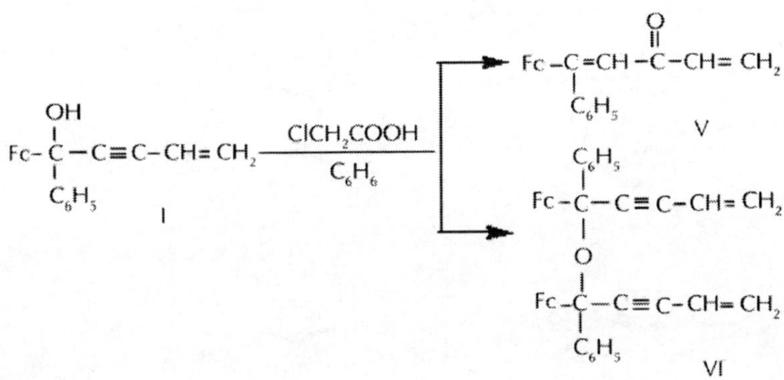

SCHEME 5 Anionotropic rearrangement of monovinylacetylenic alcohol of ferrocene series in benzene.

With participation of alcohol FK-2G mentioned reaction process is more difficult. In particular, it is needed to heat the reactionary mix at 60°C for 2 h, and yield of the product of rearrangement (Scheme 6, VII) decreases up to 43%, because at the same time as the product of rearrangement is formed, the product of intermolecular dehydratation of alcohol FK-2G –ferrocenofan (Scheme 6, VIII) is also formed with yield of up to 15%. Course of reaction was supervised by method of the TLC.

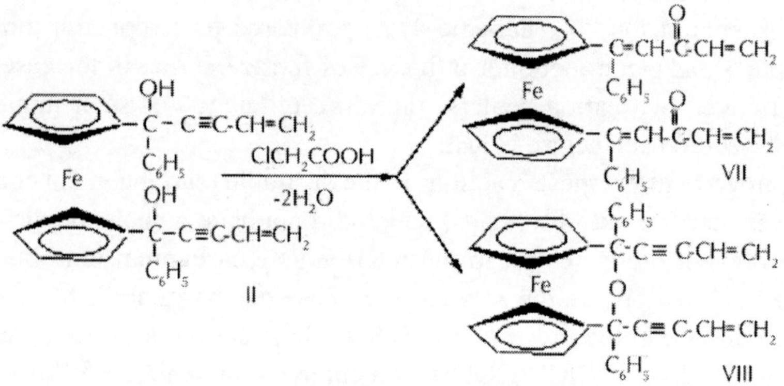

SCHEME 6 Anion-tropic rearrangement of divinylacetylene alcohol of ferrocene series in benzene.

Reaction of formation of the ferrocene-diketone (Scheme 6, VII), obviously, proceeds by following mechanism (Scheme 7):

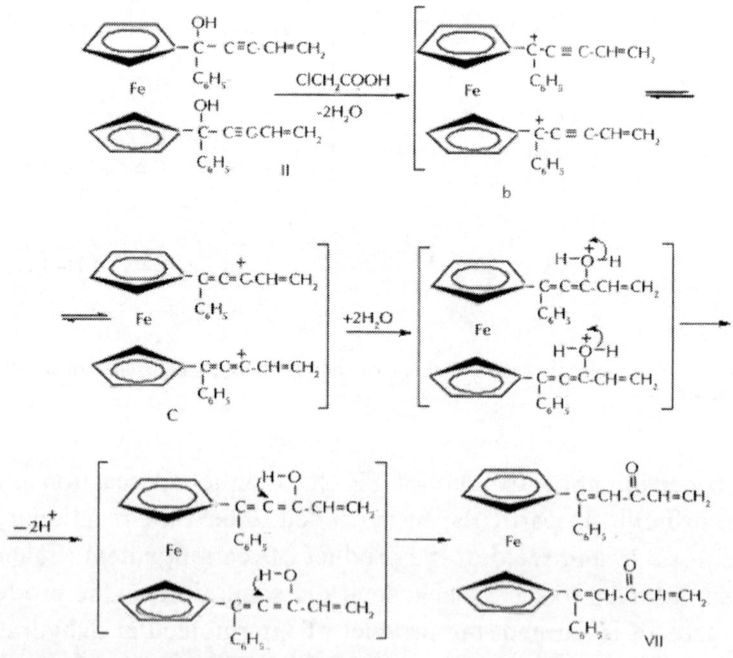

SCHEME 7 Probable mechanism of formation of the diketone VII.

The cationic centers (carbonium ion "b," Scheme 7), appear on carbon atoms neighboring the ferrocene ring at the first stage of reaction turned into the allenic cation "c" via redistribution of the bonds. It is evident that in benzene, as opposed to nucleophilic solvent, there is no nucleophilic centre; Thus carboneum ion "b," not having an opportunity to stabilize by attacking the nucleophilic centre, turns into carboneum ion "c." At the next stage, it combines with water and results the intermediate oxonium-type ion, deprotonation and isomerization of which gives unsaturated ketone VII (Scheme 7).

Assumed mechanism of formation of ferrocenophane VIII can be presented as follows (Scheme 8):

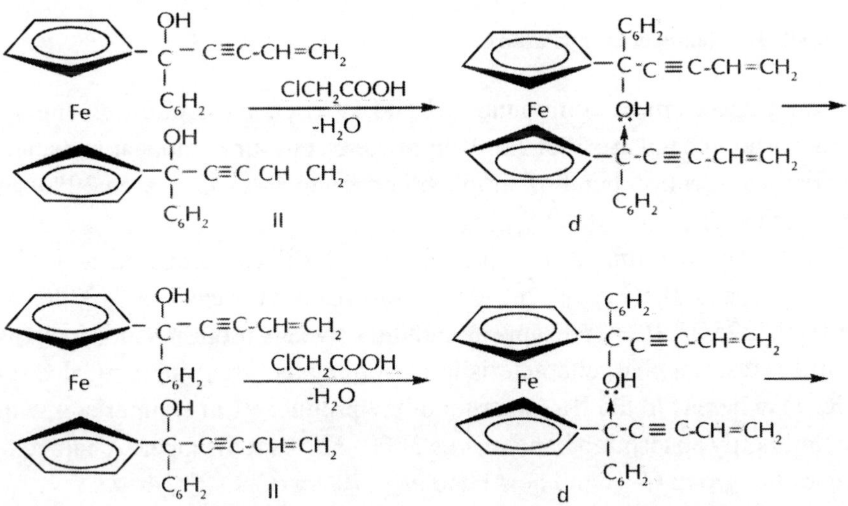

SCHEME 8 Probable mechanism of formation of the ferrocenophane VI.

As seen from the schematic, the carbonium ion "d" is stabilized not by turning into the allenic cation and then adding water, but by intramolecular attack on the carbonium ion with the oxygen atom.

In our opinion, the reaction (5) of anionotropic rearrangement of monovinylacetylenic alcohol of ferrocene series FK-1G proceeds in similar scheme.

Individuality of received substances was confirmed by method of the Thin-Layer-Chromatography – TLC.

The structure and composition of obtained compounds were established based on data of element analysis and IR-spectra. For additional identification of ketones V and VII (Scheme 5 and Scheme 7, correspondingly). their interaction with 2,4-dinitrophenylhydrazin were carried out. There were allocated the appropriate hydrazons IX and X (Scheme 9).

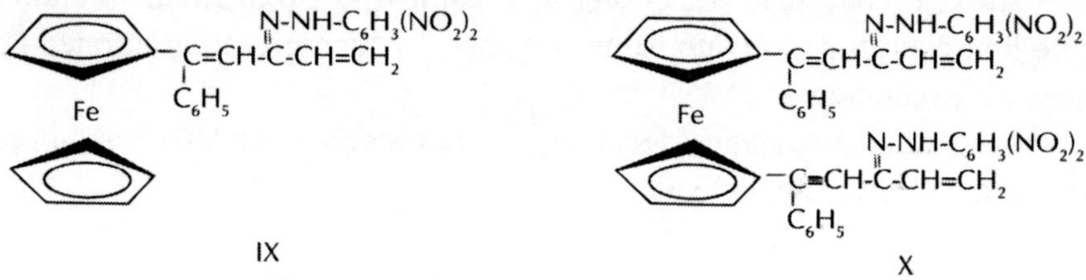

IX X

SCHEME 9 Dinitrophenyl hydrazons.

In IR spectrum of compounds III and IV (Scheme 3 and Scheme 4), the absorption band at 3600–3200/sm of the OH group disappear and characteristic absorption band of simple ether group C–O–C in area 1080/sm occur. (Fig. 4)

In the IR spectrum of compounds V and VII as compared to initial alcohols I and II, (Fig. 3) an absorption band in areas 3600–3200/sm (O-H) и 2225/sm (C≡C) disappear and intensive absorption bands in areas 1650,1750/sm appear, characteristic of conjugated grouping C=CH-C=O (Fig. 5), whereas in the IR-spectrum of the product VI in comparison with alcohol I only absorption band in area 3600–3200/sm disappear, characteristic of the group O–H and new band at 1270/sm (C–O–C) arises.

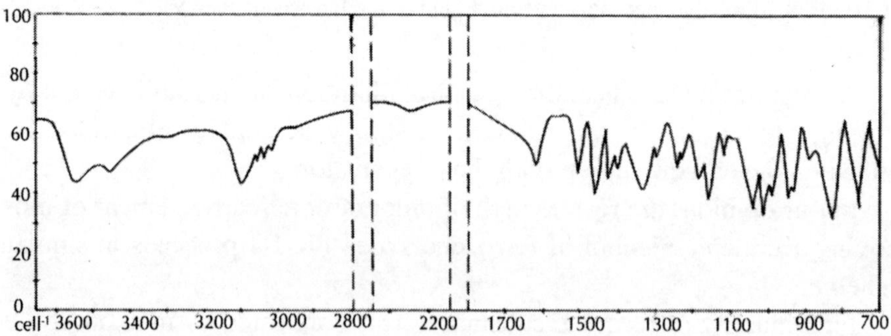

FIGURE 3 IR spectrum of feroceneacetilene alcohol FK-1G.

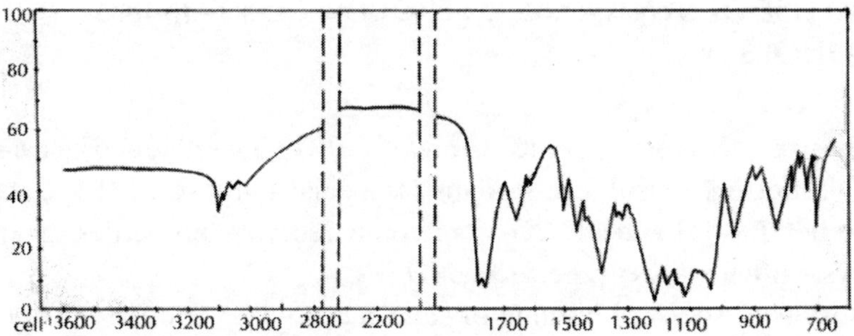

FIGURE 4 IR spectrum of compound III (Scheme 3).

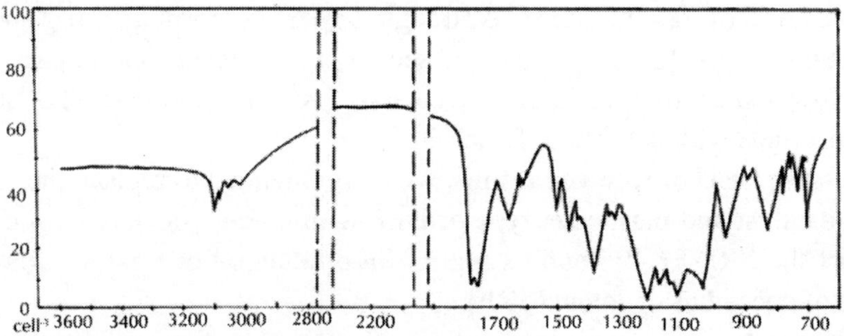

FIGURE 5 IR spectrum of unsaturated ketone VII.

In IR spectrum of hydrazones – IX and X the bands, characteristic for group –CH=CH-C=O at 1750, 1650/sm disappear and weak band in area 1630/sm (C=N) occur. In IR spectrum of ferrocenophane – VIII (Fig. 6), there are characteristic absorption bands in areas 1610/sm (C=C), 3110/sm (C-H ferrocene), 1280/sm (C-O-C cycl.) and absent absorption bands of HO-group (3600–3200/sm).

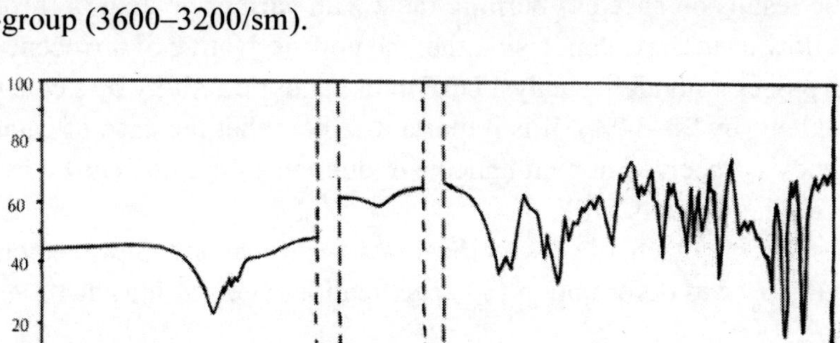

FIGURE 6 IR spectrum ferrocenophane VIII.

17.4 USE OF SYNTHESIZED FEROCENECONTAINING ALCOHOLS

Synthesized alcohols – I – FK-1G and II – FK-2G were tested as catalysts and plasticizers of solid propellant of special purpose (SRF). Catalytic properties FK-1G and FK-2G were compared with properties of known catalyst-softener - diethylferrocene (DEF).

Results of speed of burning of solid fuels, with various content of the catalysts, indicate that FK-1G surpasses considerably in its catalytic efficiency the standard catalyst and plasticizer type of DEF. In addition, the introduction of FK-1G improves strength (σ_p) and deformation (ε_p) specification of fuels. Explosive risk and sensitivity to mechanical impacts (to the blow and friction) of fuels with catalyst FK-1G is on the level of standard of fuels with catalyst-softener DEF.

On the basis of received results, we recommended to replace the standard catalyst and plasticizer type of DEF with a new effective burn catalyst of the SRF-FK-1G and to apply it in production of mixture of solid fuels of special assignment (SRF).

For evaluation of efficiency of FK-2G action as the burn catalyst of solid rocket fuels (MSRF) mixture by standard technology, there were made fuels on the basis of isoprenic rubber type of SCI-NL (type LK-12) and polybutadienic rubber with terminal carboxylic groups – SCB-KTP (type NK-I). The catalyst FK-2G was entered in to composition fuels at over 100%, and its efficiency was compared to known standard catalyst – diethylferrocene (DEF).

The results on speed of burning fuels with various content of the catalysts allow us to draw conclusion that the new derivative of ferrocene FK-2G surpasses a standard catalyst DEF in its catalytic activity by a considerable extent (by 28–34%). It is important to note that increase of catalytic efficiency is received at simultaneous reduction of the content Fe in FK-2G as compared with DEF.

Results on mechanical specifications testify about improvement of strength (σ_p) and deformation (ε_r) specifications fuels at introduction FK-2G.

Explosive risk and sensitivity to mechanical impacts (to the blow and friction) fuels with catalyst FK-2G is on the level of standard fuels with catalyst-softener - DEF.

Comparing results of test FK-2G with data of test FK-1G as catalyst of burning of solid fuels special assignment, a conclusion can be drawn that introduction of the second vinylacetylenic grouping in the molecule FK-1G improves a catalytic activity of received at the same time FK-2G.

On the basis of testing results of, it is possible to recommend FK-2G [I,I′-bis-(I-phenyl-I-oxy-2-pentyn-4-enyl)] as new effective catalyst of solid fuels special assignment.

17.5 CONCLUSIONS

The interaction of mono- and dibenzoilferrocene with lithium derivative of vinylacetylene has been studied. A new mono- and bisvinylacetylenic carbinols of ferrocene series – FK-1G and FK-1G were synthesized.

Technical and preparative method of synthesis of those carbinols was elaborated.

Catalytic and plasticizer properties of synthesized ferrocene-acetylene carbinols were investigated. The results of the carried out investigation showed that FK-1G and FK-2G considerable (by 28–34%) predominate other burn catalysts and plasticizers type of diethyl- ferrocene (DEF).

An anionotropic rearrangement of synthesized ferrocene containing carbinols mono- and bis-vinylacetylenicseries- FK-1G and FK-2G was studied under different conditions:

a) It has been established that in the nucleophilic solvents (methanol and other alcohols), the reaction proceeds anomalously – in the presence of $HgSO_4$ methylation of hydroxy group takes place and the corresponding methyl ether is obtained. It must be noted that in a non-polar solvent (for example benzene), in the other similar conditions, the aforementioned reaction does not proceed.
b) In the presence other catalytic system (for example $ClCH_2COOH$) both an anionotropic rearrangement (predominantly by forming an unsaturated ketone in the cases of FK-1G and of FK-2G) and intermolecular (in the case of FK-1G) as well as inner-molecular (in

the case of FK-2G) dehydratation of the synthesized alcohol took place.

KEYWORDS

- **Buchner filter**
- **Ferrocene**
- **polybutadienic rubber**
- **Specord 75 IR**
- **yellow crystals**

REFERENCES

1. Asin, M. B. In *Propellant fuel (toplivo)*. M., Мир, 1975.
2. Perevalova, E. G.; Reshetova, M. D.; Grandberg, K. I. Methods of elementorganic chemistry. Organoferrum compounds (Zhelezoorganicheskie soedinenia). Ferrocene. «Nauka», Moscow, 1983; 544 p.
3. Seils, D. S. Pat. USA 3447981. Firm propellant fuel and method of regulation of speed of his burning with use of the derivatives ferrocene.
4. Haskins, S. B.; Tomas, A. D. Pat. USA 3781179. Mixed firm propellant fuel with bi-functional additive – hardener and modifier of ballistic properties.
5. Haskins, S. B.; Tomas, A. D. Pat. USA 3770786. Ferrocenyloxybutenes.
6. Morits, V. P. Pat. USA 3968127. Preparation of the derivatives ferrocene.
7. Arendel, V. F. Pat. USA 4023994. Firm propellant fuel with ferrocenecontaining softener.
8. Gigineishvili, E. E.; Asatiani, L. P.; Phanjikidze, A. G. Ferrocenecontaining mono- and bisvinylacetylene alcohols – new effective catalysts of burning and softeners of special assignment. Message II. *Chemistry*, 360, Tbilisi State University Press, 2005; p. 60.
9. Gigineishvili, E. E.; Asatiani, L. P.; Phanjikidze, A.G. Ferrocenecontaining mono- and bis-vinylacetylene alcohols – new effective catalysts of burning and softeners of special assignment. Message IV. *Chem. Georgia J.*, **2006**, *6(2)*, 141–143.
10. Nazarov, I. N. Selected works (Izbrannie trudi). Academy of Sciences (USSR), Moscow, 1961; 690 p.
11. Nazarov, I. N. Chemistry of vinylethinylcarbinoles. *Uspekhi Khimii*, **1945**, *14*, 3–41.
12. Richards, S. H.; Hill, E. A. α-metallocenyl Carbonium Jons. *J. Amer. Chem. Soc.*, **1959**, *81*, 3484–3485.

13. Gigineishvili, E. E.; Asatiani, L. P.; Egorov, P. K.; Mazalov, I. A. Author certificate of USSR, 224808. 198.
14. Rausch, M., Vogel, M., Rosenberg, H. derivations of Ferrocene. Some reduction products of benzoylferrocene and 1,1[1]-dibenzoylferrocene. *J. Org. Chem.*, **1975**, *22*, 903–906.
15. Nesmeyanov, A. N.; Perevalova, E. G.; Gubin, S. P.; Grandberg, K. I.; Koslovski, A. C. Electronic properties of the Ferrocenyl as a Substituent. *Tetrahedron Lett.*, **1966**, 2381–2387.
16. Nesmeyanov, A. N.; Anisimov, K. N.; Kolobova, N. E.; Magomedov, G. K. Anionotropic rearrangament of tertiary α-acetylene alcohols of cyclopentadienylmanganumtricarbonile. *J. Org. Chem.*, **1967**, 1149–1153.
17. Hill, E. A.; Richards, T. H. Carbonium ion stabilization by metallocene: nuclei II. α-Metallocenylcarbonium ions. *J. Amer. Chem. Soc.*, **1961**, *83*, 3840–3846.
18. Cais, M. E.; Dannerberg, T. T.; Eisenstandt, A.; Levenberg, M. I.; Richards, T. H. Nuclear Magnetic Resonance Spectra of Ferrocenyl Carbonium ions. *Tetrahedron Lett.*, **1966**, 1695–1701.

CHAPTER 18

THE CHANGES OF DYNAMIC LIPID STRUCTURE OF MEMBRANES UPON THE EFFECT OF OXAZOLES IN VITRO

E. L. MALTSEVA, V. V. BELOV, and N. P. PALMINA

CONTENTS

18.1 INTRODUCTION

2,5-diphenyl-1,5-oxazole (DPhO) is a well known component of scintillation liquid for measuring of radioactivity [1, 2], fluorescence and chemiluminescence detection of oxazole-labeled amines and thiols [3, 4]. DPhO was used as probe to determine a level of mono-oxygenation in liver as well [5]. Moreover, the oxazoles showed a number of biochemical and biophysical properties, for example, phototoxicity [6], inhibitory activity (phospholipase C) and others [7]. Additionally, it was found that DPhO and one of its derivative – iod-methylate (IM-DPhO) protected mice on radiation, and this effect depends on both the dose of g-radiation and concentration of the oxasoles injected into the mice before irradiation [8, 9].

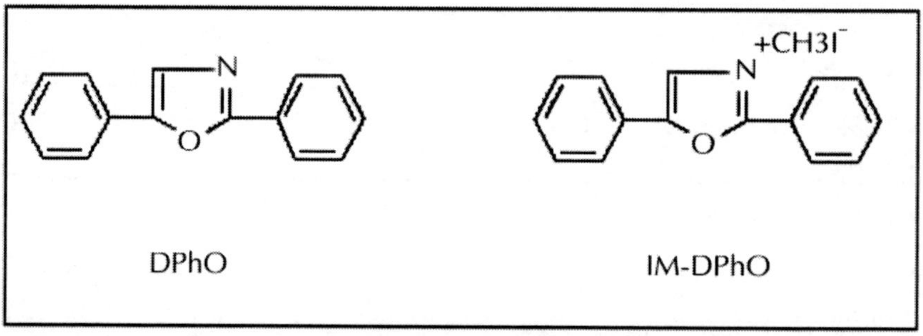

Due to the fact that irradiation effects on cells and protection from it are accompanied by the change in the lipid composition and structure of membranes, microviscosity, in particular [10,11], we proposed that the oxazoles can change the dynamic lipid structure of cell membranes as radioprotector.

Therefore, a purpose of the present work was to study the effect of DPhO and IM-DPhO in a wide range of concentration (10^{-2}–10^{-14} mol/l) on the endoplasmic reticulum membranes (microsomes) isolated from Balb-line mice. Electron paramagnetic resonance (EPR) technique and spin-probe method were used to study the dynamic structure of deep hydrophobic and surface lipid regions of microsomal membranes. We suggested the different effects of DPhO and IM-DPhO on the membrane lipids structure, because iod-methylate derivative is charged.

18.2 EXPERIMENTAL METHODS

Balb-line mice available from the "Stolbovaya" nursery station were decapitated and used to obtain membrane preparations. Membranes of endoplasmic reticulum (microsomes) were isolated from liver cells of Balb mice as described in [12] with some modifications. Membrane isolation was produced at $t = 0 \pm 4$°C. The steps were following: 1) A 2.5 g of liver portion was replaced in 40 ml of medium containing 0.25 mg sucrose, 10mM Tris-HCl and 1mM EDTA (SET); 2) Liver washed and cut very small by scissors, then in Potter homogenizer. 3) Centrifugation of homogenate for 15 min at 1500 g on centrifuge K-24 (Germany), repeated two times; 4) First supernatant centrifuged for 20 min at 2000 g at the same centrifuge, the sediment thrown away (removal of heavy fractions); 5) Second supernatant dissolved in SET was centrifuged for 90 min at 10.5000 g on the centrifuge L-65 "Beckman" (Austria); 6) The last sediment containing microsomes was suspended in SET and poured in epindorphs.

Protein level determined by Lowry method [13] using bovine serum albumin "Serva" (Germany) for calibration on the automatic spectrophotometer DU-50 "Beckman" (Austria) to measure an absorption. All membrane samples prepared finally contained 4 mg/ml of protein stored in refrigerator at t = –70°C.

DPhO from "Serva" (Germany) and its iod-methilate derivative IM-DPhO synthesized in the State Research Institute of Organic Chemistry and Technology (Moscow, Russia) were used as effectors.

The initial solution of DPhO – 1mol/l and IM-DPhO – 10^{-2} mol/l (maximal solubility) dissolved in ethanol and the concentrations ($10^{-(n+1)}$ M) were prepared by following: 20 μl (10^{-n} mol/l of DPhO or IM-DPhO) + 180 μl of ethanol. These solutions of oxazoles (20 ml) added to the samples – suspension of membranes (200 ml) in a wide range of concentrations of oxazoles from 10^{-2} to 10^{-14} mol/l. The final acting concentrations of DPhO and IM-DPhO in the membranes were in 100× less.

The dynamic lipid structure of membranes was studied by EPR – spectroscopy by use of computerized radio-spectrometer Bruker 200D (Germany) and spin – probes: two nitroxyl radicals: 5- and 16-doxyl-stearic acids (5- and 16-DSA) from "Sigma" (USA). The radicals prepared in concentration

10^{-2} mol/l in ethanol and added to membrane suspension in a final concentration 10^{-4} mol/l using Hamilton syringe: 200 μl membrane suspension + 2 μl spin-probe, which was incubated for 30 min.

TABLE 1 The parameters of EPR spectra registration of spin-probes (5- and 16-DSA).

Parameters of EPR Spectra	5-DSA	16-DSA
Central magnetic field	3,500 Gs	3,370 Gs
Assessment of magnetic field	300 Gs	200 Gs
Gain of magnetic field	1×10^5	8.0×10^4
Modulation of magnetic field	3.0 Gs	4.0 Gs
Frequency of UHF-radiation	9.6 GHz	9.6 GHz
Power of UHF-radiation	20 MW	20 MW
Time constant	500 ms	200 ms
Time of scanning	500 s	500 s
Temperature	20°C	20°C

Electron paramagnetic resonance (EPR) – technique by use "Bruker-200D" (Germany) computerized spectrometer with special computer programs has been applied to registration the spectra. A basic EPR theory applicable to study a lipid structure of membranes described in [14, 15]. The parameters of registration of EPR-spectra are shown on the table.

A value of lipid microviscosity was estimated by a rotational correlation time (τ) of 16-DSA calculated from the EPR spectra obtained using formula for fast motion range ($6 \times 10^{-11} < \tau < 3 \times 10^{-9}$ s) of the nitroxyl radicals:

$$\tau = 6.65\, \Delta H_{+1} / [(I_{+1}/I_{-1})^{1/2} - 1] \times 10^{-10}\ s,$$

where I_{+1} and I_{-1} – the resonance heights of low- and high- field components of EPR spectra correspondingly, ΔH_{+1} – the resonance width of low-field component (Fig. 1, right side).

An order parameter (S) related with the order degree of long axis of 5-DSA and molecular motion of lipid matrix has been calculated using formula:

$$S = 1.66\,(H_{//} - H_{I}) / (H_{//} + 2H_{I}),$$

where $2H_{//}$ – the resonance width between low- and high-field (external) extremes of EPR-spectra and $2H_{I}$ – between internal extremes (Fig.1, left side).

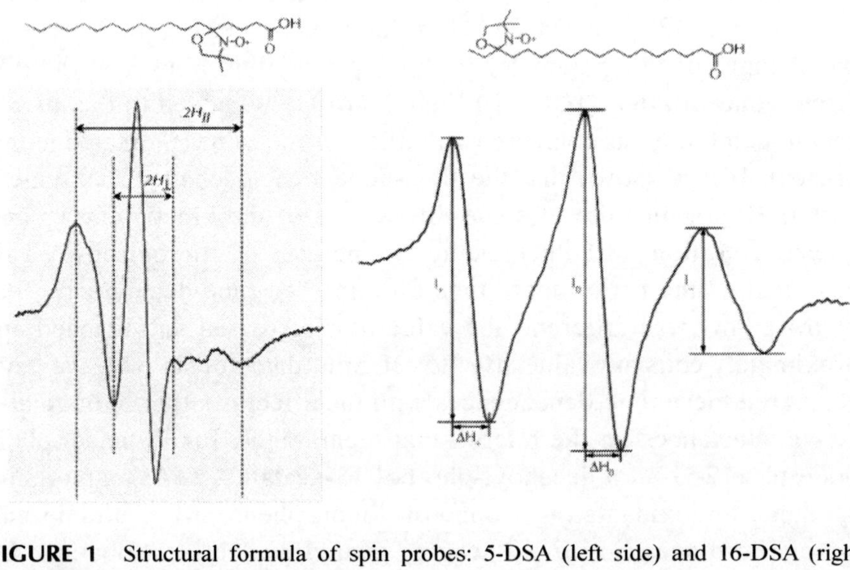

FIGURE 1 Structural formula of spin probes: 5-DSA (left side) and 16-DSA (right side), EPR spectra in surface and deep hydrophobic regions of lipid bylaer of microsomal membranes correspondingly.

The values of S and τ were processed by methods of parametric and nonparametric statistics with 95% statistic reliability by special computer programs EXEL and STATISTICA. The data shown in the Figures were obtained not less than in 3 independent experiments and 3 parallel measurements of membrane samples at each concentration of DPhO and IM-DPhO.

18.3 RESULTS AND DISCUSION

The spin probes 5- and 16-DSA are localized in the different regions of membrane lipids: 5-DSA in surface lipids at depth of ~ $8A^0$ and hydrophobic nitroxyl radical 16-DSA penetrates in lipid bilayer to the depth more than $20A^0$ [14, 16]. A motion of 16-DSA can be characterized by rotation correlation time (τ) by which a microviscosity value is estimated and anisotropic rotation of 5-DSA by order parameter (S) [15, 16].

The first step was to study a time-dependence of τ of 16-DSA upon the effect of DPhO or IM-DPhO on the membrane. Figure 2 presents the changes of this parameter in the depth of lipid bilayer at the different concentration (10^{-2} – 10^{-11} mol/l) of DPhO added to the microsomal membranes as compared with the control value (untreated membranes). It was shown that the time-dependences changed extremely within 30 min in spite of some differences of the kinetic curves and all concentrations of DPhO resulted to increase of microviscosity values for the time period more than 60 min. The time-dependent effect of IM-DPhO was different: the value of τ increased and reached approximately constant value after 15–20 min (data not shown). We have compared these time-dependences with the effect of other biologically active substances on the microsomal membranes, for instance, phorbol ester (12-O-tetra-decanoyl-phorbol-13-acetate – TPA), natural and synthetic antioxidants (a-tocopherol, ionol, phenozan), d protein and steroid hormones (insulin, glucagon, guanyl nucleotides) obtained by us previously [17–20]. It should be noticed that extremely changes of microviscosity value at different concentrations of DPhO are similar to the effect of TPA (for 30 min), but its increased level 0.5–2 hr – to antioxidants and hormones. Thus, the effect of oxazoles on the membrane lipids is different.

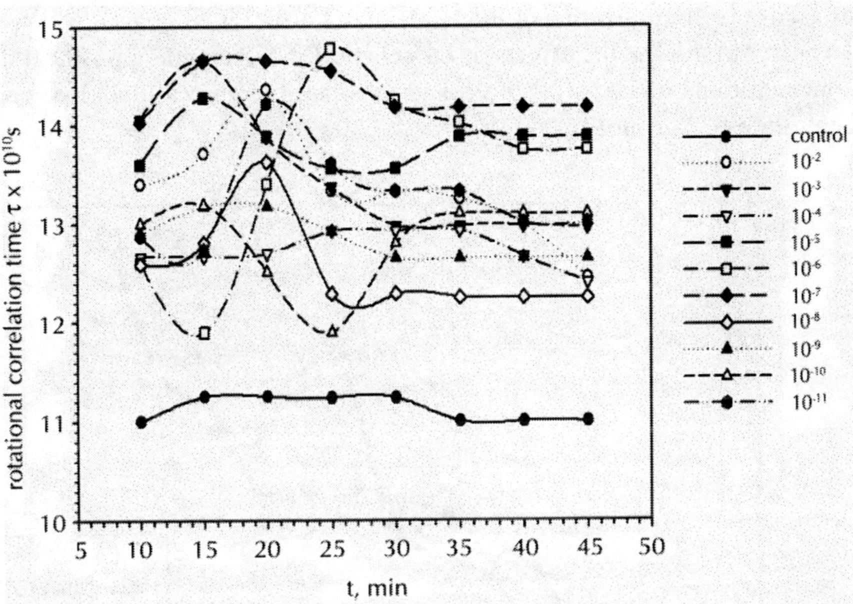

FIGURE 2 The changes of microviscosity value (rotational correlation time – τ of 16-DSA) depending on the time in the depth of membrane lipids at different concentrations of DPhO (10^{-2}–10^{-11} M).

We used the maximum increase of τ value at each concentration of DPhO shown in Fig. 2 to obtain the concentration τ – dependence, which is presented in Fig. 3 (curve 1). It was found that the microviscosity value decreased at high concentration range of 10^{-2} M –10^{-5} mol/l of DPhO, extremely increased from 10^{-6} M to 10^{-7} mol/l and than decreased up to concentration of 10^{-14} mol/l. Thus, the highest concentration – 10^{-2} mol/l is more effective and increased microviscosity value at 25%, the second maximum (effect about of 15%) was observed at concentration 10^{-7} mol/l of DPhO. It should be take into consideration that effective concentrations in membranes are in 100× less than added to the membranes.

The dependence of τ 16-DSA on the concentrations of IM-DPhO is presented in Fig. 3 (curve 2). In the case of IM-DPhO the most increase of microviscosity value is about 13% at the concentration of 10^{-4} mol/l, and the constant value observed within concentration range 10^{-5}–10^{-9} mol/l (effect is about 10%). A comparison of the effects of two oxazoles on the lipid membrane microviscosity results to a conclusion that the effect of DPhO is more significant than IM-DPhO at the high concentration of 10^{-2}–10^{-3} mol/l, but the effects of DPhO and IM-DPhO are opposite at the concentrations of 10^{-5}–10^{-7} mol/l and decreased to zero at the same concentrations (10^{-12} mol/l).

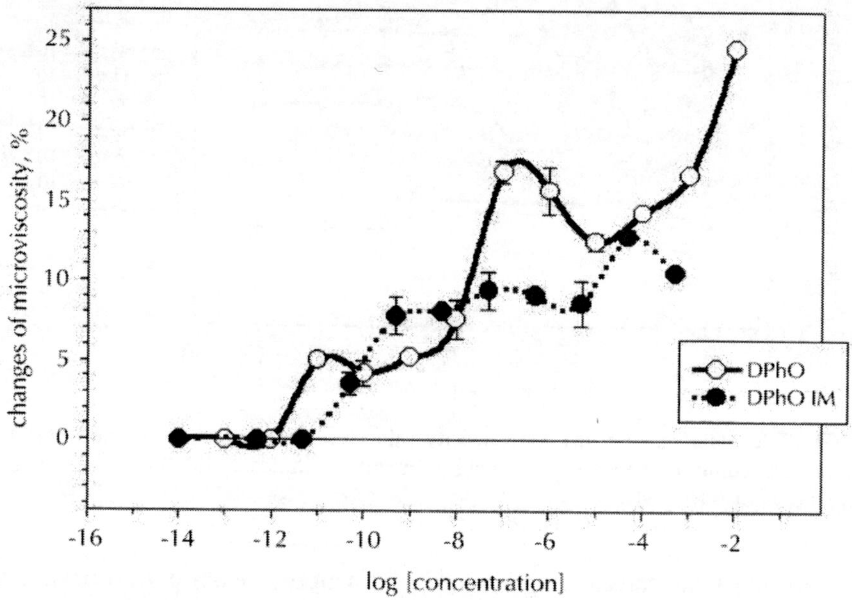

FIGURE 3 The effect (expressed in %) of two oxazoles on the lipid microviscosity (τ of 16-DSA) of microsomal membranes depending on the concentrations of DPhO (a) and IM-DPhO (b).

The increase of lipid microviscosity about 15–25% in cooperative structure of biological membranes can lead to the significantly changes, for example, activity of membrane-bound enzymes and receptors [21–24].

The order parameter – S of spin-probe – 5-DSA localized in the surface membrane lipids estimates a rigidity of these regions. In our experiments the membranes treated by DPhO and IM-DPhO shown the similar time-dependences of order parameter, which reached maximum after 10 min. The effect of the oxazoles supported for 40 min (data not shown). The concentration dependences of order parameter are presented in Fig. 4. In general, the quantitative changes of parameter S *in vitro* and *in vivo* in 10 times less than value of τ. For example, the change by 1–3% of parameter S is physiologically important for membrane function [25, 26].

Figure 4a shows that parameter S diminished from 3.5% to 0 while concentration of DPhO decreased from 10^{-3} mol/l to 10^{-8} mol/l. A comparison of the dependences of τ-value and parameter S in the membrane lipids depending on the concentration of DPhO indicate of their similarity and non-monotonous decrease within the concentration range from 10^{-3} to 10^{-8} mol/l with intermediate increase at 10^{-6}–10^{-7} mol/l. But the structural changes in the deep hydrophobic lipid regions of membrane are more prolonged on the concentration scale upon the effect of DPhO (Fig. 3).

Figure 4b presents the concentration dependence of the order parameter for IM-DPhO. In this case we have obtained the bimodal curve with two maximum at 10^{-4} and 10^{-9} mol/l, where the effect is the same, and co-called "dead zone" at 5×10^{-7}–5×10^{-9} mol/l, within the effect equals to zero. It should be noticed that a polymodal type of "dose-effect" dependences has been observed for a number of biological active substances effective at ultra-low concentrations [27]. Usually, the bimodal curve is characterized by second maximum at the effective ultra-low doses – less than 10^{-11}–10^{-12} mol/l [28]. In our experiments the second maximum of parameter S observed at 10^{-9} mol/l IM-DPhO, but as mentioned above the effective concentration of substances in the membranes is in 100 times less than added ones. Thus, the second maximum observed at 10^{-11} mol/l of IM-DPhO and it can be regarded as substances effective at the ultra-low concentrations.

In the present paper we have found that a chemical modification of DPhO by iod-methyl group resulted in different effects of active oxazoles on the dynamic lipid structure of membranes, at the surface membrane lipids, in particular. The effects of DPhO and its derivative IM-DPhO were opposite at the concentration of 10^{-5}–10^{-7} mol/l.

Regarding radio-protector properties of these compounds it can be pointed that, first of all, the survival curves of mice protected by IM-DPO and g-irradiated by different doses were described by a bimodal type of "dose-effect" dependences [8, 9], which can be related to the changes of lipid rigidity (parameter S) of membranes by IM-DPO. The increase of lipid microviscosity of membranes at 10^{-7} mol/l of DPhO and 10^{-5} mol/l of IM-DPhO corresponds to the maxima on the survival curves of mice protected by these concentrations of oxasoles at the average doses of irradiation (4 and 8 Gy).

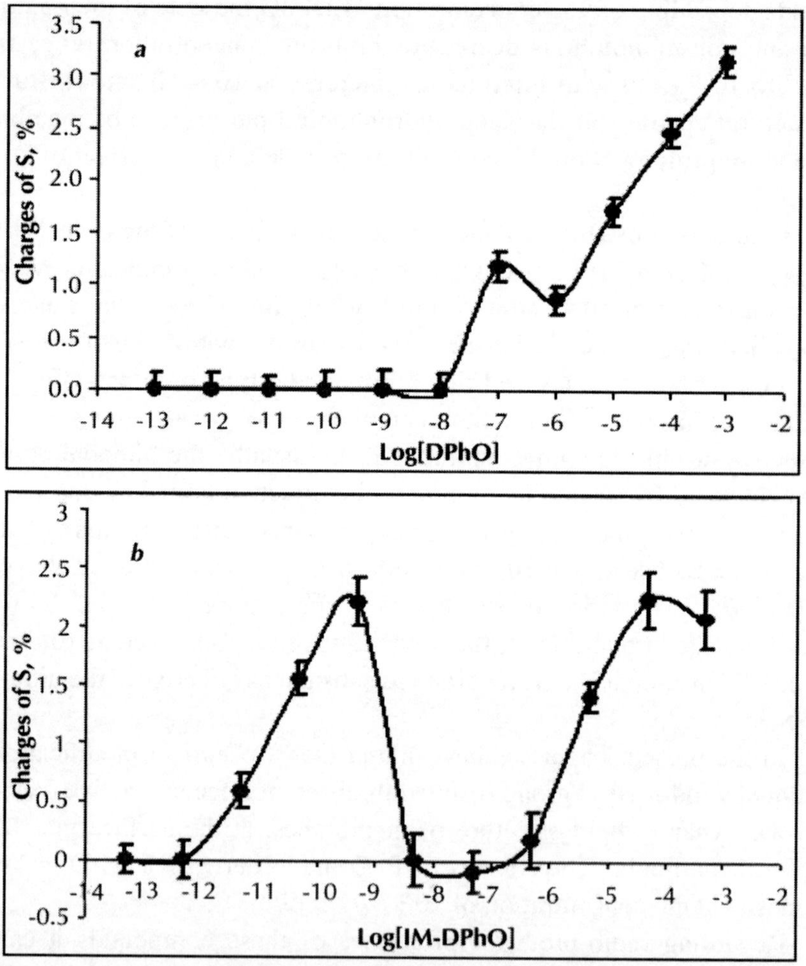

FIGURE 4 The changes of order parameter S of 5-DSA (expressed in %) depending on the concentrations of DPhO (a) and IM-DPhO (b).

18.4 CONCLUSIONS

1. By spin probes 5- and 16-DSA it was found that DPhO and its iod-methylate derivate (IM-DPhO) increased the microviscosity (τ) in the depth hydrophobic regions and rigidity (S) of surface membrane lipids depending on the concentration by nonlinear manner.
2. It was shown that IM-DPhO, but not DPhO changed the order parameter (S) of surface membrane lipids by bimodal type depending on the concentration with two maxima (10^{-4} and 10^{-9} mol/l) and "dead-zone," which is typically to the effects of substances at ultra-low concentrations.

KEYWORDS

- **2,5-diphenyl-1,5-oxazole**
- **Active oxazoles**
- **Balb-line mice**
- **Dead-zone**
- **Iod-methylate derivative**
- **Microviscosity**

REFERENCES

1. Chapman, B.; Sutherland, A. J. *Chem. Commun.*, **2003**, *21,* 84–85.
2. Culliford, S. J.; McCauley, P.; Sutherland, A. J.; McCair, M.; Sutherland, J.; Blackburn, J.; Kozlowski, R. Z. *Biochem. Biophys. Res. Commun.*, **2002**, *296(4),* 857–863.
3. DiCesare, N.; Lakowicz, J. R. *Chem. Commun.,* **2001**, *19,* 2022–2033.
4. Toyo'oka, T.; Chokshi, H. P.; Givens, R. S.; Carlson, R. G.; Lunte, S. M.; Kuwana, T. *Biomed. Chromatogr.*, **1993**, *4,* 208–216.
5. Mutch, E.; Woodhouse, K. W.; Williams, F. M.; Lambert, D.; James, O. F.; Rawlins, M. D. *Xenobiotica.,* **1985**, *7,* 599–603.
6. Kagan, J.; Kolyvas, C. P.; Jaworski, J. A.; Kagan, E.D .; Kagan, I. A.; Zang, L. H. *Photochem .Photobiol.,* **1984**, *4,* 479–483.
7. Vogt, A.;. Pestell, K. E.; Day, B. W.; Lazo, J. S.; Wipf, P. *Mol. Cancer. Ther.*, **2002**, *11,* 885–892.

8. Zhizhina, G. P.; Zavarykina, T. M.; Burlakova, E. B.; Noskov, V. G.; Dukhovich, F. S. *Radiats. Biol. Radioecol.*, **2005**, *45(1)*, 56–62.
9. Shishkina, L. N.; Smotryaeva, M. A.; Kozlov, M. V.; Kushnireva, E. V.; Dukhovich, F. S.; Noskov, N. G. *Radiats. Biol. Radioecol.*, **2005**, *45(5)*, 610–615.
10. Maltseva, E. L.; Palmina, N. P. *Appl. Radiat. Isotopes,* **1996**, *47(11–12),* 1683–1687.
11. Zima, G. V.; Dreval, V. I. *Radiat. Biol. Radioecol.*, **2000**, *40(3)*, 261–265.
12. Hostetler, K.Y.; Zenner, H.P. Morris. *Biochem. Biophys. Acta.*, **1976**, *441*, 231–235.
13. Lowry, O.; Rosebrough, N.; Farr, A.; Randall, R. *J. Biol. Chem.*, **1951**, *193*, 265–269.
14. Griffith, O. H.; Jost, P. C. Lipid spin labels in biological membranes. In *Spin Labeling Theory and Application;* Berliner, L. J., Ed.; New York–San Francisco–London, Acad. Press, **1979**, 489–569.
15. Likhtenshtein, G. I. In *Method of spin labels in molecular biology.* Nauka: Moscow, **1974**, p.255.
16. Kuznetsov, A. V. In *Method of Spin Probe*. Nauka: Moscow, **1976**, 210 p.
17. Palmina, N. P.; Bogdanova, N. G.; Maltseva, E. L.; Pinzar, E. I. *Biol. Membr,* **1999**, *9(8)*, 810–820.
18. Zlatanov, I.; Maltseva, E. L.; Borovok, N. V.; Spassov, V. *Intern. J. Biochem.*, **1993**, *25(7)*, 971–977.
19. Belov, V. V.; Mal'tseva, E. L.; Pal'mina, N. P. *Biofizika.*, **2007**, *52(1)*, 75–83.
20. Palmina, N. P.; Chasovskaya, T. E.; Belov, V. V.; Maltseva, E. L. *Doklady Biochem. Biophy.*, **2012**, *443*, 100–104.
21. Hershkowitz, M.; Heron, D.; Samuel, D.; Shinitzky, M. *Prog. Brain. Res.*, **1982**, *56*, 419–422.
22. Tsakiris, S. *Naturforsh.*, **1984**. *39*, 1196–1202.
23. Maltseva, E. L.; Belokoneva, O. S.; Zaitsev, S. V.; Varfolomeev, S. D. *Biol. Membr.*, **1991**, *8(8),* 830–836.
24. Maltseva, E. L.; Borovok, N. V.; Zlatanov, I. *Membr. Cell. Biol.*, **1996**, *9(6)*, 621–630.
25. Kury, P. G.; Ramwell, P. W.; McConnell, H. M. *Biochem. Biophys. Res. Commun.*, **1974**, *57*, 726–730.
26. Huetis, W. H.; McConnell, H. M. *Biochem. Biohys. Res. Commun.*, **1974**, *57*, 26–730.
27. Burlakova, E. B.; Konradov, A. A.; Maltseva, E. L. *J. Adv. Chem. Phy.*, **2003**, *2(2)*, 140–162.
28. Burlakova, E. B. *Russ. Chem. J.*, **1999**, *18(5)*, 3–11.

CHAPTER 19

MAGNETIC NANOPARTICLES

A. V. BYCHKOVA, M. A. ROSENFELD, V. B. LEONOVA, O. N. SOROKINA, and A. L. KOVARSKI

CONTENTS

19.1 INTRODUCTION

Magnetic nanoparticles (MNPs) have many applications in different areas of biology and medicine. MNPs are used for hyperthermia, magnetic resonance imaging, immunoassay, purification of biologic fluids, cell and molecular separation, tissue engineering [1–6]. The design of magnetically targeted nanosystems (MNSs) for a smart delivery of drugs to target cells is a promising direction of nanobiotechnology. They traditionally consist on one or more magnetic cores and biological or synthetic molecules, which serve as a basis for polyfunctional coatings on MNPs surface. The coatings of MNSs should meet several important requirements [7]. They should be biocompatible, protect magnetic cores from influence of biological liquids, prevent MNSs agglomeration in dispersion, provide MNSs localization in biological targets and homogenity of MNSs sizes. The coatings must be fixed on MNPs surface and contain therapeutic products (drugs or genes) and biovectors for recognition by biological systems. The model which is often used when MNSs are developed is presented in Fig. 1.

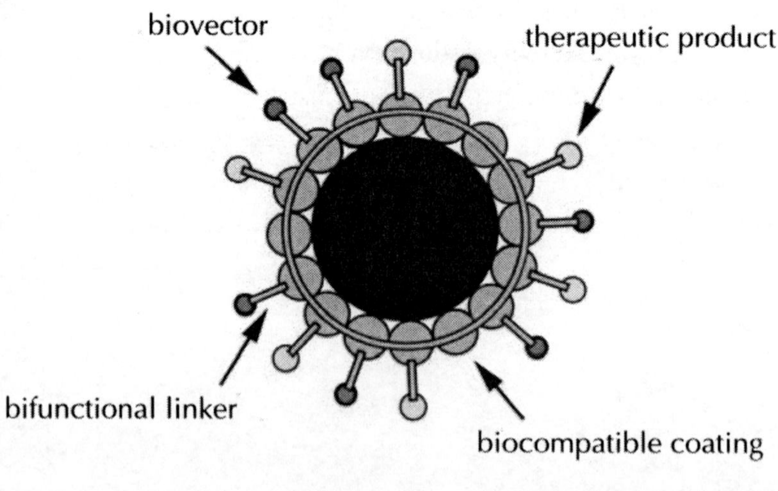

FIGURE 1 The classical scheme of magnetically targeted nanosystem for a smart delivery of therapeutic products.

Proteins are promising materials for creation of coatings on MNPs for biology and medicine. When proteins are used as components of coatings it is of the first importance that they keep their functional activity [8]. Protein binding on MNPs surface is a difficult scientific task. Traditionally bifunctional linkers (glutaraldehyde [9, 10], carbodiimide [11, 12]) are used for protein cross-linking on the surface of MNPs and modification of coatings by therapeutic products and biovectors. Authors of the study [9] modified MNPs surface with aminosilanes and performed protein molecules attachment using glutaraldehyde. In the issue [10] bovine serum albumin (BSA) was adsorbed on MNPs surface in the presence of carbodiimide. These works revealed several disadvantages of this way of protein fixing which make it unpromising. Some of them are clusters formation as a result of linking of protein molecules adsorbed on different MNPs, desorption of proteins from MNSs surface as a result of incomplete linking, uncontrollable linking of proteins in solution (Fig. 2). The creation of stable protein coatings with retention of native properties of molecules still is an important biomedical problem.

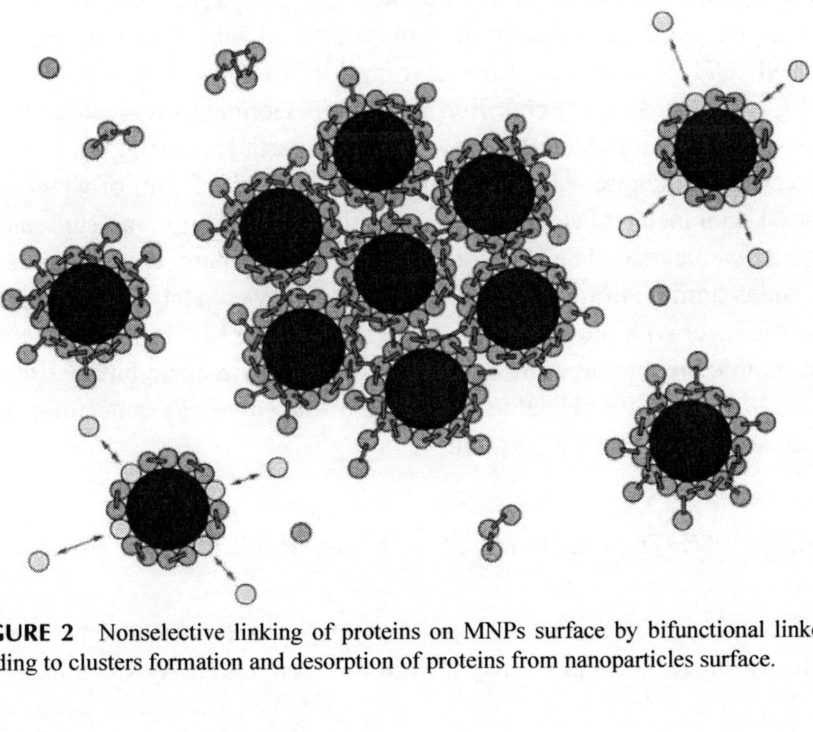

FIGURE 2 Nonselective linking of proteins on MNPs surface by bifunctional linkers leading to clusters formation and desorption of proteins from nanoparticles surface.

It is known that proteins can be chemically modified in the presence of free radicals with formation of cross-links [13]. The goals of the work were to create stable protein coating on the surface of individual MNPs using a fundamentally novel approach based on the ability of proteins to form interchain covalent bonds under the action of free radicals and estimate activity of proteins in the coating.

19.2 MATERIALS AND METHODS

19.2.1 MAGNETIC SORBENT SYNTHESIS

Nanoparticles of magnetite Fe_3O_4 were synthesised by co-precipitation of ferrous and ferric salts in water solution at 4°C and *in the alkaline medium*:

$$Fe^{2+} + 2Fe^{3+} + 8OH^- \rightarrow Fe_3O_4\downarrow + 4H_2O$$

A 1.4 g of $FeSO_4{\cdot}7H_2O$ and 2.4 g of $FeCl_3{\cdot}6H_2O$ ("Vekton," Russia) were dissolved in 50 ml of distilled water so that molar ratio of Fe^{2+}/Fe^{3+} was equal to 1:2. After filtration of the solution 10 ml of 25 mass percentage NH_4OH ("Chimmed," Russia) was added to it on a magnetic stirrer. A 2.4 g of PEG 2 kDa ("Ferak Berlin GmbH," Germany) was added previously in order to reduce the growth of nanoparticles during the reaction. After the precipitate was formed the solution (with 150 ml of water) was placed on a magnet. Magnetic particles precipitated on it and supernatant liquid was deleted. The procedure of particles washing was repeated for 15 times until neutral pH was obtained. MNPs were stabilized by double electric layer with the use of US-disperser ("MELFIZ," Russia). To create the double electric layer 30 ml of 0.1 M phosphate-citric buffer solution (0.05 M NaCl) with pH value of 4 was introduced. MNPs concentration in hydrosol was equal to 37 mg/ml.

19.2.2 PROTEIN COATINGS FORMATION

Bovine serum albumin ("Sigma-Aldrich," USA) and thrombin with activity of 92 units per 1 mg ("Sigma-Aldrich," USA) were used for protein coating formation. Several types of reaction mixtures were created: "A1-MNP-0," "A2-MNP-0," "A1-MNP-1," "A2-MNP-1," "A2-MNP-1-acid," "T1-MNP-0," "T1-MNP-0" and "T1-0-0." All of them contained:

1) 2.80 ml of protein solution ("A1" or "A2" means that there is BSA solution with concentration of 1 mg/ml or 2 mg/ml in 0.05 M phosphate buffer with pH 6.5 (0.15 M NaCl) in the reaction mixture; "T1" means that there is thrombin solution with concentration of 1 mg/ml in 0.15 M NaCl with pH 7.3),
2) 0.35 ml of 0.1 M phosphate-citric buffer solution (0.05 M NaCl) or MNPs hydrosol ("MNP" in the name of reaction mixture means that it contains MNPs),
3) 0.05 ml of distilled water or 3 mass % H_2O_2 solution ("0" or "1" in the reaction mixture names correspondingly).

Hydrogen peroxide interacts with ferrous ion on MNPs surface with formation of hydroxyl-radicals by Fenton reaction:

$$Fe^{2+} + H_2O_2 \rightarrow Fe^{3+} + OH^{\cdot} + OH^{-}$$

"A2-MNP-1-acid" is a reaction mixture, containing 10 μl of ascorbic acid with concentration of 152 mg/ml. Ascorbic acid is known to form free radicals in reaction with H_2O_2 and generate free radicals in solution but not only on MNPs surface.

The sizes of MNPs, proteins and MNPs in adsorption layer were analyzed using dynamic light scattering (Zetasizer Nano S "Malvern," England) with detection angle of 173° at temperature 25°C.

19.2.3 STUDY OF PROTEINS ADSORPTION ON MNPS

The study of proteins adsorption on MNPs was performed using ESR-spectroscopy of spin labels. The stable nitroxide radical used as spin label is presented in Fig. 3. Spin labels technique allows studying adsorption of macromolecules on nano-sized magnetic particles in dispersion without complicated separation processes of solution components [14]. The principle of quantitative evaluation of adsorption is the following. Influence of local fields of MNPs on spectra of radicals in solution depends on the distance between MNPs and radicals [14–16]. If this distance is lower than 40 nm for magnetite nanoparticles with the average size of 17 nm [17] ESR spectra lines of the radicals broaden strongly and their intensity decreases to zero. The decreasing of the spectrum intensity is proportional to the part of radicals, which are located inside the layer of 40 nm in thickness around MNP. The same happens with spin labels covalently bound to protein macromolecules. An intensity of spin labels spectra decreases as a result of adsorption of macromolecules on MNPs (Fig. 4). We have shown that spin labels technique can be used for the study of adsorption value, adsorption kinetics, calculation of average number of molecules in adsorption layer and adsorption layer thickness, concurrent adsorption of macromolecules [18–20].

1

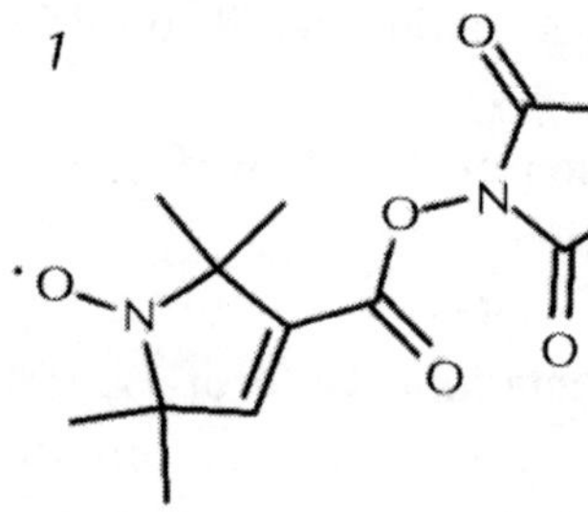

2

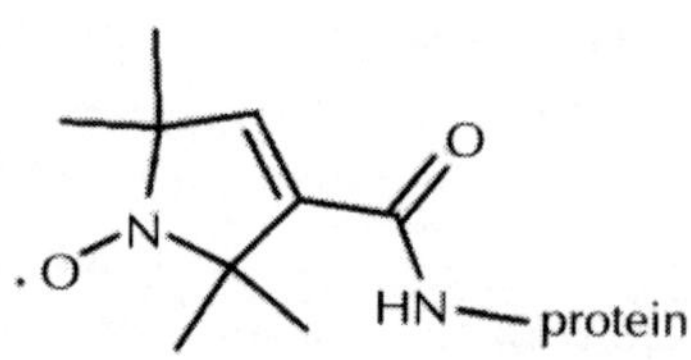

FIGURE 3 The stable nitroxide radical used for labelling of macromolecules containing aminogroups (1) and spin label attached to protein macromolecule (2).

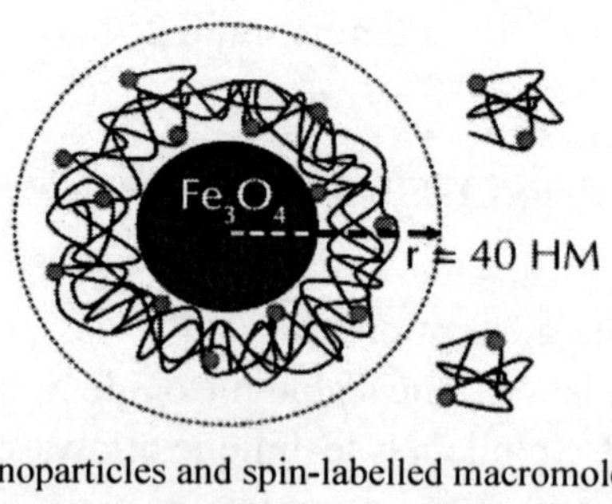

FIGURE 4 Magnetic nanoparticles and spin-labelled macromolecules in solution.

The reaction between the radical and protein macromolecules was conducted at room temperature. A 25 μl of radical solution in 96% ethanol with concentration of 2.57 mg/ml was added to 1 ml of protein solution.

The solution was incubated for 6 hr and dialyzed. The portion of adsorbed protein was calculated from intensity of the low-field line of nitroxide radical triplet I_{+1}.

The method of ferromagnetic resonance was also used to study adsorption layer formation.

The spectra of the radicals and magnetic nanoparticles were recorded at room temperature using Bruker EMX 8/2.7 X-band spectrometer at a microwave power of 5 mW, modulation frequency 100 kHz and amplitude 1 G. The first derivative of the resonance absorption curve was detected. The samples were placed into the cavity of the spectrometer in a quartz flat cell. Magnesium oxide powder containing Mn^{2+} ions was used as an external standard in ESR experiments. Average amount of spin labels on protein macromolecules reached 1 per 4–5 albumin macromolecules and 1 per 2–3 thrombin macromolecules. Rotational correlation times of labels were evaluated as well as a fraction of labels with slow motion ($\tau > 1$ ns).

19.2.4 COATING STABILITY ANALYSIS AND ANALYSIS OF SELECTIVITY OF FREE RADICAL PROCESS

In our previous works it was shown that fibrinogen (FG) adsorbed on MNPs surface forms thick coating and micron-sized structures [18]. Also FG demonstrates an ability to replace BSA previously adsorbed on MNPs surface. This was proved by complex study of systems containing MNPs, spin-labelled BSA and FG with spin labels technique and ferromagnetic resonance [20]. The property of FG to replace BSA from MNPs surface was used in this work for estimating BSA coating stability. A 0.25 ml of FG ("Sigma-Aldrich," USA) solution with concentration of 4 mg/ml in 0.05 M phosphate buffer with pH 6.5 was added to 1 ml of the samples "A1-MNP-0," "A2-MNP-0," "A1-MNP-1," "A2-MNP-1." The clusters formation was observed by dynamic light scattering.

The samples "A2-MNP-0," "A2-MNP-1," "T1-MNP-0," "T1-MNP-1" were centrifugated at 120,000 g during 1 h on «Beckman Coulter» (Austria). On these conditions MNPs precipitate, but macromolecules physically adsorbed on MNPs remain in supernatant liquid. The precipitates containing MNPs and protein fixed on MNPs surface were dissolved in buffer solution with subsequent evaluation of the amount of protein by Bradford colorimetric method [21]. Spectrophotometer CF-2000 (OKB "Spectr," Russia) was used.

Free radical modification of proteins in supernatant liquids of "A2-MNP-0," "A2-MNP-1" and the additional sample "A2-MNP-1-acid" were analyzed by IR-spectroscopy using FTIR-spectrometer Tenzor 27 ("Bruker," Germany) with DTGS-detector with 2 cm^{-1} resolution. Comparison of "A2-MNP-0," "A2-MNP-1" and "A2-MNP-1-acid" helps to reveal the selectivity of free radical process in "A2-MNP-1."

19.2.5 ENZYME ACTIVITY ESTIMATION

Estimation of enzyme activity of protein fixed on MNPs surface was performed on the example of thrombin. This protein is a key enzyme of blood clotting system, which catalyzes the process of conversion of fibrinogen to fibrin. Thrombin may lose its activity as a result of free radical modification and the rate of the enzyme reaction may decrease. So estimation of enzyme activity of thrombin cross-linked on MNPs surface during free radical modification was performed by comparison of the rates of conversion of fibrinogen to fibrin under the influence of thrombin contained in reaction mixtures. 0.15 ml of the samples "T1-MNP-0," "T1-MNP-1" and "T1-0-0" was added to 1.4 ml of FG solution with concentration of 4 mg/ml. Kinetics of fibrin formation was studied by Rayleigh light scattering on spectrometer 4400 ("Malvern," England) with multibit 64-channel correlator.

19.3 RESULTS AND DISCUSSION

ESR spectra of spin labels covalently bound to BSA and thrombin macromolecules (Fig. 5) allow obtaining information about their microenvironment. The spectrum of spin labels bound to BSA is a superposition of narrow and wide lines characterized by rotational correlation times of 10^{-9} s and $2 \cdot 10^{-8}$ s, respectively. This is an evidence of existence of two main regions of spin labels localisation on BSA macromolecules [22]. The portion of labels with slow motion is about 70%. So a considerable part of labels are situated in internal areas of macromolecules with high microviscosity. The labels covalently bound to thrombin macromolecules are characterized by one rotational correlation time of 0.26 ns. These labels are situated in areas with equal microviscosity.

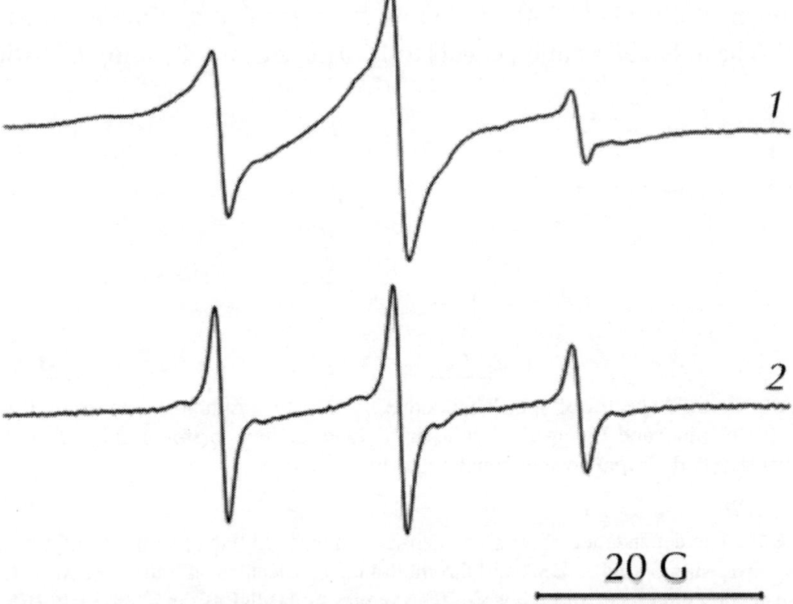

FIGURE 5 ESR spectra of spin labels on BSA (1) and thrombin (2) at 25°C.

The signal intensity of spin-labelled macromolecules decreased after introduction of magnetic nanoparticles into the solution *that testifies to the* protein adsorption on MNPs (Fig. 6). Spectra of the samples "A1-MNP-0" and "T1-MNP-0" consist of nitroxide radical triplet, the third line of sextet of Mn^{2+} (the external standard) and ferromagnetic resonance spectrum of MNPs. Rotational correlation time of spin labels does not change after MNPs addition. The dependences of spectra lines intensity for spin-labelled BSA and thrombin in the presence of MNPs on incubation time are shown in Table 1. Signal intensity of spin-labelled BSA changes insignificantly. These changes correspond to adsorption of approximately 12%

of BSA after the sample incubation for 100 min. The study of adsorption kinetics allows establishing that adsorption equilibrium in "T1-MNP-0" takes place when the incubation time equals to 80 min and ~41% of thrombin is adsorbed. The value of adsorption *A* may be estimated using the data on the portion of macromolecules adsorbed and specific surface area calculated from MNPs density (5,200 mg/m^3), concentration and size. Therefore, BSA adsorption equals to 0.35 mg/m^2 after 100 min incubation. The dependence of thrombin adsorption value on incubation time is shown in Fig. 7. Thrombin adsorption equals to 1.20 mg/m^2 after 80 min incubation.

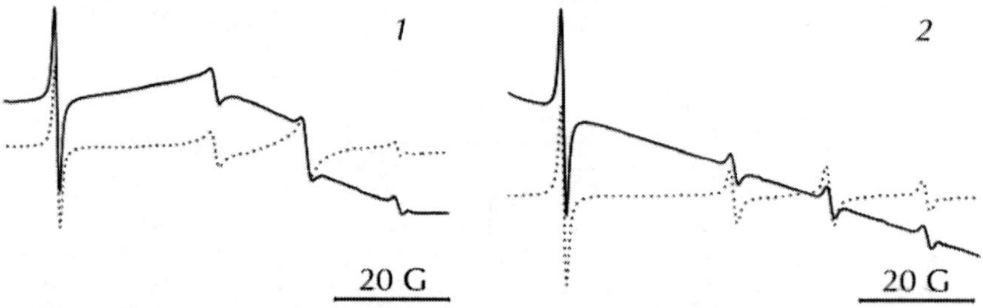

FIGURE 6 ESR spectra of spin labels on BSA (*1*) and thrombin (2) macromolecules before (dotted line) and 75 min after (solid line) MNPs addition to protein solution at 25°C. External standard – MgO powder containing Mn^{2+}.

TABLE 1 The dependence of relative intensity of low-field line of triplet I_{+1} of nitroxide radical covalently bound to BSA and thrombin macromolecules, and the portion *N* of the protein adsorbed on incubation time *t* of the samples "A1-MNP-0" and "T1-MNP-0."

	Spin-labelled BSA		Spin-labelled Thrombin	
***t*, min.**	**I_{+1}, rel. units**	***N*, %**	**I_{+1}, rel. units**	***N*, %**
0	0.230 ± 0.012	0 ± 5	0.25 ± 0.01	0 ± 4
15	-	-	0.17 ± 0.01	32 ± 4
35	0.205 ± 0.012	9 ± 5	0.16 ± 0.01	36 ± 4
75	0.207 ± 0.012	10 ± 5	0.15 ± 0.01	40 ± 4
95	-	-	0.15 ± 0.01	40 ± 4
120	0.200 ± 0.012	13 ± 5	0.14 ± 0.01	44 ± 4

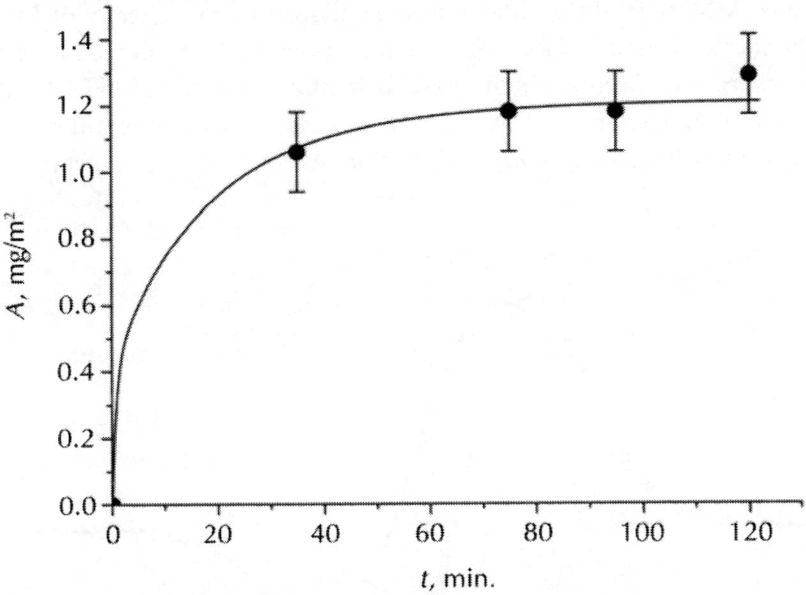

FIGURE 7 Kinetics of thrombin adsorption on magnetite nanoparticles at 25°C. Concentration of thrombin in the sample is 0.9 mg/ml, MNPs – 4.0 mg/ml.

The FMR spectra of the samples "A1-MNP-0," "T1-MNP-0" and MNPs are characterized by different position in magnetic field (Fig. 8). The centre of the spectrum of MNPs is 3254 G, while the centre of "A1-MNP-0" and "T1-MNP-0" spectra is 3253 G and 3449 G, respectively. Resonance conditions for magnetic nanoparticles in magnetic field of spectrometer include a parameter of the shift of FMR spectrum $|M_1| = \frac{3}{2}|H_1|$, where H_1 is a local field created by MNPs in linear aggregates, which form in spectrometer field. $H_1 = 2\sum_1^{\infty}\frac{2\mu}{(nD)^3}$, where D is a distance between MNPs in linear aggregates, μ is MNPs magnetic moment, n is a number of MNPs in aggregate [23]. Coating formation and the thickness of adsorption layer influence on the distance between nanoparticles decrease dipole interactions and particles ability to aggregate. As a result the centre of FMR spectrum moves to higher fields. This phenomenon of FMR spectrum centre shift we observed in the system "A1-MNP-0" after FG addition [20]. The spectrum of MNPs with thick coating becomes similar to FMR spectra of isolated MNPs. So the similar centre positions of FMR spectra of MNPs without coating and MNPs in BSA coating point to a very thin coating and low adsorption of protein in this case. In contrast according to FMR centre position the thrombin coating on MNPs is thicker than albumin coating. *This result is consistent with the data* obtained by ESR spectroscopy.

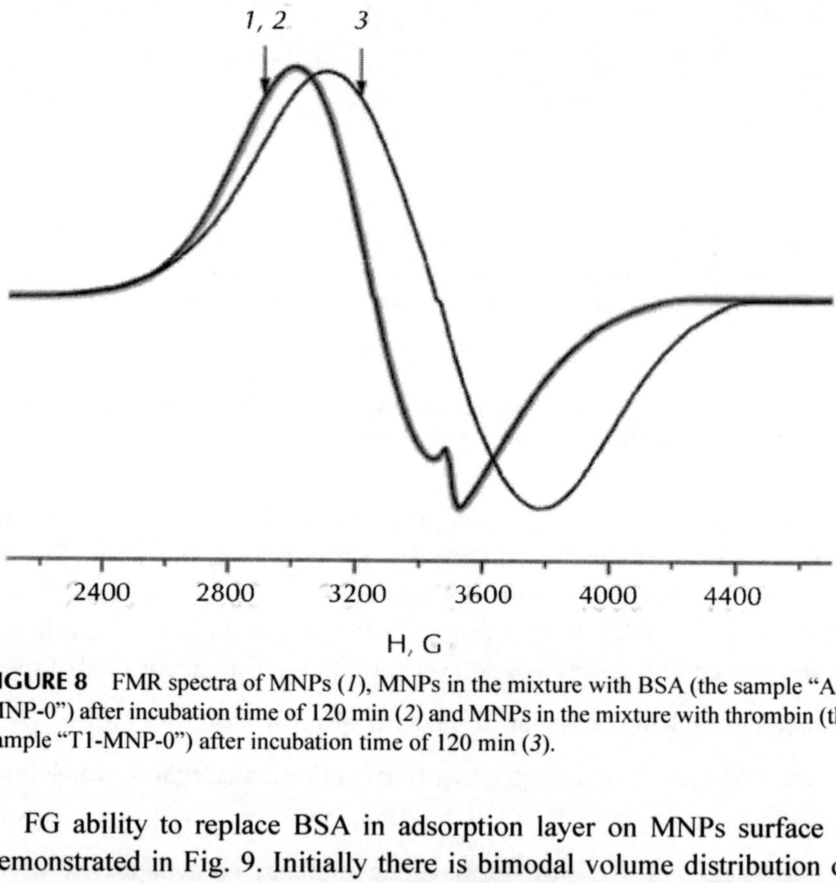

FIGURE 8 FMR spectra of MNPs (*1*), MNPs in the mixture with BSA (the sample "A1-MNP-0") after incubation time of 120 min (*2*) and MNPs in the mixture with thrombin (the sample "T1-MNP-0") after incubation time of 120 min (*3*).

FG ability to replace BSA in adsorption layer on MNPs surface is demonstrated in Fig. 9. Initially there is bimodal volume distribution of particles over sizes in the sample "A2-MNP-0" that can be explained by existence of free (unadsorbed) BSA and MNPs in BSA coating. After FG addition the distribution changes. Micron-sized clusters form in the sample that proves FG adsorption on MNPs [18]. In the case of "A2-MNP-1" volume distribution is also bimodal. The peak of MNPs in BSA coating is characterized by particle size of maximal contribution to the distribution of ~23 nm. This size is identical to MNPs in BSA coating in the sample "A2-MNP-0." It proves that H_2O_2 addition does not lead to uncontrollable linking of protein macromolecules in solution or cluster formation. Since MNPs size is 17 nm, the thickness of adsorption layer on MNPs is approximately 3 nm.

After FG addition to "A2-MNP-1" micron-sized clusters do not form. So adsorption BSA layer formed in the presence of H_2O_2 keeps stability. This stability can be explained by formation of covalent bonds between protein macromolecules [13] in adsorption layer as a result of free radicals generation on MNPs surface. Stability of BSA coating on MNPs was demonstrated for the samples "A1-MNP-1" and "A2-MNP-1" incubated for more than 100 min before FG addition. Clusters are shown to appear if the incubation time is insufficient.

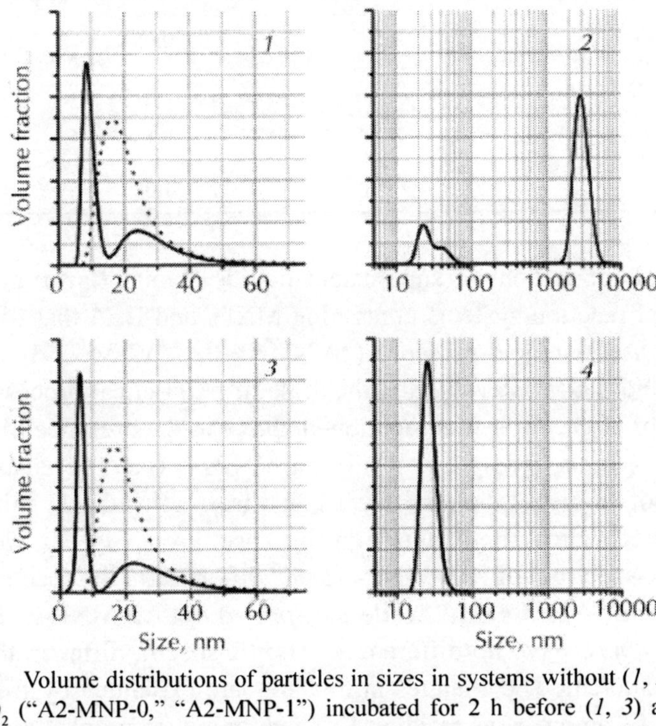

FIGURE 9 Volume distributions of particles in sizes in systems without (*1*, *2*) and with (*3*, *4*) H_2O_2 ("A2-MNP-0," "A2-MNP-1") incubated for 2 h before (*1*, *3*) and 20 min after (*2*, *4*) FG addition. Dotted line is the volume distribution of nanoparticles in sizes in dispersion.

The precipitates obtained by ultracentrifugation of "A2-MNP-0," "A2-MNP-1," "T1-MNP-0" and "T1-MNP-1" were dissolved in buffer solution. The amount of protein in precipitates was evaluated by Bradford colorimetric method (Table 2). The results showed that precipitates of systems with H_2O_2 contained more protein than the same systems without H_2O_2. Therefore in the samples containing H_2O_2 the significant part of protein molecules does not leave MNPs surface when centrifuged while in the samples "A2-MNP-0" and "T1-MNP-0" the most of protein molecules leaves the surface. This indicates the stability of adsorption layer formed in the presence of free radical generation initiator and proves cross-links formation.

TABLE 2 The amount of protein in precipitates after centrifugation of the samples "A2-MNP-0," "A2-MNP-1," "T1-MNP-0" and "T1-MNP-1" of 3.2 ml in volume.

Sample Name	Amount of Protein in Precipitates, mg
"A2-MNP-0"	0.05
"A2-MNP-1"	0.45
"T1-MNP-0"	0.15
"T1-MNP-1"	1.05

Analysis of content of supernatant liquids obtained after ultracentrifugation of reaction systems containing MNPs and BSA that differed by H_2O_2 and ascorbic acid presence ("A2-MNP-0," "A2-MNP-1" and "A2-MNP-1-acid") allows evaluating the scale of free radical processes in the presence of H_2O_2. As it was mentioned above in the presence of ascorbic acid free radicals generate not only on MNPs surface but also in solution. So both molecules on the surface and free molecules in solution can undergo free radical modification in this case. From Fig. 10, we can see that the IR-spectrum of "A2-MNP-1-acid" differs from the spectra of "A2-MNP-0" and "A2-MNP-1," while *the spectra of* "A2-MNP-0" and "A2-MNP-1" *almost* have no differences. The IR-spectra differ in the region of 1,200–800/cm. The changes in this area are explained by free radical oxidation of amino acid residues of methionine, tryptophane, histidine, cysteine, and phenylalanine. These residues are sulfur-containing and cyclic ones which are the most sensitive to free radical oxidation [13, 24]. The absence of differences in "A2-MNP-0" and "A2-MNP-1" proves that cross-linking of protein molecules in the presence of H_2O_2 is selective and takes place only on MNPs surfaces.

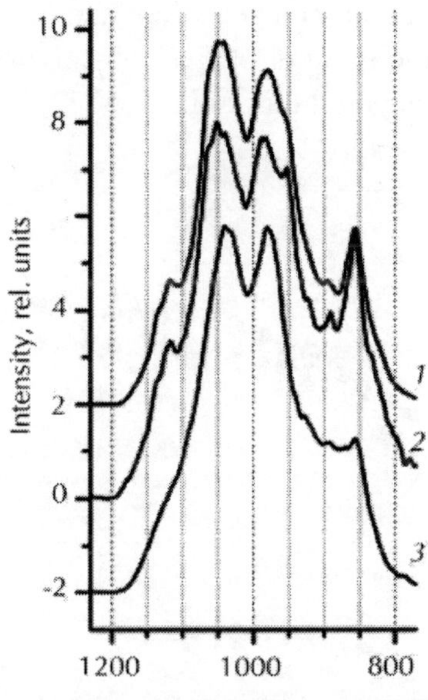

FIGURE 10 IR-spectra of supernatant solutions obtained after centrifugation of the samples "A2-MNP-0" (*1*), "A2-MNP-1" (*2*) and "A2-MNP-1-acid" (*3*).

When proteins are used as components of coating on MNPs for biology and medicine their functional activity retaining is very important. Proteins fixed on MNPs can lose their activity as a result of adsorption on MNPs or free radical modification, which is cross-linking, and oxidation but it was shown that they do not lose it. Estimation of enzyme activity of thrombin cross-linked on MNPs surface was performed by comparison of the rates of conversion of fibrinogen to fibrin under the influence of thrombin contained in reaction mixtures "T1-MNP-0," "T1-MNP-1" and "T1-0-0" (Fig. 11). The curves for the samples containing thrombin and MNPs that differ by the presence of H2O2 had no fundamental differences that illustrate preservation of enzyme activity of thrombin during free radical cross-linking on MNPs surface. Fibrin gel was formed during ~15 min in both cases. Rayleigh light scattering intensity was low when "T1-0-0" was used and small fibrin particles were formed in this case. The reason of this phenomenon is autolysis (self-digestion) of thrombin. Enzyme activity of thrombin, one of serine proteinases, decreases spontaneously in solution [25]. So the proteins can keep their activity longer when adsorbed on MNPs. This way, the method of free radical cross-linking of proteins seems promising for enzyme immobilization.

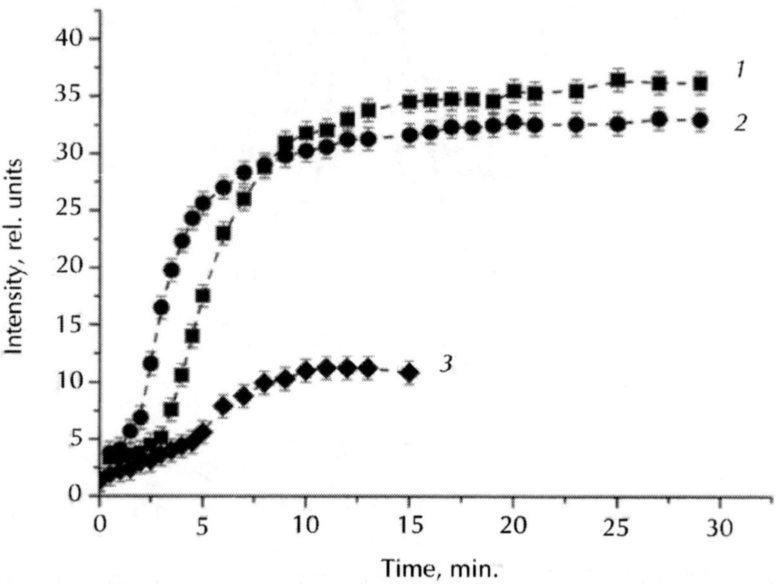

FIGURE 11 Kinetics curves of growth of Rayleigh light scattering intensity in the process of fibrin gel formation in the presence of "T1-MNP-0" (*1*), "T1-MNP-1" (*2*) and "T1-0-0" (*3*).

19.4 CONCLUSION

The novel method of fixation of proteins on MNPs proposed in the work was successfully realized on the example of albumin and thrombin. The blood plasma proteins are characterized by a high biocompatibility and allow decreasing toxicity of nanoparticles administered into organism. The method is based on the ability of proteins to form interchain covalent bonds under the action of free radicals. The reaction mixture for stable coatings obtaining should consist on protein solution, nanoparticles containing metals of variable valence (for example, Fe, Cu, Cr) and water-soluble initiator of free radicals generation. In this work albumin and thrombin were used for coating being formed on magnetite nanoparticles. Hydrogen peroxide served as initiator. By the set of physical (ESR-spectroscopy, ferromagnetic resonance, dynamic and Rayleigh light scattering, IR-spectroscopy) and biochemical methods it was proved that the coatings obtained are stable and formed on individual nanoparticles because free radical processes are localized strictly in the adsorption layer. The free radical linking of thrombin on the surface of nanoparticles has been shown to almost completely keep native properties of the protein molecules. Since the method provides enzyme activity and formation of thin stable protein layers on individual nanopaticles it can be successfully used for various biomedical goals concerning a smart delivery of therapeutic products and biologically active substances (including enzymes). It reveals principally novel technologies of one-step creation of biocompatible magnetically targeted nanosystems with multiprotein polyfunctional coatings which meet all the requirements and contain both biovectors and therapeutic products (Fig. 12).

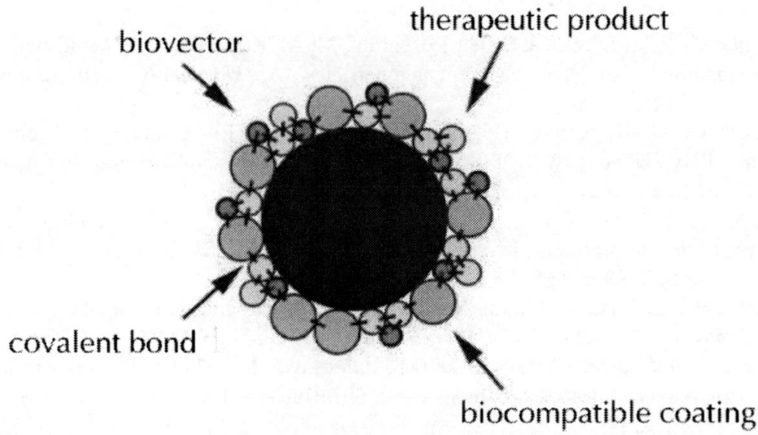

FIGURE 12 The scheme of magnetically targeted nanosystem for a smart delivery of therapeutic products based on the free radical protein cross-linking.

KEYWORDS

- **Uncontrollable linking**
- **Incomplete linking**
- **US-disperser**
- **Fibrin gel**
- **Rayleigh light scattering**

REFERENCES

1. Gupta, A. K.; Gupta, M. Synthesis and surface engineering of iron oxide nanoparticles for biomedical applications. *Biomaterials,* **2005**, *26*, 3995–4021.
2. Vatta, L. L.; Sanderson, D. R.; Koch, K. R. Magnetic nanoparticles: Properties and potential applications. *Pure Appl. Chem.*, **2006**, *78*, 1793–1801.
3. Lu, A. H.; Salabas, E. L.; Schűth, F. Magnetic Nanoparticles: Synthesis, Protection, Functionalization, and Application. *Angew. Chem. Int. Ed.*, **2007**, *46*, 1222–1244.
4. Laurent, S.; Forge, D.; Port, M.; Roch, A.; Robic, C.; Elst, L. V.; Muller, R. N. Magnetic Iron Oxide Nanoparticles: Synthesis, Stabilization, Vectorization, Physicochemical Characterizations, and Biological Applications. *Chem. Rev.*, **2008**, *108*, 2064–2110.
5. Pershina A. G.; Sazonov A. E.; Milto I. V. Application of magnetic nanoparticles in biomedicine. *B. Siberian Med.* [in Russian], **2008**, *2*, 70–78.
6. Trahms, L. Biomedical applications of magnetic nanoparticles. *Lect. Notes Phys.*, **2009**, *763*, 327–358.
7. Bychkova, A. V.; Sorokina, O. N.; Rosenfeld, M. A.; Kovarski, A. L. Multifunctional biocompatible coatings on magnetic nanoparticles. *Uspekhi Khimii (Russian Journal)*, in press (2012).
8. Koneracka, M.; Kopcansky, P.; Antalik, M.; Timko, M.; Ramchand, C. N.; Lobo, D.; Mehta, R. V.; Upadhyay, R. V. Immobilization of proteins and enzymes to fine magnetic particles. *J. Magn. Magn. Mater.*, **1999**, *201*, 427–430.
9. Xu, L.; Kim, M.-J.; Kim, K.-D.; Choa, Y.-H.; Kim, H.-T. Surface modified Fe_3O_4 nanoparticles as a protein delivery vehicle. *Colloids and Surfaces A: Physicochem. Eng. Aspects*, **2009**, *350*, 8–12.
10. Peng, Z. G.; Hidajat, K.; Uddin, M. S. Adsorption of bovine serum albumin on nanosized magnetic particles. *J. Colloid Interf. Sci.*, **2004**, *271*, 277–283.
11. Šafărík, I.; Ptáčková, L.; Koneracká, M.; Šafăríková, M.; Timko, M.; Kopčanský, P. Determination of selected xenobiotics with ferrofluid-modified trypsin. *Biotech. Lett.*, **2002**, *24*, 355–358.
12. Li, F.-Q.; Su, H.; Wang, J.; Liu, J.-Y.; Zhu, Q.-G.; Fei, Y.-B.; Pan, Y.-H.; Hu, J.-H. Preparation and characterization of sodium ferulate entrapped bovine serum albumin nanoparticles for liver targeting. *Int. J. Pharm.*, **2008**, *349*, 274–282.

13. Stadtman, E. R.; Levine, R. L. Free radical-mediated oxidation of free amino acids and amino acid residues in protein. *Amino Acids*, **2003**, *25*, 207–218.
14. Bychkova, A. V.; Sorokina, O. N.; Shapiro, A. B.; Tikhonov, A. P.; Kovarski, A. L. Spin Labels in the Investigation of Macromolecules Adsorption on Magnetic Nanoparticles. *Open Colloid Sci. J.*, **2009**, *2*, 15–19.
15. Abragam, A. *The Principles of Nuclear Magnetism*, Oxford University Press: New York, 1961.
16. Noginova, N.; Chen, F.; Weaver, T.; Giannelis, E. P.; Bourlinos, A. B.; Atsarkin, V. A. Magnetic resonance in nanoparticles: between ferro- and paramagnetism. *J. Phys.: Cond. Matter*, **2007**, *19*, 246208–246222.
17. Sorokina, O. N.; Kovarski, A. L.; Bychkova, A. V. Application of paramagnetic sensors technique for the investigation of the systems containing magnetic particles. In *Progress in Nanoparticles research*; Frisiras, C.T., Ed.; Nova Science Publishers: New York, 2008; 91–102.
18. Bychkova, A. V.; Sorokina, O. N.; Kovarskii, A. L.; Shapiro, A. B.; Leonova, V. B.; Rozenfel'd, M. A. Interaction of fibrinogen with magnetite nanoparticles. *Biophysics (Russ. J.)*, **2010**, *55(4)*, 544–549.
19. Bychkova, A. V.; Sorokina, O. N.; Kovarski, A. L.; Shapiro, A. B.; Rosenfeld, M. A. The Investigation of Polyethyleneimine Adsorption on Magnetite Nanoparticles by Spin Labels Technique. *Nanosci. Nanotechnol. Lett. (ESR in Small Systems)*, **2011**, *3*, 591–593.
20. Bychkova, A. V.; Rosenfeld, M. A.; Leonova, V. B.; Lomakin, S. M.; Sorokina, O. N.; Kovarski, A. L. Surface modification of magnetite nanoparticles with serum albumin in dispersions by free radical cross-linking method. *Russ. colloid J.*, **2012**.
21. Bradford, M. M. A rapid and sensitive method for the quantitation of microgram quantities of *protein utilizing* the *principle* of *protein-dye binding. Anal. Biochem.*, **1976**, *72*, 248–254.
22. Antsiferova, L. I.; Vasserman, A. M.; Ivanova, A. N.; Lifshits, V. A.; Nazemets, N. S. Atlas of Electron Paramagnetic Resonance Spectra of Spin Labels and Probes [in Russian], Nauka, Moscow, 1977.
23. Dolotov, S. V.; Roldughin, V. I. Simulation of ESR spectra of metal nanoparticle aggregates. *Russ. Colloid J.*, **2007**, *69*, 9–12.
24. Smith, C. E.; Stack, M. S.; Johnson, D. A. Ozone effects on inhibitors of human neutrophil proteinases. *Arch. Biochem. Biophys.*, **1987**, *253*, 146–155.
25. Blomback, B. Fibrinogen and fibrin – proteins with complex roles in hemostasis and thrombosis. *Thromb. Res.*, **1996**, *83*, 1–75.

CHAPTER 20

HEAT RESISTANCE OF COPOLYMERS OF VINYL ACETATE AS A COMPONENT OF BIODEGRADABLE MATERIALS

E. V. BELOVA, P. V. PANTYUKHOV, and V. S. LITVISHKO

CONTENTS

20.1 INTRODUCTION

Heat resistance of copolymers of vinyl acetate dibytyl maleate was investigated with the method of differential thermal, thermo gravimetric and spectrometric analysis. Studies show that thermooxidative transformation of copolymers is their intensification with the increase of the content of dibytyl maleate.

20.2 AIM AND BACKGROUND

Copolymers of vinyl acetate are known as a component of biodegradable materials [1]. For copolymer of vinyl acetate with dibytyl maleate there is data on thermal stability in absence of oxygen [2]. The purpose of work – to study thermooxidative decomposition.

Copolymers of vinyl acetate (VA) are used as a polymeric matrix in the creation of biodegradable composite materials. During the manufacture of these materials in the temperature reaches 200–250°C. Obviously, the prerequisite for the creation of such materials is the thermal resistance of copolymers of vinyl acetate as a component of biodegradable materials.

20.3 EXPERIMENTAL METHODS

In the paper studied copolymers VA with dibytyl maleate (DBM). Copolymers were characterized by molecular mass (MM) within 81,000–83,000 c.u. MM was estimated using the intrinsic viscosity acetone solutions at 20°C and calculated theoretically using ratios Van Krevelen.

The definition of a general nature thermooxidative the collapse of the copolymers carried out by the method of differential thermal analysis (DTA) and thermo gravimetric analysis (TGA).

In Fig. 1 presents data analysis for alcohol and homopolymer vinyl acetate. As can be seen from the curves of the DTA, for PVA in the range of 150–250°C, there has been an exothermic region, is not accompanied by a noticeable loss of weight. When the temperature 230–310°C is fixed acute endothermic peak, which corresponds to the maximum speed of the weight loss at 300°C.

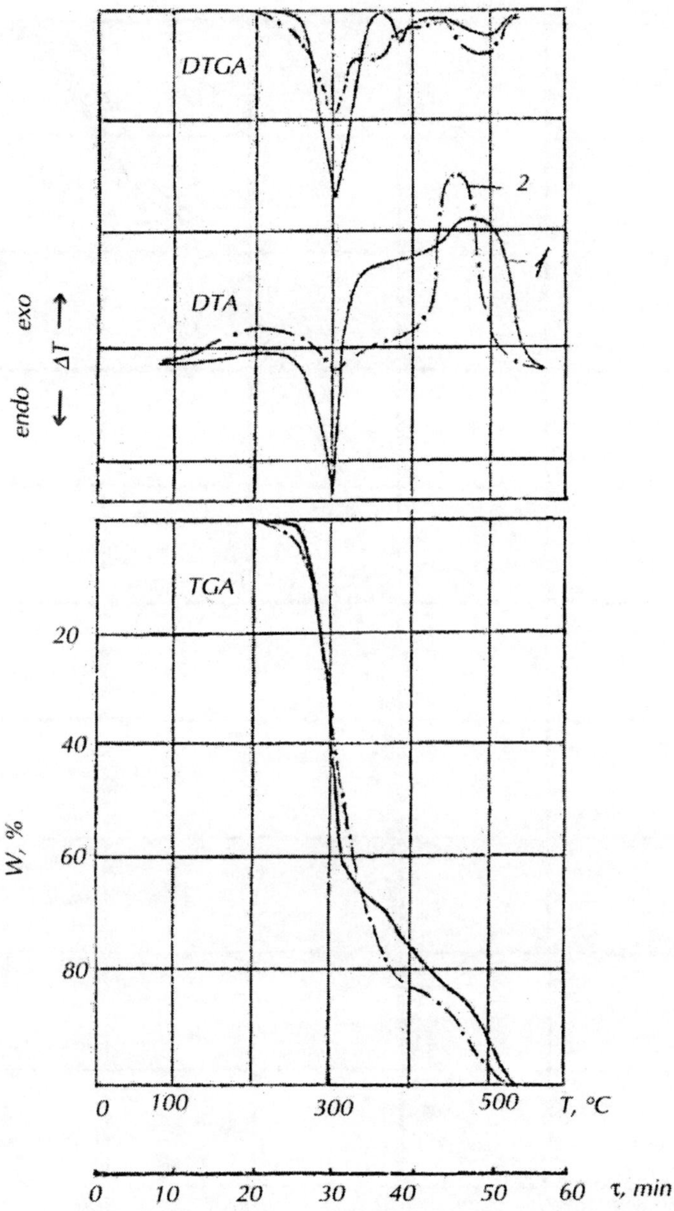

FIGURE 1 Curves TGA, DTGA, the DTA polyvinyl acetate (1) and copolymer VA with DMB (2) the mass content DMB 40% in air atmosphere.

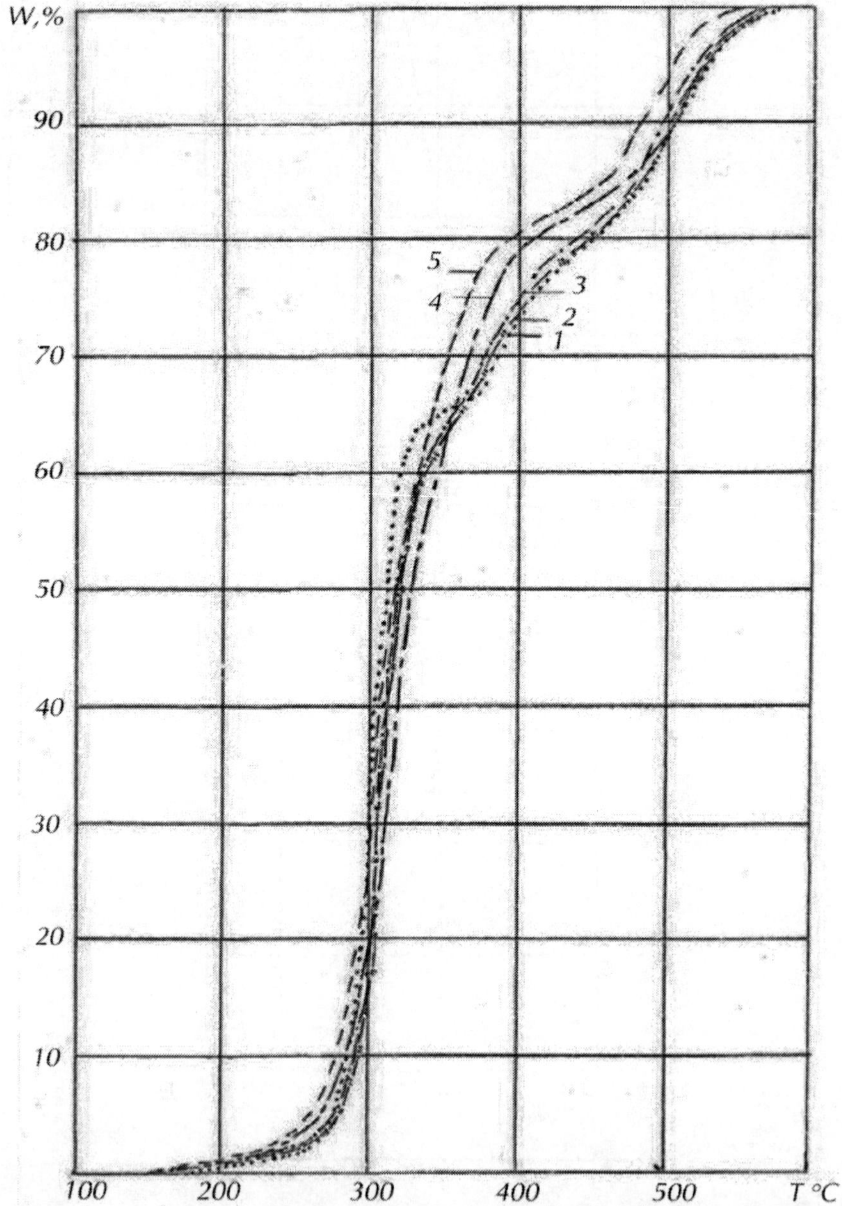

FIGURE 2 The dependence of the weight loss VA copolymer with DMB in mass content DMB: 1 – 0%; 2 – 10%; 3 – 20%; 4 – 30%; 5 – 40% in an atmosphere of air temperature.

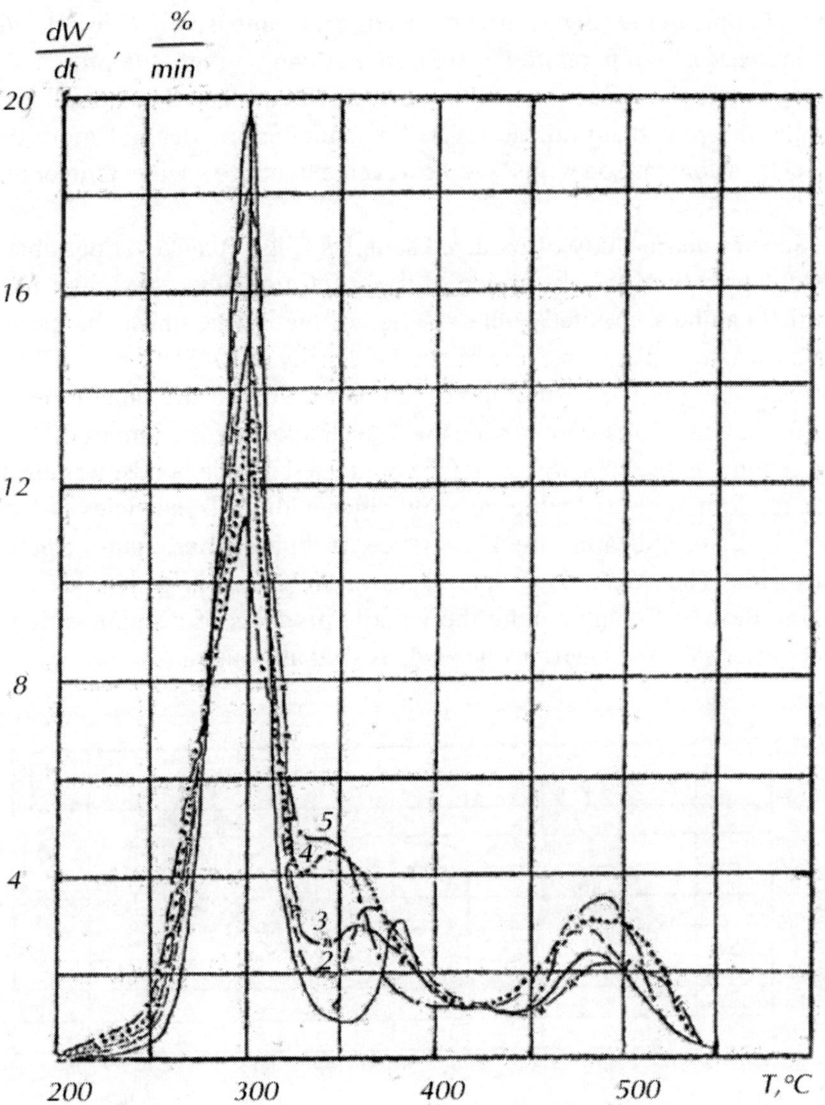

FIGURE 3 The dependence of the rate of loss of weight copolymer VA with DMB in mass content DMB: 1 – 0%; 2 – 10%; 3 – 20%; 4 – 30%; 5 – 40% in an atmosphere of air temperature.

Introduction DMB leads to significant changes in processes thermo-oxidative decomposition: the weight loss for copolymers is registered at lower temperatures (Fig. 2) and proceeds more intensively (Fig. 3). With the increase of temperature up to 300°C, as can be seen, the presence of DMB causes the reduction of the speed of weight loss. The speed of the smaller, more content maleic links. The endothermic peak of more than offset by a comparison with PVA, however, extends to a wider temperature field (200–350°C).

Spectrometric study of oxidized samples (Fig. 4) makes it possible to find the growth of the absorption of the spectrum with a maximum 1710/cm that can be associated with oxidation of methylene links. On spectrograms also took place decrease of optical density at 1025/cm and 1170/cm, corresponding to the destruction of VA and DMB. At the same time was the formation of absorption bands with characteristic maximum of 1600/cm, relating to the appearance of the conjugated double bonds. It is clearly seen the formation of double carbonyl highs with the frequencies of 1855, and 1780/cm, indicating the appearance of the polymer chain of maleic anhydride. The qualitative experiences of the chromatographic determination allowed finding among the volatile products of decomposition of copolymers VA and DMB acetic acid, as well as boutin.

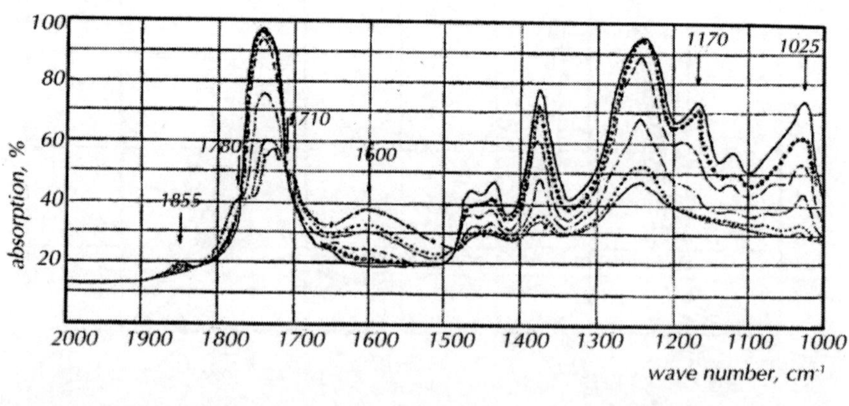

FIGURE 4 IR absorption spectra of VA with DMB in mass content DMB 40% of the original (__) of the sample and the oxidized when 230°C within 60 min. (ooo); 120 min. (- - -); 240 min. (-◦-); 360 min. (◦◦◦); 480 min. (-”-)

20.4 RESULTS AND DISCUSSION

These data allows to draw a conclusion about the fact that the decomposition of copolymers in the low temperature stage is caused by the processes of deacetylation with the formation of polienovykh circuits and decomposition of the links of the DMB for oet connection:

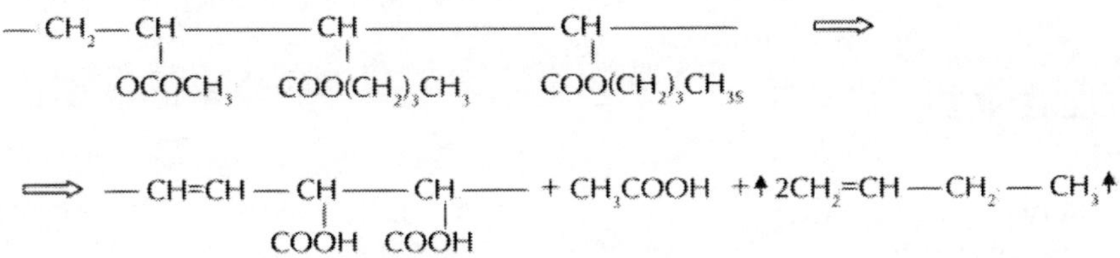

The presence in the polymer chain of maleic anhydride, apparently, is connected with dehydration maleic acid, formed after splitting butene:

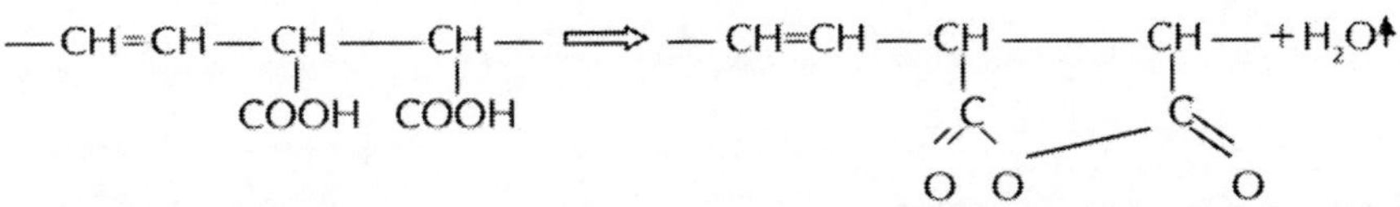

Data TGA, processed by the method of Reich, shows that during this stage of the process is the least resistance to thermal oxidation have copolymers with the highest content of DBM, resulting in a decrease in the activation energy of the process of decay from 98.5 KJ/mol (homopolymer VA) to 76.7 KJ/mol (copolymer with 40% of the masses DMB) in the temperature range of 220–250°C.

20.5 CONCLUSION

Thus, studies show that thermooxidative transformation of copolymers VA and DMB associated with oxidative reactions of methene groups, as well as the collapse of the VA and DMB – links to difficult – etheric relations. A distinctive feature of the thermooxidative processes is their intensification with the increase of the content of DMB in copolymer.

KEYWORDS

- **Dibytyl maleate**
- **Oet connection**
- **Polienovykh circuits**
- **Thermooxidative transformation**

REFERENCES

1. Suvorova, A. I.; Tyukova, I. S.; Trufanova, E. I. "Biodegradable starch-based polymeric materials," *Russ. Chem. Rev*, **2000**, *69(5)*, 451–459.
2. Afanasev, A. G.; Lugova, L. I. "Thermal stability of copolymers of vinyl acetate with dibytyl maleate," *Plasticheskie Massy*, **1984**, *6*, 62–63.

CHAPTER 21

DEGRADATION AND STABILIZATION OF POLYVINYLCHLORIDE, KINETICS AND MECHANISM

G. E. ZAIKOV, M. I. ARTSIS, L. A. ZIMINA, and A. D'AMORE

CONTENTS

21.1 INTRODUCTION

In both plasticized (semi-rigid and flexible) PVC materials as well as PVC in solutions, the rate of their thermal degradation and effective stabilization are caused by essentially different fundamental phenomena in comparison to aging of PVC in absence of the solvent.

Both structure and macromolecular dynamics render the significant influence on its stability, i.e., chemical nature of the solvent (plasticizer), its basicity, specific and non-specific solvation, degree of PVC in a solution (solubility), segmental mobility of macromolecules, thermodynamic properties of the solvent (plasticizer), formation of associates, aggregates, etc.

The chemical stabilization of PVC plays a less significant role. The effect of above factors on stability (behavior) of semi-rigid and flexible PVC will be done on quantitative level.

It will be described effect of "echo"-type of stabilization on the stability of PVC in the presence of plasticizers. If we would like to have stable material from PVC we should make stabilization of plasticizers as more reactive chemical compounds.

At PVC's degradation in solution, one of the basic reasons of change of the process kinetic parameters is the nucleophilic activation of PVC's dehydrochlorination reaction. The process is described by E_2 mechanism [1–3]. Thus, there is a linear dependence between PVC's thermal dehydrochlorination rate and parameter of solvent's relative basicity *B*/cm (Fig. 1) [1–3]. (The value B/cm is evaluated by shift of a characteristic band OH of phenol at $\lambda = 3600$/cm in an IR-spectrum at interaction with the solvent [4]). It is essentially important that the rate of PVC's dehydrochlorination in the solvents with relative basicity $B > 50$/cm was always above, than the rate of PVC's dehydrochlorination without the solvent, while when $B <$ 50/cm, PVC's desintegration rate was always less, than at it's destruction without the solvent. The revealed dependence $V_{HCl} = f(B)$ is described by the equation:

$$V_{HCl}^{*} = V_{HCl} + k(B - 50) \quad (1)$$

An inhibition of PVC's disintegration in the solvents with basicity $B <$ 50/cm is very interesting and practically important phenomenon. It has received the name "solvatational" stabilization of PVC. Let's notice, however, that ignoring of the fact that PVC solutions even at low concentration (2 wt.%) do not represent solutions with isolated macromolecules but rather with structured systems, results that in a number of cases a deviation from linear dependence of PVC dehydrochlorination rate of the solvent basicity *B*/cm is observed. In particular, an abnormal behavior of PVC is observed at destruction in certain ester-type solvents (plasticizers) (Fig. 1, points 25–28) that apparently caused by structural changes of macromolecules. This was never before taken into account at work with PVC in solutions.

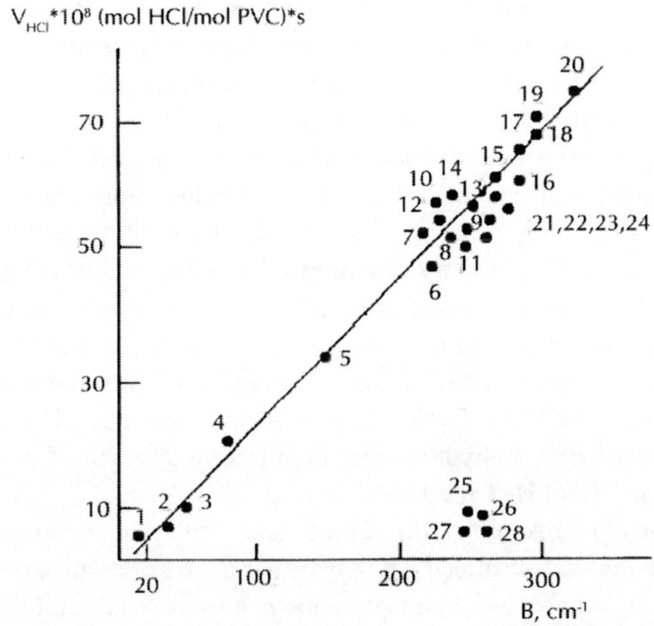

FIGURE 1 Influence of the solvent's basicity on the rate of thermal dehydrichlorination in solution: 1 – *n*-dichlorobenzene, 2 – *o*-dichlorobenzene, 3 – naphthalene, 4 – nitrobenzene, 5 – acetophenone, 6 – benzonitrile, 7 – di-*n*-(chlorophenyl-chloropropyl) phosphate, 8 – triphenylphosphite, 9 – phenyl-*bis*-(β-chloroethyl) phosphate, 10 – tri-(*n*-chlorophenyl phosphate), 11 – 2-ehtylhexylphenyl phosphate, 12 – tricresyl phosphate, 13 – cyclohexanone, 14 – phenyl-*bis*-(β-chloropropyl) phosphate, 15 – tri-β-chloroethyl phosphate, 16 – tri-β-chloropropyl phosphate, 17 – di-2-(ethylhexyl) phosphate, 18 – 2-ethylhexylnonyl phosphate, 19 – tri-(2-ethylhexyl) phosphate, 20 – tributyl phosphate, 21, 25 – dibutyl phthalate, 22, 26 – di-2-ethylhexyl adipinate, 23, 27 – dioctyl phthalate, 24, 28 – dibutyl sebacinate. Concentration of PVC in solution: 1–24 – 0.2 wt.%, 25 – 28 – 2wt.%; 423 K, under nitrogen.

It was revealed quite unexpectedly that not only interaction "polymer–solvent", but also interaction "polymer–polymer" in solutions provide significant influence on rate of PVC disintegration. It's known that structure and properties of the appropriate structural levels depend from conformational and configurational nature of macromolecules, including a supermolecular structure of the polymer, which in turn determines all basic (both physical and chemical) characteristic of polymer.

"Polymer–polymer" interaction results to formation of structures on a supermolecular level. In particular, as getting more concentrated the PVC-solvent system consistently passes a number of stages from isolated PVC macromolecules in a solution (infinitely diluted solution) to associates and aggregates from macromolecules in a solution. At the further increase of PVC concentration in a solution formation of spatial fluctuational net with structure similar to a structure of polymer in the block occurs.

When polymer's concentration in a solution increases, the rate of PVC's dehydrochlorination reaction changes as well, and various character of influence of the solvent on a PVC disintegration rate in solution is observed depending on a numerical value of basicity parameter B/cm [5–10]. If the relative basicity of employed solvents was $B > 50$/cm, the polymer's degradation rate decreases when its concentration increases. If a basicity of the employed solvents was $B < 50$/cm, the polymer's degradation rate increases with increased concentration of a polymer. In all cases the rate of HCl elimination from a polymer has a trend in a limit to reach values of PVC dehydrochlorination rate in absence of the solvent $V_{HCl}^{PVC} = 5*10^{-8}$ (mol HCL/mol PVC)/s. (Fig. 2).

Equation (1) turns into an Eq. (2) if to take into account that the PVC's degradation rate is determined not only by parameter of relative basicity of the solvent B, but also by its concentration in a solution (C, mol PVC/l), as well as by degree of "polymer–polymer" interaction (degree of macromolecules structurizarion in a solution $\Delta C = /C–C_0/$, where C_0 – concentration of a beginning of PVC macromolecules association in a solution):

$$V_{HCl} = V_{HCl}^{0} + A_1 /(C + /\Delta C/ + d_1)(B - 50) \qquad (2)$$

Here factor $A_1 = (0.8 \pm 0.2) \times 10^{-9}$ (mol HCL/mol PVC)/s; d_1 – dimensionless factor reflecting interaction "polymer–solvent" ($d_1 = 0.5 \pm 0.25$).

The deviation from the moment of a beginning of macromolecules association in a solution is taken on the absolute value, since it can change in both directions to more concentrated and more diluted solutions of a polymer.

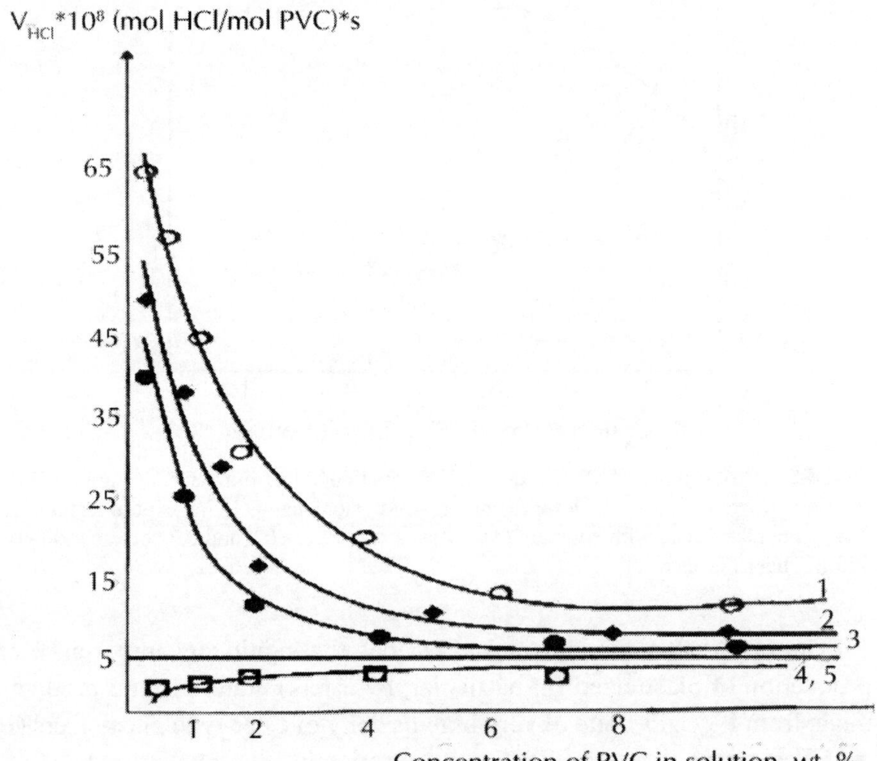

FIGURE 2 A change of PVC's dehydrochlorination rate of its concentration in a solution: 1 – cyclohexanol, 2 – cyclohexanone, 3 – benzyl alcohol, 4 – 1, 2, 3 – trichloropropane, 5 – *o*-dichlorobenzene, 6 – no solvent; 423 K, under nitrogen.

Equation (2) well describes a change of PVC's thermal dehydrochlorination rate of its concentration in a solution in view of parameter of relative basicity of the solvent B, irrespective of the chosen solvent (Fig. 3).

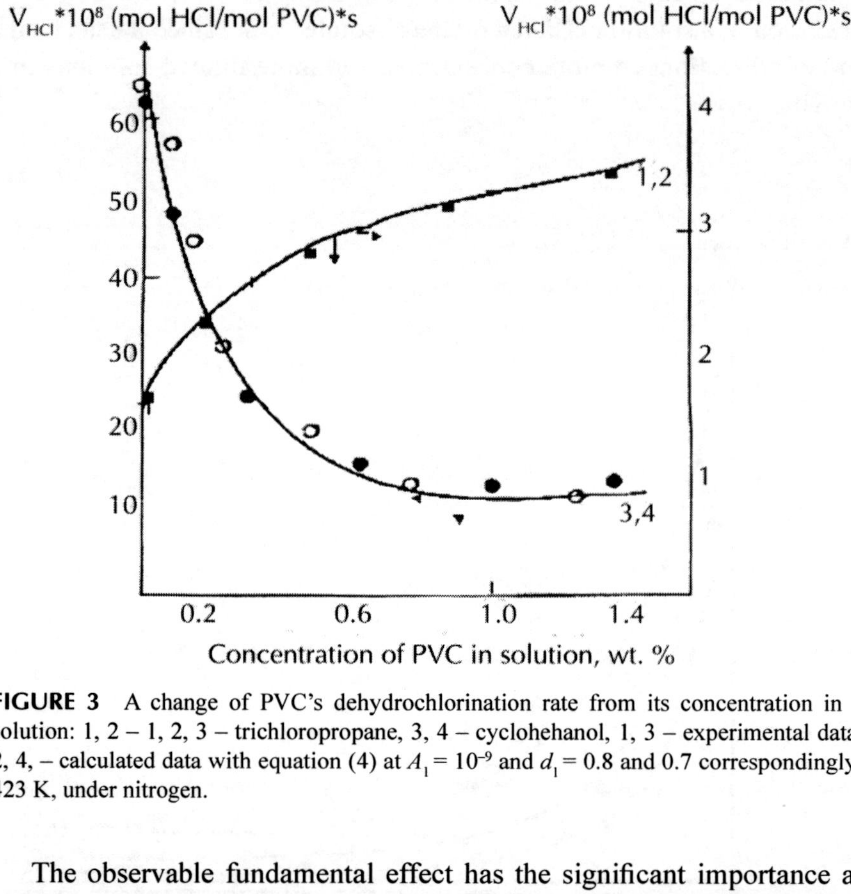

FIGURE 3 A change of PVC's dehydrochlorination rate from its concentration in a solution: 1, 2 – 1, 2, 3 – trichloropropane, 3, 4 – cyclohehanol, 1, 3 – experimental data, 2, 4, – calculated data with equation (4) at $A_1 = 10^{-9}$ and $d_1 = 0.8$ and 0.7 correspondingly; 423 K, under nitrogen.

The observable fundamental effect has the significant importance at production of plasticized (in particular, by esters) materials and products made from PVC. Despite of very high basicity of ester-type plasticizers (B = 150/cm) in an interval of PVC concentration in solutions more than 2%, a noticeable reduction of PVC degradation rate is observed (Fig. 1, curve, points 25–28), i.e., on essence, stabilization of PVC occurs. This effect is caused by formation of dense globules, associates, etc. in the system PVC – plasticizer. Practically this allows creating economic formulations of plasticized materials from PVC with the very little contents of metal-containing stabilizers – HCl acceptors, or without their use at all.

Temperature is very important on formation of the heterophase system. Even at low concentration of PVC in ester-type plasticizer (e.g., in dioctyl phthalate at $C > 0.1$ mol/l) the true solutions are formed only at temperatures above 400 K. Globular structure of suspension PVC and formation of associates retain at temperatures up to 430–445 K. In other words, PVC at plasticization is capable to keep its structural individuality on a supermolecular level, which is formed during polymer's synthesis. Specifically in these conditions the ester-type plasticizer behaves not as a highly-basic solvent, but as a stabilizer at PVC's thermo-degradation due to formation of associates etc. This leads to a reduction of stabilizer's amount, extension of exploitation time of materials and products, etc.

It is necessary to note that the change of PVC's degradation rate at association of macromolecules is the general phenomenon and does not depend on how it was achieved. In particular, similar (as well as at concentration of PVC solutions, Figs. 2, and 3), character of change of PVC dehydrochlorination rate in a solution is observed, if the change of PVC's structural – physical condition in a solution is reached upon addition even chemically inert non-solvents, for example, hexane, decane, undecane, polyolefines, polyethylene wax, etc. [6, 9–12] (Fig. 4). It is interesting to note that the degree of relative change of PVC disintegration rate under action of the second inert polymer (non-solvent) is much higher, than at concentration of a PVC solution, especially in case of use of the low-basic solvents (trichloropropane, dichlorobenzene – a result of formation of more dense formations on a supermolecular level, corresponding associates and aggregates, thanks to whom there is a significant change of a PVC destruction rate. The more contents of non-solvent (including an inert polymer) in a blend and lower thermodynamic compatibility of components in a solution, the more structural formation takes place in a solution, including one at the presence of a polymer blends (associates, aggregates). Formation of a fluctuational net with participation of macromolecules is probable. Since the reason of change of PVC thermal dehydrochlorination rate in case of its blends with chemically inert thermodynamically incompatible polymers is the same, as at concentration of a PVC solution (structural-chemical changes of a polymer in a solution), the parameters determining the rate of PVC disintegration will be similar, obviously. Therefore, at consideration of PVC thermal destruction a concentration of the second polymer in a blend with PVC, as well as a degree of its thermodynamic affinity to PVC have to be taken in account in addition to an influence of polymer's concentration in a solution, basicity of the solvent *B*/cm and forces of interaction "polymer–solvent". In view of these factors the Eq. (2) turns into an Eq. (3):

$$V_{HCl} = V_{HCl}^{0} + (A_1 / c + /(c + |\Delta c| + d_1 + \alpha n))(B - 50) + A_1 / B)(d_2 \alpha^n / c) \quad (3)$$

where α is the fraction of the second polymer, varying from 0 to 0.99; n is the dimensionless parameter describing a degree of thermodynamic affinity of PVC to the second polymer and varying from zero (in a case of a complete thermodynamic compatibility of the components) up to certain value equals ~10 (in a case of a complete thermodynamic incompatibility of the polymers). Dimensionless coefficient d_2 reflects interaction of the second polymer with the solvent. At destruction of PVC in a blend with poly(ethylene) in a solution of dichlorobenzene, trichloropropane, and cyclohexanol it equals 2.5 ± 0.1.

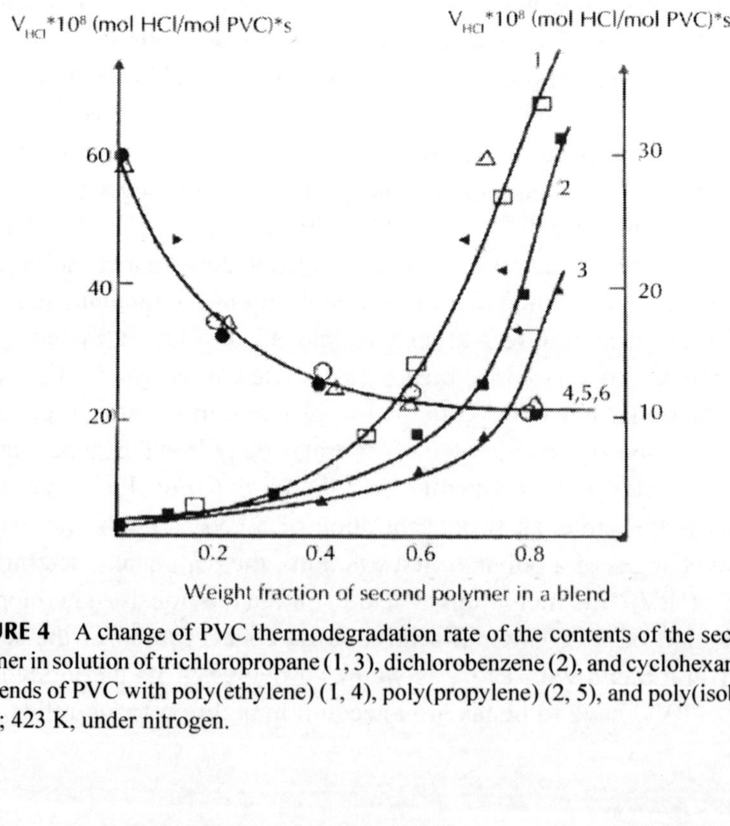

FIGURE 4 A change of PVC thermodegradation rate of the contents of the secong inert polymer in solution of trichloropropane (1, 3), dichlorobenzene (2), and cyclohexanol (4–6) for blends of PVC with poly(ethylene) (1, 4), poly(propylene) (2, 5), and poly(isobutylene) (3, 6); 423 K, under nitrogen.

Observable changes of PVC thermal disintegration rate under action of second thermodynamically incompatible with PVC polymer (or owing to an increase of PVC concentration in a solution) are caused by a displacement of the solvent from macromolecular globules of PVC with transformation to the structure, which it has in absence of the solvent. This evokes unexpected effect of "the solvent action" (a delay or an acceleration depending on the solvent's basicity *B*/cm) in relation to PVC's thermal disintegration. The solvent's displacement, which accelerates PVC's disintegration ($B > 50$/cm), results to easing of its interaction with PVC and leads to a delay of process of HCl elimination from macromolecules, i.e., to stabilization. This occurs both in a case of concentration of PVC's solutions, and in case of addition of second polymer, which is thermodynamically incompatible with PVC. In the solvents slowing down PVC's disintegration ($B > 50$/cm) by virtue of low nucleophilicity, an effect of the solvent displacement and the easing of its influence on PVC results in an opposite result – an increase of HCl elimination rate from PVC upon of increase of its concentration in a solution or at use of chemically inert non-solvent. It is obvious that irrespective of the fact how to make changes in PVC's structure in a solution – by increase of its concentration in a solution or by addition of second thermodynamically incompatible with PVC chemically inert non-solvent – the varying structural – physical condition of the polymer results in a noticeable change of its thermal dehydrochlorination rate in a solution. These effects are caused by structural – physical changes in system polymer – solvent, and previously unknown phenomena can be classified as structural – physical stabilization (in case of a reduction of gross – rate of PVC disintegration in highly-basic solvents at $B >$ 50/cm) and, respectively, structural – physical antistabilization (in case of increase of gross – rate of PVC disintegration in low-basic solvents with $B > 50$/cm).

21.2 "ECHO" - STABILIZATION OF PVC

At last, it is necessary to specify to one more appreciable achievement in the field of aging and stabilization of PVC in a solution. In real conditions the basic reason of the sharp accelerated aging of plasticized materials and products is the oxidation of the solvent by oxygen of air (Fig. 5, curve 3).

$$RO_2^{\bullet} + RH \xrightarrow{K_6} ROOH + R\cdot$$
$$ROOH \xrightarrow{K_6} RO^{\bullet} + HO^{\bullet}$$
$$RO_2^{\bullet} + RO_2^{\bullet} \xrightarrow{K_6} \text{inactive products}$$

Peroxides, formed at oxidation of ester-type plasticizers, initiate disintegration of macromolecules. In these conditions the rate of PVC destruction can increase in two and more orders of magnitude and is determined by oxidizing stability of the solvent to oxygen – parameter $K_{ef} = K_2 \times K_3^{0.5} \times K_6^{-0.5}$. Then higher an oxidizing stability of the solvent (in particular, ester-type plasticizer), at which's presence a thermooxidative disintegration of PVC occurs, then lower its degradation rate and longer an exploitation time of semi-rigid and flexible materials on a basis of PVC [13–18]. An inhibition of process of the solvent's oxidation (including plasticizers) due to of incorporation of stabilizers – antioxidants or their synergistic compositions slows a thermo-oxidative disintegration of PVC in a solution (Fig. 5, curve 5).

At effective inhibition of the ester-type plasticizers' oxidation by oxygen of air the rate of PVC thermo-oxidative destruction in their concentrated solutions is getting closer to the rate of polymer's disintegration, what is characteristic for its thermal destruction at plasticizer's (solvent's) presence, i.e., slower, than PVC's desintegration without a solvent. This occurs due to a structural – physical stabilization. In these cases an inhibition of reaction of the solvent's oxidation at use of "echo" – type stabilizers – antioxidants causes PVC's stabilization (Fig. 5, curve 5). This fundamental phenomenon of PVC's stabilization in a solution at its thermo-oxidative destruction has received the name of an "echo" – stabilization of PVC [2, 15, 16].

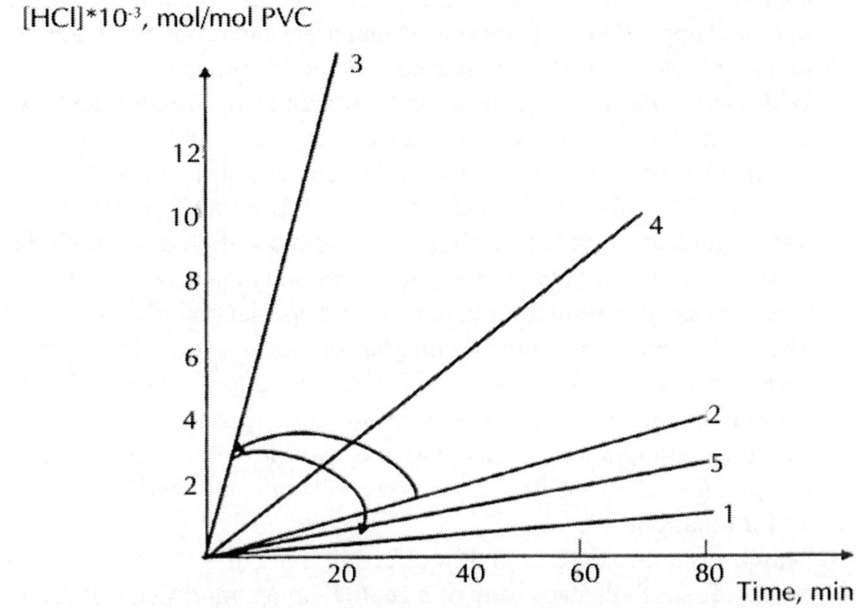

FIGURE 5 "Echo"-stabilization of PVC. Elimination of HCl during thermo- (argon) (1, 2) and thermo-oxidative (air) (3 – 5) destruction of PVC in solution of dioctyl sebacinate: 1 – 4 unstabilized PVC, 5 – PVC, stabilized with diphenylpropane (0.02 wt. %) – "echo" stabilization; 2, 4 – PVC with no solvent; 448 K.

21.3 TASKS FOR A FUTURE [26–47]

Thus, a creation of high-quality and economic semi-rigid and flexible materials and products on a basis of PVC, including ones where solvents are employed, require the specific approach, essentially differing from principles of manufacture of rigid materials and products from PVC. In particular, account and use of the fundamental phenomena: solvatational, structural–physical and "echo" – stabilization of a polymer in a solution.

As to paramount tasks of fundamental and applied research in the field of PVC's manufacture and processing in the beginning of XXI century, obviously they are following:

- Manufacture of an industrial PVC, not containing of labile groups in a backbone. It will provide drastic increase of an intrinsic stability of polymeric products, possibility of PVC processing with the minimal contents or in total absence of stabilizers and other chemicals – additives and opportunity of creation of materials and products on a PVC basis with the essentially increased service life-time;

- Wide use of the latest achievements in area of destruction and stabilization of PVC, both at presence and absence the solvents. First of all, phenomena of chemical, solvatational, structural–physical, self- and "echo" – stabilization of PVC will allow to create rigid, semi-rigid and flexible (plasticized) materials and products with the minimal contents of chemicals – additives and increased life-time of their service at exploitation in natural and special conditions;
- The use of non-toxic, non-flammable products which do not emit toxic and other poison gaseous and liquid products at elevated temperature at manufacture of materials and products from PVC;
- Complete slimination of all toxic and even low-toxic (particularly compounds based on Pb, Cd, Ba, etc.) chemicals – additives from all formulations;
- Search of non-toxic and highly effective inorganic chemicals – additives, first of all, stabilizers of a zeolite – type, modified clays, etc.

At the same time new "surprises" may be expected, which undoubtedly will be presented us by this outstanding polymer – puzzle, a plastic's "working horse" for many decades. Certainly it will give new stimulus in development of scientific bases and practical development with opening of new pathways, conducting to essential delay of PVC's ageing in natural and special conditions at reduction of amounts of the appropriate chemicals – additives, down to their complete elimination [18–25].

KEYWORDS

- **Echo-type**
- **Ester-type**
- **Globular structure**
- **Heterophase**
- **Solvatational**

REFERENCES

1. Minsker, K. S.; Abdullin, M. I.; Manushin, V. I.; Malyshev, L. N.; Arzhakov, S. A. *Dokl. Akad. Nauk SSSR,* **1978**, *242(2)*, 366–368.
2. Minsker, K. S. In *Polymer Yearbook*; Pethrick, A., Ed.; Acad. Publ.: Harwood, 1994; pp. 229–241.
3. Minsker, K. S.; Kulish, E. I.; Zaikov, G. E. *Vysokomol. Soedin.*, **1993**, *35B(6)*, 316–317.
4. Palm, V. A. *Osnovy kolichestvennoi teorii organicheskikh reaktsii*; Khimiya: Leningrad, 1977; p. 114.
5. Kolesov, S. V.; Kulish, E. I.; Minsker, K. S. *Vysokomol. Soedin.*, **1994**, *36B(8)*, 1383–1384.
6. Kolesov, S. V.; Kulish, E. I.; Zaikov, G. E.; Minsker, K. S. *Russian Polym. News,* **1997**, *2(4)*, 6–9.
7. Kulish, E. I.; Kolesov, S. V.; Minsker, K. S. *Bashkirskii Khimicheskii Zhurnal,* **1998**, *56(2)*, 35–37.
8. Kulish, E. I.; Kolesov, S. V.; Minsker, K. S.; Zaikov, G. E. *Vysokomol. Soedin.*, **1998**, *40A(8)*, 1309–1313.
9. Kolesov, S. V.; Kulish, E. I.; Zaikov, G. E.; Minsker, K. S. *J. Appl. Polym. Sci.*, **1999**, *73(1)*, 85–89.
10. Kulish, E. I.; Kolesov, S. V.; Minsker, K. S.; Zaikov, G. E. *Chem. Phys. Report,* **1999**, *18(4)*, 705–711.
11. Kulish, E. I.; Kolesov, S. V.; Akhmetkhanov, R. M.; Minsker, K. S. *Vysokomol. Soedin.,* **1993**, *35B(4)*, 205–208.
12. Kulish, E. I.; Kolesov, S. V.; Minsker, K. S.; Zaikov, G. E. *Int. J. Polym. Mater.*, **1994**, *24(1–4)*, 123–129.
13. Martemyanov, V. S.; Abdullin, M. I.; Orlova, T. E.; Minsker, K. S. *Neftekhimiya,* **1981**, *21(1)*, 123–129.
14. Minsker, K. A.; Abdullin, M. I.; Zueva, N. P.; Martemyanov, V. S.; Teplov, B. F. *Plast. Massy,* **1981**, *9*, 33–34.
15. Minsker, K. S.; Abdullin, M. I. *Dokl. Akad. Nauk SSSR,* **1982**, *263(1)*, 140–143.
16. Minsker, K. S.; Zaikov, G. E. *Chemistry of chlorine-containing polymers: synthesis, degradation, stabilization*; NOVA Science: Huntington, 1999; p 295.
17. Minsker, K. S.; Lisitsky, V. V.; Zaikov, G. E. *Vysokomol. Soedin.*, **1981**, *23A(3)*, 498–512.
18. Minsker, K. S.; Kolesov, S. V.; Yanborisov, V. M.; Berlin, Al. Al.; Zaikov, G. E. *Vysokomol. Soedin.*, **1984**, *26A(5)*, 883–899.
19. Minsker, K. S.; Lisitsky, V. V.; Zaikov, G. E. *J. Vinyl Technol.*, **1980**, *2(4)*, 77–86.
20. Minsker, K. S.; Kolesov, S. V.; Zaikov, G. E. *Degradation and Stabilization of Vinyl Chloride Based Polymers*; Pergamon Press, 1988.
21. Minsker, K. S.; Kolesov, S. V.; Zaikov, G. E. *J. Vinyl Technol.*, **1980**, *2(3)*, 141–151.
22. Minsker, K. S.; Kolesov, S. V.; Zaikov, G. E. *Vysokomol. Soedin.*, **1987**, *23A(3)*, 498–512.
23. Minsker, K. S.; Kolesov, S. V.; Yanborisov, V. M.; Adler, M. E.; Zaikov, G. E. *Dokl. Akad. Nauk SSSR,* **1983**, *268(6)*, 1415–1419.

24. Kolesov, S. V.; Minsker, K. S.; Yanborisov, V. M.; Zaikov, G. E.; Du-Jong, K.; Akhmetkhanov, R. M. *Plast. Massy,* **1983**, *12*, 39–41.
25. Kolesov, S. V.; Steklova, A. M.; Zaikov, G. E.; Minsker, K. S. *Vysokomol. Soedin.*, **1986**, *28A(9)*, 1885–1890.
26. Zaikov, G. E.; Buchachenko, A. L.; Ivanov, V. B. In *Aging of Polymers, Polymer Blends and Polymer Composites*, Nova Science Publ.: New York, 2002; *1*, 258 pp.
27. Zaikov, G. E.; Buchachenko, A. L.; Ivanov, V. B. In *Aging of Polymers, Polymer Blends and Polymer Composites*, Nova Science Publ.: New York, 2002; *2*, 253 pp.
28. Zaikov, G. E.; Buchachenko, A. L.; Ivanov, V. B. In *Polymer Aging at the Cutting Edge*, New York, Nova Science Publ., 2002, 176 pp.
29. Semenov, S. A.; Gumargalieva, K. Z.; Zaikov, G. E. In *Biodegradation and Durability of Materials Under the Effect of Microorganisms*", VSP International Science Publ.: Utrecht, 2003; 199 pp.
30. Rakovsky, S. K.; Zaikov, G. E. In *Interaction of Ozone with Chemical Compounds. New Frontiers*, Rapra Technology: London, 2009.
31. In *Chemical Reactions in Gas, Liquid and Solid Phases: Synthesis, Properties and Application*, Zaikov, G. E.; Kozlowski, R. M., Eds. Nova Science Publishers: New York, 2010.
32. Davydov, E. Ya.; Pariiskii, G. B.; Gaponova, I. S.; Pokholok, T. V.; Zaikov, G. E. In *Interaction of Polymers with Polluted Atmosphere. Nitrogen Oxides*, Rapra Technology: London, 2009.
33. In *Monomers, Oligomers, Polymers, Composites and Nanocomposites Research. Synthesis, Properties and Applications*, Pethrick, R. A.; Petkov, P.; Zaikov, G. E.; Rakovsky, S. K. *Polymer Yearbook*, 2nd Edition; 2012; *23*, 481 pp.
34. In *Polymer Yearbook – 2011. Polymers, Composites and Nanocomposites. Yesterday, Today, Perspectives*, Zaikov, G. E.; Sirghie, C.; Kozlowski, R. M. Nova Science Publishers: New York, 2012; 254 pp.
35. Zaikov, G. E.; Kozlowski, R. M. In *Chemical Reactions in Gas, Liquid and Solid Phases: Synthesis, Properties and Application,* Nova Science Publishers: New York, 2012; 282.
36. Stoyanov, O. V.; Kubica, S.; Zaikov, G. E. In *Polymer Material Science and Nanochemistry*, Institute for Engineering of Polymer Materials and Dyes Publishing House, Torun (Poland), 2012.
37. In *Thr Problems of Nanochemistry for the Creation of New Materials*, Lipanov, A. M.; Kodolov, V. I.; Kubica, S.; Zaikov, G. E. Institute for Engineering of Polymer Materials and Dyes Publishing House: Torun, Poland, 2012; 252 pp.
38. "Unique properties of polymers and composites some examples and perspectives," In *Today and Tomorrow*; Bubnov, Y. N.; Vasnev, V. A.; Askadskii, A. A.; Zaikov, G. E. Eds.; Nova Science Publishers: New York, 2012; *1*, 278 pp.
39. "Unique properties of polymers and composites Some examples and perspectives," In *Tomorrow and Perspectives*, Bubnov, Y. N.; Vasnev, V. A.; Askadskii, A. A.; Zaikov, G. E. Eds.; Nova Science Publishers: New York, 2012; *2*.
40. "Kinetics, catalysis and mechanism of chemical reactions. From pure to applied science," In *Today and Tomorrow*, Islamova, R. M.; Kolesov, S. V.; Gennady, E., Eds.; Nova Science Publishers: New York, 2012; *1*, 312 pp.

41. "Kinetics, catalysis and mechanism of chemical reactions. From pure to applied science." In *Tomorrow and Perspectives*, Islamova, R. M.; Kolesov, S. V.; Gennady, E., Eds.; Nova Science Publishers: New York, 2012; *2*, 444 pp.
42. Kubica, C.; Zaikov, G. E.; Liu, L. In *Biochemical Physics and Biodeterioration. New Horizons*, Institute for Engineering of Polymer Materials and Dyes Publishing House: Torun, Poland, 2012; 292 pp.
43. Kubica, C.; Zaikov, G. E.; Pekhtasheva, E. L. In *Biodamages and their Sources for Some Materials*, Institute for Engineering of Polymer Materials and Dyes Publishing House: Torun, Poland, 2012; 248 pp.
44. Kubica, C.; Zaikov, G. E.; Stoyanov, O. V. In *Polymers, Composites and Nanocomposites*, Institute for Engineering of Polymer Materials and Dyes Publishing House: Torun, Poland, 2012; 398 pp.
45. Turovsky, À. À.; Bazylyak, L. I.; Kytsya, À. R.; Turovsky, N. À.; Zaikov, G. E. In *Non-Valency Interaction in Organic Peroxides Homolysis Reactions*, Nova Science Publishers: New York, 2012; 250 pp.
46. Pekhtasheva, E. L.; Neverov, A. N.; Zaikov, G.E. In *Biogamages and Biodegradation of Polymeric Materials*, iSmithers: London, 2012; 230 pp.
47. Mikitaev, A. K.; Kozlov, G. V.; Zaikov, G. E. In *Polymer as Natural Composites* Nova Science Publishers: New York, 2012; 198 pp.

CHAPTER 22

THE INFLUENCE OF UV AND VISIBLE LASER RADIATION ON LAYERED ORGANIC-INORGANIC NANOCOMPOSITES ZINC AND COPPER

V. T. KARPUKHIN, M. M. MALIKOV, T. I. BORODINA, G. E. VALYANO, O. A. GOLOLOBOVA, and D. A. STRIKANOV

CONTENTS

22.1 INTRODUCTION

Layered organic-inorganic hybrid nanocomposites are of significant interest to researchers in terms of their applications in science and technology. These include a broad class of chemical compounds such as, (1) – layered double hydroxides, (2) – hydroxyl double salts (HDS's) and (3) – hydroxides of metals, in which various organic anions were intercalated in the space between the layers.

The structural variability of these materials leads to the appearance of new chemical and physical properties – variable magnetism [1, 2], efficient catalysis, adsorption and high ion-exchange capacity [3]. The use of such nanocomposites promises to improve the mechanical and thermal stability of polymers in which composites have been dispersed [4], and also opens up the possibility for a creation of new optoelectronic devices (stochastic lasers, LEDs, sensors [5–7]), including a variety of diagnostic purposes [8]. Layered materials have potential application in pharmacology and medical science [9].

The chemical synthesis and analysis of properties of layered nanocomposites devoted tens publications. In the last decade for the synthesis of metal, oxide, hydroxide nanostructures used method of laser ablation of metals in a liquid medium [10–12]. However, the researches aimed at producing layered organic-inorganic composites by this flexible, simple method are not enough [13]. Practically important is also the question of the structural stability of these composites in the colloidal state under the influence, in particular, laser radiation optical range (colloidal stability of drugs, film, optical data carriers, etc.).

The problem of nanostructures modification theoretically and experimentally studied in many papers [14–16]. As objects of research were considered nanoparticles of finite size and usually spherical shape (nanoparticles of gold, platinum, copper oxide, zinc, etc.) and colloids with fractal structures. Analyzed the mechanisms of fragmentation by heating and melting of the particles, the formation of a plasma cloud, depending on the energy input. In [16], they were discussed in detail the effects photomodification and photoaggregation in fractal formations of silver colloids. However, the information for a complete solution of problem colloidal stability under the action of various types of radiation is not enough.

In this paper, the authors present the results of studying of the structural and morphological changes that occur after UV and visible laser irradiation of layered organic–inorganic composites zinc and copper synthesized by laser ablation in liquid. The synthesized materials are the second and third group composites. Their structural formulas are as follows: $(M)_2(OH)_3X \times z \times H_2O$ and $(M)(OH)_2X \times z \times H_2O$, where M – divalent metals (Zn, Cu) and X – intercalated anion – alkyl sulfate ($C_nH_{2n+1}SO_4^-$), where n = 12.

22.2 DESCRIPTION OF THE EXPERIMENT

The technique of laser ablation in liquids is described in detail in a number of original articles and reviews [10–12]. In this paper, the radiation in the visible range (λ = 510.6 and 578.2 nm) was generated by a copper vapor laser (CVL) and UV light (λ = 255, 271, and 278 nm) was obtained by nonlinear transformations radiation of 510–578 nm lines on the crystal BBO [17]. Average output power of CVL ≈ 10–12 W, the ratio of power line radiation was respectively 2:1, pulse width 20 ns, the pulse repetition frequency – 10 kHz, average power UV ≈ 0.6W. In the synthesis of nanocomposites, the laser beam was focused onto the target surface by an achromatic lens with a focal length f = 280 mm, which provided a spot size of less than 100 microns. The target was placed in a cell with deionized water or aqueous surfactant solutions. The volume of liquid in the cell was about 2 cm³. The cell was placed in the vessel with the cooling water, the temperature of which was maintained at ≈ 310 K. The vessel was mounted on a movable stage, which allowed continuous move the focal spot on the target surface. Surfactant used in the experiment – SDS ($C_{12}H_{25}SO_4Na$) is anionic surfactant. The treatment of the colloid containing nanocomposites by UV and visible light was carried out in the cell volume ≈ 2 cm³. Laser beam was focused by a lens with f = 280 mm at the center of the cell.

Optical characteristics of the obtained colloidal solution containing nanocomposites were analyzed by the method of optical absorption, with spectral range from 200 to 700 nm, on a spectrophotometer SF-46 with automatic data processing system. For registration Raman spectra was used system KSVU-23 with double monochromator. The structure and composition of the solid phase colloidal solution prepared by centrifugation at 4000-13000 rpm and dried at a temperature of 320–330 K were investigated by X-ray diffractometer DRON-2 (K_α line of copper). The shape and size of nanostructures were studied with an atomic force microscope (AFM) Solver P47-PRO (in semi-contact topography mode) and scanning electron microscopy (SEM, Hitachi S405A, 15 kV). The samples for AFM analysis were prepared by single or multiple coats of wet precipitate obtained after centrifugation on a glass slide with a follow-drying the precipitate at a temperature of 312–323 K under atmospheric conditions. The number of layers of precipitate was adjusted empirically to achieve a sufficient sharpness of X-ray diffraction patterns and AFM images.

22.3. THE EXPERIMENTAL RESULTS AND DISCUSSION

In the experiments, the authors, as noted above, used a powerful (≈10W) laser with a high pulse repetition frequency -10^4Hz. Compared with the majority of work using Nd: YAG lasers, it provides a more intense productivity of nanoparticles at a time and, in other words, the formation of dense colloid. Self-organization of a large number of nanoparticles in dense colloid leads to the formation of fractal aggregates (FA) and large fractal structures. The theory of optical properties of FA is different from the well-known Mie theory for the metal nanoparticles, since taking into account the electrodynamics interaction of neighbors of the particles that make up fractal aggregate [16]. Conclusions of this theory, which confirmed experimentally, are a broadening and shift a plasmon resonances of a fractal to the low-frequency range of spectra, i.e., the increase of the length of the long-wavelength part of the spectra in comparison with a spectra of system of individual, non-interacting particles. The frequency shift $\Delta\omega \sim \omega_p$, i.e., a mutual electrodynamics influence of neighboring particles is so high, that leads to a shift of their resonance by an amount comparable to the resonant frequency. The length of the long-wavelength range of spectra increases with the size of fractals. From the theory of optical properties of fractals, in particular, illustrated by the example of silver colloid particles, it should be the possible appearance in FA spectra of additional peaks and "dips" in the long-wavelength range of spectra, associated with changes in particle size distribution, including the presence of surfactants in the colloid. There is also photodynamic effect on the particles of fractals. In particular, this effect leads to the destruction of nanoparticles and appearance rather narrow dips – 'burn-through' – in the absorption spectra of the colloid, which are close to the resonance frequencies of the incident radiation. Also, this effect connected with photoaggregation of small particles [18].

Analysis of the absorption spectra $A(\lambda)$ of colloids zinc and copper (Fig. 1) obtained by irradiation of the target in aqueous solution SDS ($M = 0.01$) at different exposure time τ_e confirms the existence of a developed fractal structures of colloids. Increasing the exposure time leads to a significant increase in the absorption of colloid, especially in long-wave part of the spectra (Fig. 1, curve ....). In addition, in the range $\lambda \approx 600$–630 nm (for copper colloids) can see a rather narrow the peaks of 'transparency.' Impact on the original colloids UV and visible light in the range of power densities of $\approx 5.10^5$–10^7 W cm^2 almost does not change their absorption spectra.

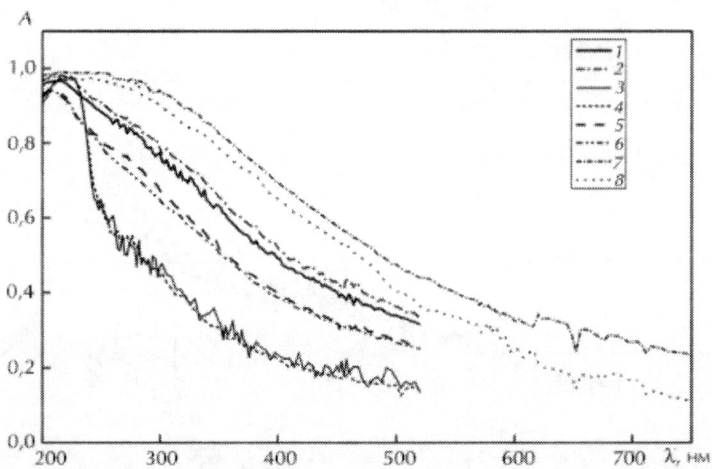

FIGURE 1 The absorption spectra of A (λ) colloidal solutions [Zn + 0.01M SDS], curves 1, 3, 5 – basic colloids, curves 2, 4, 6 – colloids after laser irradiation in the visible range; [Cu + 0.01M SDS]curve 7 – basic colloid, curve 8 – colloid after laser irradiation of UV range.

Consideration X-ray diffraction patterns, AFM and SEM images of the solid phase of colloids show the processes occurring in colloids under irradiation in more detail. Comparison of the diffraction patterns of the solid phases of colloids [$Cu_2(OH)_3$ + DS], obtained before UV irradiation (λ = 255 nm) and after (Fig. 2a), points to the destruction of these complexes and the possible emergence of other layered structures with interlayer distance d = 4.87 nm and 3.658 nm. Diffraction patterns of [$Zn(OH)_2$ + DS], obtained after UV irradiation (Fig. 2b) demonstrate the presence of the product of its decomposition – zinc carbonate $ZnCO_3$ size of about 17 nm. In addition, the composite was fragmented to X-ray amorphous state as evidenced by the presence of a strong halo of Bragg angles $2\theta = 8–15°$.

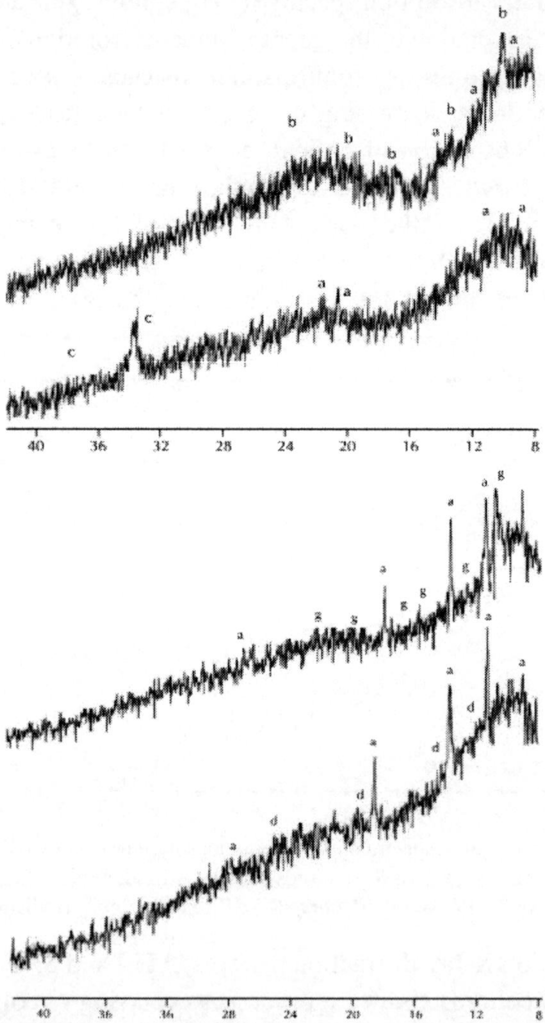

FIGURE 2 X-ray diffraction patterns of solid phase extracted from the colloidal solution of [Zn + 0.01 M SDS] (a) and [Cu + 0.01M SDS] (b), curve 1 – after laser irradiation of the UV range, curve 2 – the basic colloid. Phases: a – SDS, b – [Zn $(OH)_2$ + DS], c – $ZnCO_3$, g – layered structure with interlayer spacing d = 4.87 nm, d – layered structures with interlayer spacing d = 3.658 nm.

Figures 3 and 4 presents AFM and SEM images of the dispersed phase composites of zinc and copper, obtained before (Figs. 3a, and 4a) and after (Figs. 3b, and 4b) UV treatment ($\lambda = 255, 271$ nm). Before zinc and copper composites have the form colonies (layers) densely laid down against each other plates with linear dimensions from hundreds of nanometers to tens of microns or more, and their thickness in the transverse direction in the four – ten times less. The picture changes significantly after UV and visible ($\lambda = 510, 578$ nm) laser treatment. The fragmentation of composites structures is obvious. Length plates zinc composites rarely exceed ≈ 1 μm (150–800 nm) and a thickness of about 20–50 nm. Similar transformations are observed in composites of copper.

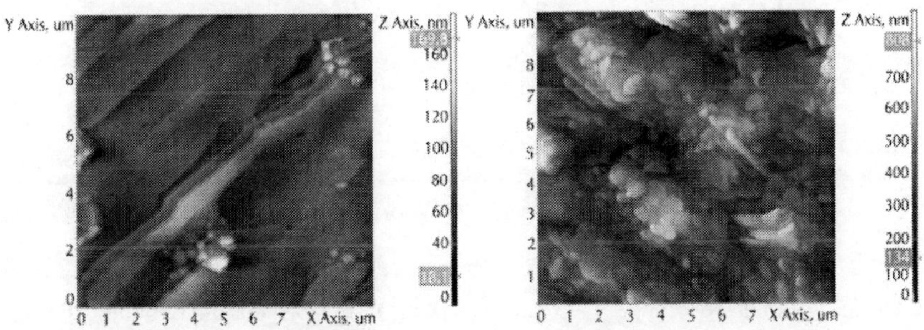

FIGURE 3 AFM images of the surface area of solid phase obtained after laser ablation of colloidal solution of [Zn + 0.01 M SDS]. Basic colloid (a) and colloid after laser irradiation UV range (b).

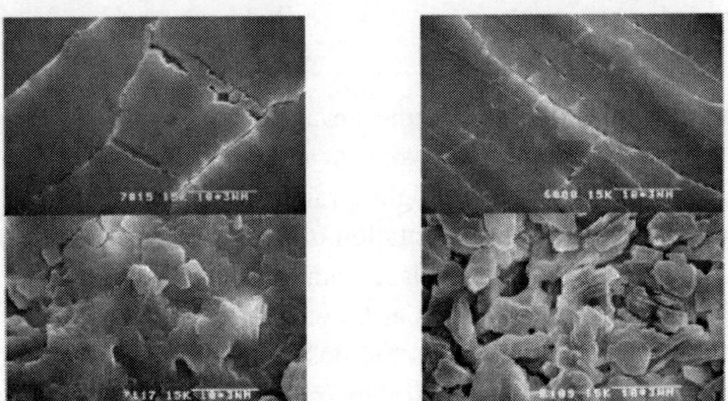

FIGURE 4 The microstructure of the surface area of colloids solid phase obtained after laser ablation of a target of zinc (a, b) and copper (c, d). Marking a line on the scale micrographs (B × C NM) stands for (B × 10^C nm). Basic colloid (a, c) and colloid after laser irradiation in the visible range (b, d).

Consideration of the Raman spectra of zinc composites after UV exposure confirms their decomposition, in particular, the appearance of zinc carbonate $ZnCO_3$ (Fig. 5).

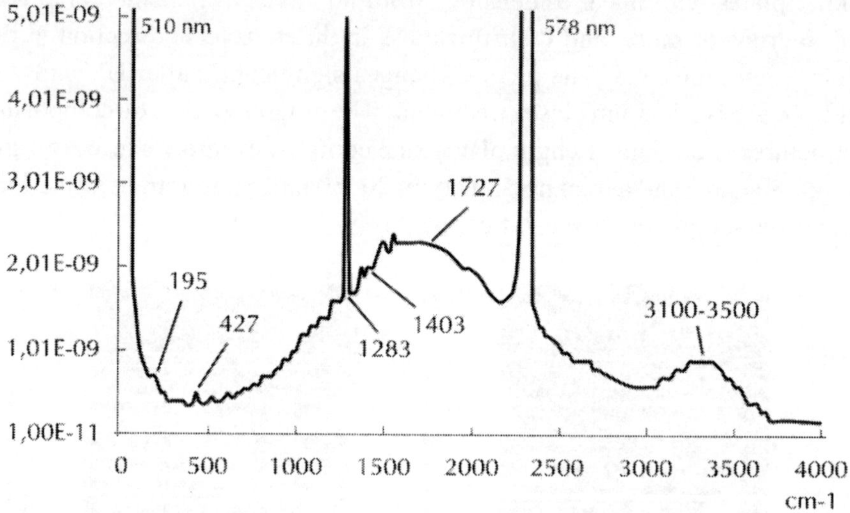

FIGURE 5 Raman spectra ($\lambda_{exc.}$ = 510nm) of [Zn $(OH)_2$ + DS] colloid obtained after laser irradiation (λ = 510, 578 nm, t_e = 1.5 h): 195/cm – $ZnCO_3$, 427/cm – $ZnCO_3$, 1283/cm – SDS, 1403/cm – $ZnCO_3$, 1727/cm – $ZnCO_3$, 3100–3500 – H_2O.

22.4 CONCLUSION

The experimental data indicate the instability of fractal structures nanocomposites at sufficiently intensive irradiation (≈ 5.10^5–10^7 Wcm^2) on them by laser UV and visible light (fragmentation, chemical changing). The main reason for the fragmentation of the colloids solid phase is heating nanoparticles of colloids by laser radiation. The estimation of the temperature of heating the particles at these power densities, made by analogy with [14, 15], demonstrates the possibility of nanoparticles melting and separation of small parts and thus fragmentation of fractal structures. Changes in the structure and composition of the composites, the appearance of new chemical compounds caused destruction a number of chemical bonds (C–H, C–O, C–C, and others, the binding energy ≈ 400 kDj/

mol) molecules of composites under the influence of sufficiently energetic UV rays and visible light.

KEYWORDS

- **Burn-through**
- **Dips**
- **Intensive irradiation**
- **Transparency**
- **UV irradiation**

REFERENCES

1. Fujita, W.; Awaga, K. "Reversible structural transformation and drastic magnetic change in a copper hydroxides intercalation compound", *J. Am. Chem. Soc.,* **1997**, *119 (19)*, 4563–4564.
2. Okazaki, M.; Toriyama, K.; Tomura, S.; Kodama, T.; Watanabe, E. "A monolayer complex of $Cu_2(OH)_3C_{12}H_{25}SO_4$ directly precipitated from an aqueous SDS solution", *Inorg. Chem.*, **2000**, *39(13)*, 2855–2860.
3. Newman, S. P.; Jones, W. "Synthesis, characterization and applications of layered double hydroxides containing organic guests", *New J. Chem.,* **1998**, *22(2)*, 105–115.
4. Kandare, E.; Chigwada, G.; Wang, D.; Wilkie, C. A.; Hossenlopp, J. M. "Nanostructured layered copper hydroxy dodecyl sulfate: A potential fire retardant for poly(vinyl ester) (PVE)", Polymer Degradation and Stability, **2006**, *91(8),* 1781–1790.
5. Usui, H.; Sasaki, T.; Koshizaki, N. "Ultraviolet emission from layered nanocomposites of $Zn(OH)_2$ and sodium dodecyl sulfate prepared by laser ablation in liquid", Applied Physics Letters, **2005**, *87(6)*, 63105–63108.
6. Karen, L.; Van der Molen, A. P.; Lagendijk, M. A. "Quantitative analysis of several random lasers", *Opt. Commun.*, **2007**, *278(1)*, 110–113.
7. Kumar, N.; Dorfman, A.; Hahm, J. "Ultrasensitive DNA Sequence Detection of Bacillus anthracis using nanoscale ZnO sensor arrays", *Nanotechnology*, **2006**, *17(12)*, 2875–2881.
8. Yu, V. A.; Yu, P. V.; Polyakov, A. F. "Effect of Particle Concentration on Fluctuation Velocity of the Disperse Phase for Turbulent Pipe Flow", *Int. J. Heat Fluid Flow*, **2000**, *21(5),* 562–567.
9. Nalawade, P.; Aware, B.; Kadam, V. J.; Hirlekar, R. S. *"Layered double hydroxides:* A review", *J. Sci. Ind. Res.,* **2009**, *68(4)*, 267–272.

10. Yang, G. W. "Laser ablation in liquids: applications in the synthesis of nanocrystals", *Prog. Mat. Sci.*, **2007**, *52(4)*, 648–698.
11. Karpuhin, V. T.; Malikov, M. M.; Val'yano, G. E.; Borodina, T. I.; Gololobova, O. A. "Investigation of the characteristics of a colloidal solution and its solid phase obtained through ablation of zinc in water by high power radiation from a copper vapor laser", *High Temp+*, **2011**, *49(5)*, 681–686.
12. Bozon-Verdura, F.; Brainer, R.; Voronov, V. V.; Kirichenko, N. A.; Simakin, A. V.; Shafeev, G. A. "Synthesis of nanoparticles by laser ablation of metals in liquids", *Kvantovaya Electron.* (Moscow), **2003**, *33(8)*, 714.
13. Liang, Ch.; Shimizu, Y.; Masuda, M.; Sasaki, T.; Koshizaki, N. "Preparation of layered zinc hydroxide/surfactant nanocomposite by pulsed-laser ablation in a liquid medium", *Chem. Mat.*, **2004**, *16(6)*, 963–965.
14. Werner, D.; Hashimoto Sh. "Improved Working Model for Interpreting the Excitation Wave length – and Fluence – Dependent Response in Pulsed Laser-Induced Size Reduction of Aqueous Gold Nanoparticles", *J. Phys. Chem. C.*, **2011**, *115,* 5063–5072.
15. Kawasaki, M. "Laser- induced Fragmentative Decomposition of Fine CuO Powder in Acetone as Highly Productive Pathway to Cu and Cu_2O Nanoparticles", *J. Phys. Chem. C.*, **2011**, *115,* 5165–5173.
16. Karpov, S. V.; Slabko, V. V. "Optical and photophysical properties of fractal-structured metal sols", *Novosibirsk: SB RAS*, **2003**, 265 p.
17. Batenin, V. M.; Karpukhin, V. T.; Malikov, M. M. "Efficient sum-frequency and second harmonic generation in a two-pass copper vapour laser amplifier", *Quantum Electron+*, 2005; *35(9)*; 844–848.
18. Karpov, S. V.; Popov, A. K.; Rautian, S. G.; Safonov, V. P.; Slabko, V. V.; Shalaev, V. M.; Shtokman, M. I. "Finding photomodification silver clusters, selective wavelength and polarization", *JETP Lett+*, **1988**, *48(10)*, 528–531.

CHAPTER 23

MODELING AND OPTIMIZATION OF THE DESIGN PARAMETERS SCRUBBER

R. R. USMANOVA and G. E. ZAIKOV

CONTENTS

23.1 INTRODUCTION

Vortex gas flow is a complex form of movement is entirely dependent on the design parameters are tightening devices. These devices determine thc aerodynamic characteristics and flow chambers: the degree of twist, hydraulic resistance, structure and uneven speed, features recirculation zones, injection capacity, turbulence intensity. Possibility of a tangential inlet gas into the unit and the formation of internal swirling flows are extremely diverse. However, despite the differences of known devices in design, size and purpose, formed in these gas streams have common patterns.

Currently used mathematical models of gas purification formed on simplified theoretical concepts of gas flow. They are not sufficiently taken into account the operational and design parameters of gas cleaning devices, as well as aero-hydrodynamic properties of gas-dispersed flows. These models cannot be used to search for the best options of integrated gas cleaning systems, as they show the properties of objects in a narrow range of parameters. We need more complete and appropriate mathematical models based on the study of the aerodynamics of gas and taking place in these events.

The use of computer technology and software to compute the hydrodynamic characteristics of eddy currents during the development and design of industrial devices. This avoids the need for costly field tests of gas purification apparatus.

Software suite is a modern ANSYS-14/CFX-modeling tool, based on the numerical solution of the equations of hydrodynamics [1, 2]. Hydrodynamic calculation makes it possible to determine the flow resistance of the device and to predict the efficiency of the separation process in the design stage.

23.2 STATEMENT OF THE PROBLEM

Mathematical modeling is always based on some physical hypothesis that simplifies considered real objects. At this stage of the development of mathematics is possible not only to describe the physical model in the form of equations and additional conditions, but also to solve. For the formulation of the problem of modeling and the subsequent investigation of the processes occurring in the vortex of centrifugal machines, you need to define the relationship between the parameters of the device is twisted and formed them flow.

In studies of vortex centrifugal devices primarily include spatial flow. Similarly it is conducted in models based on the hypothesis of a plane vortex. Motion of the gas is described by the Navier-Stokes equations. The equation is introduced with the closure of the fluctuating components of the hypothesis on the path of displacement. The values of the tangential and radial velocity are taken close to each other. Axial velocity is very small. Exclusion from consideration of the axial movement of the gas is greatly reduced, the (idealized). This is not consistent with the physical picture, in which a large place occupied by forward and backward axial currents. The results of these studies are of interest to determine the effects of the vortex flow asymmetry with respect to the axis of the camera.

Numerical analysis of the gas inside the dynamic scrubber reduces to solving the Navier-Stokes equations [3]. For the solution of equations of Navier-Stokes equations with a standard (k–ε)-turbulence model. To find the scalar parameters k and ε are two additional model equations containing empirical constants [4–6]. The computational grid was built in the grid generator ANSYS ICEM CFD. The grid consists of 1247 542 elements.

The grid was constructed on a uniform and non-uniform radius along the axial coordinate. Unevenness wondered law exponentially decreasing to the exhaust pipe and increasing the bottom unit. Denominators geometric progressions were chosen so that in the high-velocity gradient mesh was thick as possible. On solid walls slip conditions from which is vorticity. On the input and output sections of laws of changing the current function, on the axis of the unit gradient target variable–constant. Conical part determine the boundary conditions and restrictions on the size of the computational grid, i.e., the reduction of radius corresponds to a decrease of the boundary node.

General view of the computational domain is shown in Fig. 1.

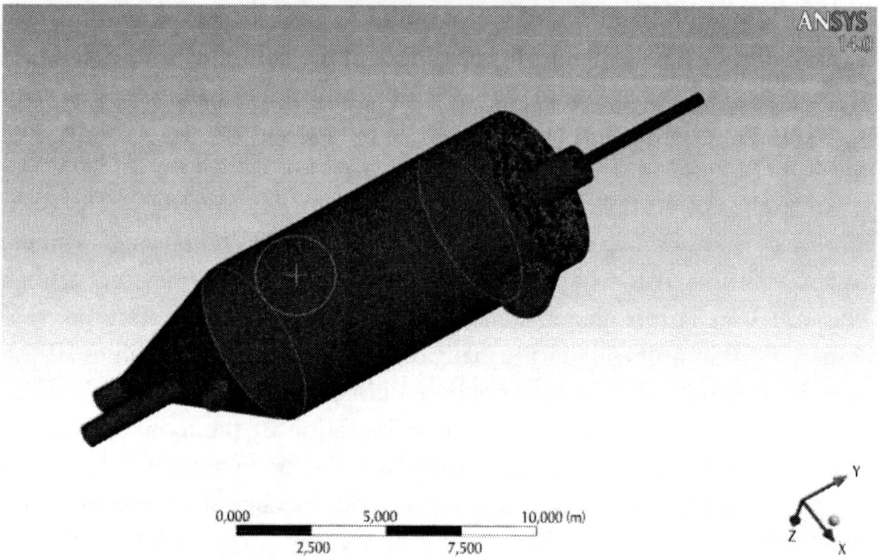

FIGURE 1 General view of the computational domain.

23.3 ANALYSIS OF THE RESULTS

During the flow of gas and dust in a dynamic scrubber is complex. This is explained by the fact that in the central part of the device is vane swirler. Analysis of fluid flow and distribution of deposited particles in a dynamic scrubber showed that the presence of turbulent diffusion particles concentrate at the vessel wall is not a thick layer, and in the form of gas and dust loosened concentrated layer (Fig. 2).

In this case, the dust is localized in the annular wall layer of a certain thickness in a spiral of dust accumulations in the form of bundles. Initiated the formation of helical bundles is dusty swirl vane. With the passage of dust in vane curved channel is the concentration of particles in the peripheral zone of the channel. Thus, after the passage of a uniform flow swirler vane is divided into a number of parallel streams with alternating then small, the large concentration of dust. The thickness and density of the boundary layer depends on the gas velocity, the angle of twist, and the character input stream to a dynamic scrubber. Higher speeds may reduce the thickness of the boundary layer, in spite of increasing the role of turbulent diffusion.

Reducing the length of the inlet pipe reduces the eccentricity of the axis of rotation of the gas flow from the geometrical axis of the apparatus. Even in this case, the center of the rotational flow does not coincide completely with the geometric axis of the machine. There is some minor eccentricity, the value of which does not exceed 6–8% of the radius of the device. The presence of such eccentricity swirling flow, there are also researchers [8].

Given that the eccentricity of the machine is small, and in its central part of the irrigation device, we consider for a gas flow as symmetric about the axis of the machine.

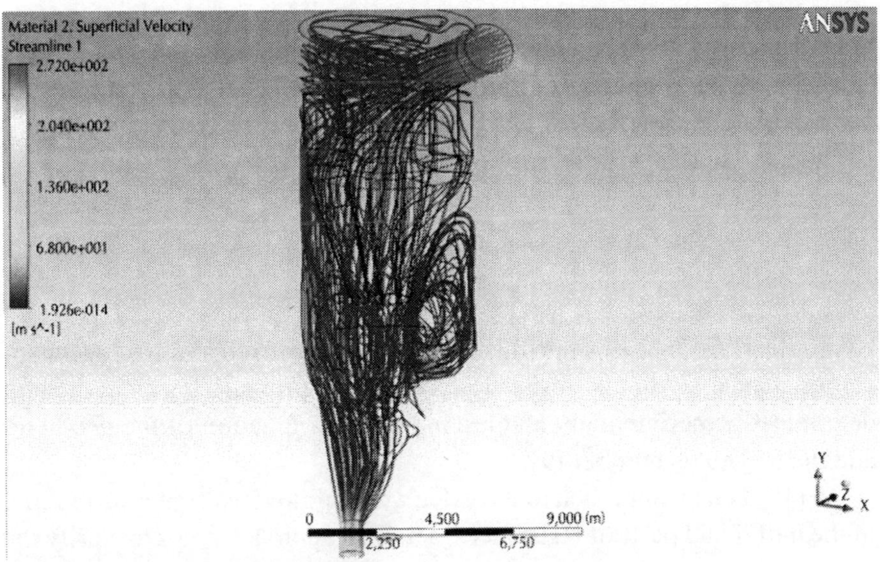

FIGURE 2 The flow pattern of gas and dust in the dynamic flow scrubber.

23.4 OUTPUT CALCULATION FORMULAS

Aerodynamics of the apparatus were conducted in the range of the Reynolds number changes from 3.5×10^4 to 15×10^4. This corresponded to an average of 5 to 25 m/s. The degree of spin flow was constant at $K = 1.5$.

A large study focused on determination of the resistance unit and study of the effect of dynamic geometry scrubber on the energy characteristics P and ξ [7, 8]. Resistance scrubber is calculated by the total pressure drop at the entrance to the unit and to exit. In our case, we write the Bernoulli equation for incompressible gas in the form of:

$$\rho\frac{\alpha_k W_1^2}{2} + P_1 + \rho_g g z_1 = \rho_g \frac{\alpha_k W_2^2}{2} + P_2 + \Delta P \qquad (1)$$

where z is the distance between the sections; P_1, P_2 is the Static pressure, Pa; W_1, W_2 is the average flow velocity in the annulus and in the exhaust pipe; α_k is the Coriolis coefficient taking into account the non-uniformity of the velocity distribution in the cross section. Ratio is the ratio of the true kinetic energy to kinetic energy of the flow, calculated at the average rate (for the turbulent regime of motion take $\alpha = 1.05–1.10$). Knowing ΔP, we can calculate the coefficient of hydraulic resistance, referred to the conditional mean in terms of the speed machine W_0.

$$\zeta_0 = \frac{2\Delta P}{\rho_g W_0^2} \qquad (2)$$

Hydraulic resistance of centrifugal machines is generally viewed as the local resistance. Hydraulic resistance coefficient, pressure losses in the unit determined experimentally and are mainly as a function of the geometry and the Reynolds number [9].

In [10, 11], an approach to the calculation of flow resistance as the sum of the individual parts of resistance tract. This approach helps to clarify the physical nature of the process, to evaluate different designs aerodynamic perfection swirlers.

Hydraulic resistance machines centrifugal type represented by the sum of the resistances of the cylindrical device, swirler and exhaust pipe. The resistance of the cylindrical part of the theoretically calculated for different distribution laws tangential velocity.

The processing of the experimental data suggested an empirical equation for calculating the coefficient of hydraulic resistance machines centrifugal type.

Found that the flow resistance of the "dry" machine obeys the square of the velocity of the gas. With the increase in the coefficient of spin ξ down. This is due to the decreasing level of the tangential component of the gas velocity in the swirler.

At some value of *K*, the coefficient of hydraulic resistance is almost independent of the flow scrubbing liquid. This is explained by the influence of two factors related to the supply of irrigating fluid dynamic scrubber.

1 side – increasing ξ due to the increase of pressure loss of the gas stream to transport liquids; with 2 sides – reducing ξ due to the decrease of the tangential velocity of the gas by the braking action of the liquid.

On this basis was constructed empirical mathematical model for calculating the coefficient of hydraulic resistance, including a formula to calculate ξ «dry» machine,

$$\xi_{\text{dry}} = \frac{1}{n}\left(\left(R_m\right)^{2n} - 1\right) + \frac{\alpha}{K^2}\cdot\left(\frac{\vartheta_{out}}{\vartheta_{in}}\right)^2 \tag{3}$$

empirical relationship for calculating the pressure drop in the gas transport liquid,

$$\xi_{tr} = 4\cdot\left(\frac{Q}{G}\right)^{0,4}\cdot\sqrt{1+\frac{1}{K^2}} \tag{4}$$

and ultimate dependence for calculating ξ irrigation apparatus,

$$\xi_{irr} = \frac{1}{n}\left(\left(R_m\right)^{2n} - 1\right) + \alpha\frac{\varepsilon^2}{K^2}\cdot\left(1+\frac{\rho_l}{\rho_g}\right)\cdot\left(\frac{\vartheta_{out}}{\vartheta_{in}}\right)^2 + 4\left(\frac{Q}{G}\right)^{0,4}\cdot\sqrt{1+\frac{1}{K^2}} \tag{5}$$

where ξ is the coefficient of hydraulic resistance; R_m is the radius of the cylindrical chamber, m; $\rho_{n,}$ ρ_l is the density of gas and liquid, kg/m³; υ_{in} υ_{ou} is the velocity of gas at the inlet and outlet of the unit, m/s; *n*, ε is the indicators vortex movement, *K* is the factor twist swirler; *Q*, *G* is the liquid and gas flow kg/m³; *α*-twist angle of the flow, °.

The resulting formula takes into account the presence of the dispersed phase and the partial loss of swirling flow.

23.5 PROCESSING OF RESULTS

Figure 3 shows that the swirl angle of the blades with $\alpha = 45°$ has the lowest power characteristics. However, the efficiency in the gas cleaning unit with swirler is reduced by 6–8% in comparison with the unit, where the blades are tilted at an angle $\alpha = 30°$. This can be explained by the decreasing flow swirling, which is characterized by a relative twist angle (90°–45°), and the lowest centrifugal forces.

Therefore, from the point of view of increasing the efficiency of gas treatment, preference should be given with the greatest Swirl swirled flow-30°.

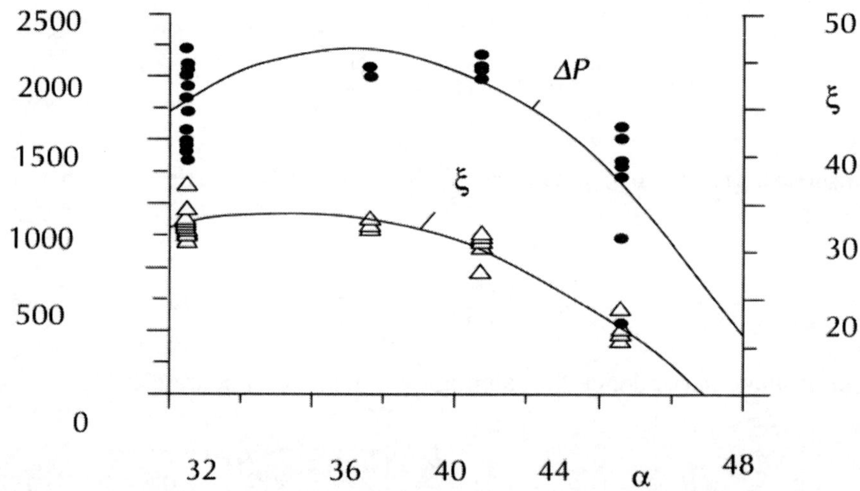

FIGURE 3 The impact of the angle α on the energy characteristics of dynamic scrubber.

Study of hydrodynamic characteristics scrubber showed that the coefficient of hydraulic resistance depends strongly on the angle of the blade swirler α. It also depends on the movement of gas-dispersed medium defined by the Reynolds number $Re = \rho D\upsilon/\mu$. As seen in Fig. 4, with increasing Reynolds number of 8×10^4 sets scaling of ξ. The exception is with the swirl angle of blades 35.5°, installation of which continues to increase hydraulic resistance.

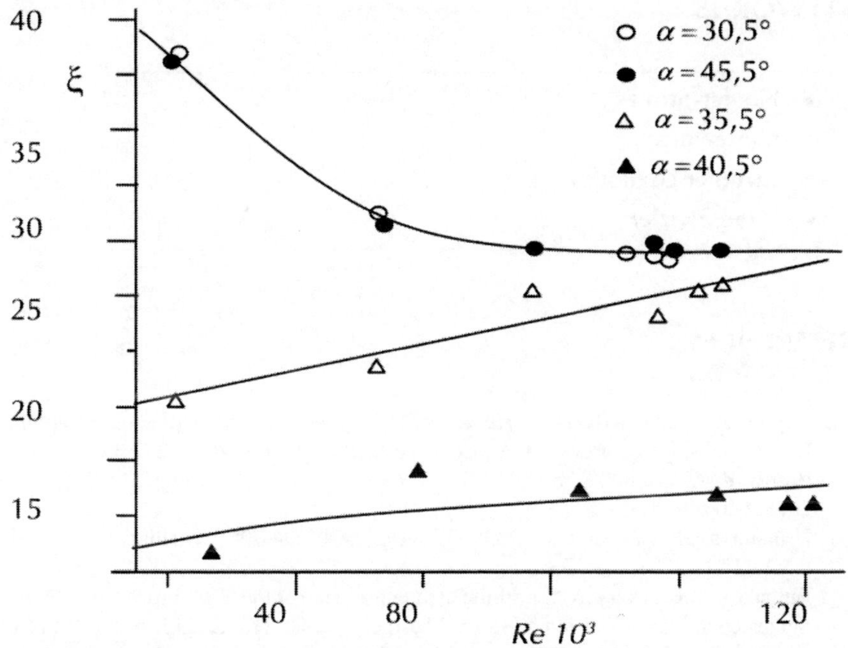

FIGURE 4 The dependence of the hydraulic resistance of the Reynolds number and the angle of the blades of the swirler.

23.6 CONCLUSION

1. The program ANSYS-14/CFX mathematical model of motion of polydispersed gas system. The character of the movement of dust particles under the influence of centrifugal force. This allowed choosing the desired hydrodynamic conditions and taking into account of the design under various conditions of scrubber.
2. Experimental study of the hydrodynamic characteristics of the device to determine the empirical constants and test the adequacy of the hydrodynamic model.
3. The empirical relationships for determining the coefficient of hydraulic resistance scrubber. They take into account the diameter of the separation chamber, the gas density and the scrubbing liquid, the angle of the blades of the swirler. The dependences obtained are suitable for the calculation of centrifugal machines of any type.

KEYWORDS

- **Navier-Stokes**
- **plane vortex**
- **turbulence model**
- **Vane Swirler**

REFERENCES

1. Kaplun, A. B. In *ANSYS in the Hands of the Engineer—A Practical Guide,* Kaplun, A. B.; Morozov, E.; Olfereva, M. A., Eds.; VNIIMP: Moscow, 2003; 272 p.
2. Basow, K. A. In *ANSYS and LMS Virtual Lab. Geometric Modeling.* M.: DMK Press, 2005; 640 p.
3. Usmanova, R. R. Patent 2339435 RF, B01 D47/06 dynamically scrubber. Opubl.27.11.2008. *Bull, 33.*
4. Bulgakov, V. K.; Potapov, I. I. Finite element scheme of the High-order for the Navier-Stokes equations. Modified by the SUPG-method, Sat. proceedings of the 16th International Conference "Mathematical Methods in Engineering and Technology". T. 1., St. Petersburg. Publishing House of the Saint-Petersburg State Tech. University Press, 2003; pp. 129–132.
5. Goncharov, A. L.; Fryazinov, I. V. On the construction of monotone of difference schemes for the Navier-Stokes equations on devyatito of point patterns, *Inst. Appl. Math. Keldysh RAN.M.*, **1986**, *93*, 14–16.
6. Goncharov, A. L.; Fryazines, I. V. Difference schemes on a nine-point template "cross" for the Navier-Stokes equations in velocity-pressure, Preprint. *Inst. Appl. Math. MV Keldysh Sci.* Moscow, **1986**, *53*, 17.
7. Usmanova, R. R.; Panov, A. K.; Minsker, K. S. Increased efficiency in flue gas kilns, *Chem. Ind. Today*, **2003**. *9*, 43–46.
8. Sazhin, B. S.; Akulich, A. V. Mathematical modeling of the gas in the separation zone of the once-through vortex apparatus based on the model of turbulence. *Theor. Found. Chem. Eng.*, **2001**, *35(5)*, 472–478.
9. Idelchik, I. E. Hydraulic resistance of a (physical-mechanical basis) M., **1954**. 316.
10. Tarasova, L.; Terekhov, M. A.; Troshkin, O. A. Hydraulic calculation of the resistance of the vortex unit, *Chem. Petrol. Mech. Eng.*, **2004**, *2*, 11–12.
11. Idelchik, I. E. Hydraulic resistance cyclones, its definition, and how to reduce the value of , *Mechanical Cleaning of Industrial Gases NIIOGAZ.* Moscow, 1974; pp. 135–159.

CHAPTER 24

EVALUATION OF BIOCOMPATIBILITY AND ANTIBACTERIAL PROPERTIES OF POLYSULPHONE/NANOSILVER COMPOSITES

M. ZIĄBKA, A. MERTAS, W. KRÓL, J. CHŁOPEK, and R. ORLICKI

CONTENTS

24.1 INTRODUCTION

This chapter concerns the evaluation of the biocompatibility and the bactericidal activity of composites made of polysulfone (PSU) modified by silver nanoparticles (PSU/Ag), obtained by extrusion and injection moulding. The antibacterial activity against the Gram-positive *Staphylococcus aureus* and the Gram-negative *Escherichia coli* was evaluated by means of the surface deposition method. With the use of the MTT and LDH tests, the degree of the composites' cytotoxicity was specified *in vitro* on human cells: osteoblasts (HTB-85 cell line, ATCC) and fibroblasts (CRL-7422 cell line, ATCC). By inductively coupled plasma mass spectrometry (ICP-MS), it was possible to evaluate the amount of the released silver ions and to determine their impact on the environment surrounding the bacteria. The materials' microstructure and the dispersion of the modifier phase were estimated with the use of electron scanning microscopy, with elemental analysis in micro-areas (SEM+EDS).

Tests of antibacterial efficacy proved that nanosilver-modified composites have a bactericidal effect against the tested bacteria. The material tests proved no cytotoxicity against osteoblasts and negligible cytotoxicity for fibroblasts. The spectrometric analysis showed that the amount of the released silver ions depend on the amount of the nanoparticles present in the polymer matrix.

The popularity of silver nanoparticles (nAg) has its effect on the continuous development of methods for their obtaining and application. Similarly to all nanometric size materials, the characteristic properties of nAg are their very small size, a big surface area and unique physicochemical properties, which make silver the antibacterial. Among various nanomaterials like copper, zinc, titanium, magnesium and gold, nAg demonstrates the highest bactericidal efficacy against bacteria, viruses and other eukaryotic microorganisms [1].

The mechanism responsible for the bactericidal activity of nAg has not been fully explained yet. However, their bactericidal efficacy is observed to be originated in the absorption of free silver ions and then with the interruption of the ATP (adenosine triphosphate), as well as the DNA replication, the creation of reactive oxygen species (ROS), and also a direct cell membrane damage [2].

A wide spectrum of activity of nAg promotes their frequent application in various aspects of antimicrobial therapy. The innovative applications of nano-silver currently include not only reduction of hospital infections, but also improvement of patient's life comfort and disposition, resulting from the reduction of surgical infections. In the future, one can thus expect nanosilver to be applied in various fields of implantational medicine as a bactericidal agent. In stomatology, the unique properties of nanosilver are used to neutralize the bacterial or fungal colonies that populate the prosthetic inlays and to eliminate the infections of mucous membrane. Nanosilver is also applied in the hygiene of the oral cavity, in toothbrushes, as an anti-caries agent, in pastes, mouthwashes, dental prosthesis disinfectant liquids, inlays and dental dressings [3]. In the recent years, a big interest was aroused by the antibacterial coatings used on implants in order to eliminate the formation of bacterial biofilms, as well as by the bioactive bactericidal coatings which simultaneously stimulate the implant's ostheointegration with the bone and improve the bactericidal activity [4, 5].

Beside the bactericidal function of nanosilver discussed in the literature, there are many factors that can cause the undesirable cytotoxicity effect of nanomaterials. The reason for silver's cytotoxicity should also be found in the formation of ROS, which damage the DNA and RNA of the cells by way of apoptosis. Due to the strong reduction-oxidation function (redox) of precious metals, nAg increase the ROS concentration and induce oxidation stress in the cells, initiating the peroxidation process, which causes a deformation of the cell membrane's structure, the latter's depolarization and the inhibition of the membrane enzymes' activity. This leads to the loss in the integrity of the cell membrane and a disorder of the oxidative phosphorylation in mitochondria [6], which in consequence, contributes to cell degradation and leads to numerous diseases and accelerated ageing of the body [7].

Taking into consideration the differences in the effect of nanomaterials on living cells, before their clinical application, an assessment of their biocompatibility should be performed. This chapter presented the obtaining and testing of bactericidal composite materials confronted with Gram-positive and Gram-negative bacteria and then, the assessment of their biocompatibility under *in vitro* conditions.

24.2 PREPARATION OF SAMPLES

The polymer PSU and composite PSU/Ag materials for the biological tests were obtained in the process of extrusion and injection moulding with the use of a vertical screw injection moulder with three heating zones, produced by Multiplas. The polymer matrix was a commercial polysulfone by Sigma Aldrich, with the molecular mass of 22,000 g/mol and the melting point of 343°C. As the bactericidal addition, silver nano-powder by Nano-Amor was used, with the purity of 99.9%, the mean particle size of 80 nm and the density of 10.49 g/cm^3. Nanopowder of the following weight fractions was introduced into the dried polymer granulates: 1%, 1.5% and 2%, thus obtaining composite mixtures. Next, with the use of the screw injection moulder, the homogenization process of the mixtures was performed. The injection parameters of particular compositions were selected and adapted to the parameters recommended by the polysulfone's producer. The process of melting and injection was conducted at the temperatures within the range of 343–350°C, the stream flow of 40–50% and the pressure of 80 kg/cm^2. The samples assigned for the tests were obtained in the form of squares sized 10 mm × 10 mm and disks, 11 mm in diameter, and next, they were sterilized with the use of low temperature plasma (apparatus Sterrad 120) and with the application of hydrogen peroxide vapor in a double cycle (2 × 45 min).

24.3 *IN VITRO* TESTS

The antibacterial activity of PSU and the PSU/Ag composites was examined with the use of Gram-positive *Staphyloccocus aureus* (ATCC 25923) and Gram-negative *E. coli* (ATCC 25922) bacteria standard strains. PSU and the PSU/Ag materials were introduced into a bacterial suspension in tryptonic water, with the colony-forming bacteria density of 1.5×10^5 CFU/ml (CFU–colony forming units) of *S. aureus* or *Escherichia coli*, respectively. Then the suspensions were incubated under static conditions at 37°C for 17 h. The respective bacteria cells of *S. aureus* or *E. coli* in tryptonic water were used as the positive controls (blank). PSU in tryptonic water, PSU/Ag in tryptonic water and pure tryptonic water were used as the negative controls. After incubation, the volume of 5 μl of each specimen was put on 5% sheep blood agar plates. The plates were incubated at 37°C for 24 h. Two specimens were tested for each group of the studied materials. The antibacterial efficiency of the tested materials was determined according to Xiaoyi Xu [8].

The cytotoxicity tests were performed with the use of the direct contact method on the extracts obtained by an 8-d incubation of the polymer PSU and PSU/Ag composite samples, placed at the bottom of the well of a 24-well culture plate, in 2 ml of culture medium. The incubation was conducted at 37°C in air atmosphere with a 5% content of CO_2 and 100% of relative air humidity. Next, 0.2 ml of the obtained extract and its fourfold dilution in a proper culture medium was dosed for the cultures of human osteoblasts (HTB-85 cell line, ATCC, USA) and human fibroblasts (CRL-7422 cell line, ATCC, USA) adhered to the bottom of the well of a 96-well culture plate. In the case of the control test, the extract of the examined samples was replaced by the proper volume of culture medium. The plates were incubated at 37°C in air atmosphere with a 5% content of CO_2 and 100% of relative air humidity. The incubation time for two parallelly conducted experiments was 24 h and 48 h. The cytotoxicity of PSU and the PSU/Ag composites was measured by 3-[4,5-dimethylthiazol-2-yl]-2,5 diphenyltetrazolium (MTT) assay [9, 10] and by lactate dehydrogenase (LDH) activity measured with the use of a commercial cytotoxicity assay kit (Roche Diagnostics GmbH, Mannheim, Germany), [11]. Each of the indications was repeated three times.

In the MTT assay, after a 24 or 48 h incubation of the cultured cells, the culture medium was removed and 20 μl of the MTT solution prepared at 5 mg/ml (Sigma Chemical Company, MO, USA) was added to each well for 4 h. The resulting crystals were dissolved in dimethyl sulfoxide (DMSO). The controls included native cells and the medium itself. The spectrophotometric absorbance of each well was measured with the use of a microplate reader (ELx 800, Bio-Tek Instruments, Winooski, VT, USA) at 550 nm. The cytotoxicity percentage was calculated by the formula: percent cytotoxicity (cell death) = (1 − [absorbance of experimental wells/absorbance of control wells]) × 100%.

Lactate dehydrogenase (LDH) is a stable cytosolic enzyme that is released from the necrotic cells. The LDH activity in the culture supernatants was measured with a coupled enzymatic assay, resulting in the conversion of the tetrazolium salt into a red formazan product. The supernatants were removed, and the activity of LDH released from the cells was measured in this sample of the culture medium. The maximal release was obtained after treating the control cells with 20% Triton X-100 (Sigma Chemical Company, St. Louis, MO, USA) for 10 min at room temperature. The cytotoxicity percentage was expressed with the use of the formula: (sample value/maximal release) × 100%.

24.4 INDUCTIVELY COUPLED PLASMA MASS SPECTROMETRY (ICP-MS)

The amount of the silver ions Ag^+ was determined by the technique of mass spectrometry with ionization, in inductively coupled plasma (ICP-MS). To that end, the samples with the contact surface of 1 cm^2 were immerged in water of ultra-high purity (UHQ H_2O) at 37°C. The analysis of the ion amount was performed after 1, 4, 6, 12 and 24 wk of the samples' incubation under static conditions. The test included three representative samples from each series.

24.5 SCANNING ELECTRON MICROSCOPY – MEASUREMENTS

The morphology of the surface of the polysulfone PSU and the composites PSU/Ag was examined with the use of scanning electron microscope (SEM) Jeol JSM 5400, with a unit for the chemical analysis in micro-areas EDS (Link ISIS series 300) in the secondary electron detection system (SE).

24.6 RESULTS AND DISCUSION

The tests assessing the antibacterial efficacy showed the antibacterial properties function of nAg against the model Gram-positive and Gram-negative bacterial strains. After 24 h of incubation, a significant decrease of the amount of *S. aureus* bacteria was observed, and also a slightly lower decrease of the amount of the *E. coli* bacteria (Table 1). This proves a higher resistance of the latter to the operation of nanosilver. The bactericidal performance is more effective in the case of the Gram-positive bacteria, such as *S. aureus* (that is cocci) than in the case of the Gram-negative *E. coli*, which is explained by the differences in the structure of the cell walls of those bacteria. Also noted was the lack of bacterial growth for the negative controls and a confluent growth of the bacterial colony for the positive controls (blank). Together with the increasing nanosilver addition, the antibacterial efficacy of the composites rises.

TABLE 1 Antibacterial efficacy of polysulfone (PSU) and its composites (PSU/Ag) against *S. aureus* and *E. coli* standard strains.

Material	***S. aureus*** **ATCC 25923**		***E. coli*** **ATCC 25922**	
	(the initial density of bacterial suspensions was 1.5 × 10⁵ CFU/ml)			
	CFU/ml after 17-h incubation with material	**Antibacterial efficacy ABE (%)**	**CFU/ml after 17-h incubation with material**	**Antibacterial efficacy ABE (%)**
PSU	5.95×10^7	0.0	1.67×10^8	0.0
PSU/1%wt.Ag	3.85×10^7	34.8	1.45×10^8	9.4
PSU/1.5%wt.Ag	2.1×10^7	64.4	1.35×10^8	15.6
PSU/2%wt.Ag	2.55×10^4	99.9	1.1×10^8	31.2
BLANK	5.9×10^7	–	1.6×10^8	–

The effect of the nAg contained in the tested composites on the presence of the *S. aureus* bacterial strain is graphically presented in Fig. 1. One can observe a decreasing number of bacterial colonies together with the increasing content of nAg in the tested composites.

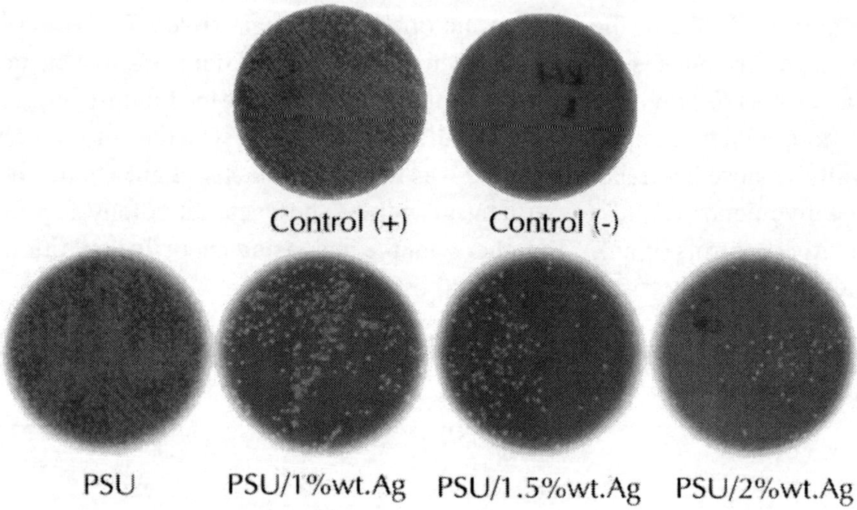

FIGURE 1 The 5% sheep blood agar plates with colonies of *S. aureus* ATCC 25923 standard strain from bacterial suspensions after incubation with tested materials (PSU and PSU/Ag composites) and controls.

The results of the MTT determination of the cytotoxicity of PSU and its composites with nanosilver, for human osteoblasts and fibroblasts, are shown in Table 2. The introduction of nanosilver into the polysulfone's matrix has a positive effect on the behaviour of the composites in the biological environment. Both after 24 and 48-h incubation of osteoblasts with an extract of the tested materials, the observed cytotoxic effect (CT) did not exceed 5%, which points to the fact that none of the tested materials is cytotoxic for human osteoblasts. Weakening cytotoxic effect of the tested composites together with the increasing content of nanosilver was also observed and the results suggests that the proposed amounts of the added nAg are not cytotoxic. In case of the tests performed on diluted extracts, the cytotoxicity was insignificantly low, both after 24 and 48-h incubation of the tested materials with osteoblasts.

TABLE 2 Cytototoxic effect (CT,%) of tested materials (extracts of PSU and PSU/Ag composites) in contact with human osteoblasts and human fibroblasts measured by MTT assay.

Cytotoxicity CT (%) – for human osteoblasts (HTB-85 cell line, ATCC)				
Material	**Undiluted extracts**		**Fourfold diluted extracts**	
	after 24h	**after 48h**	**after 24h**	**after 48h**
PSU	3.0±2.01	2.3±2.11	0.07±0.12	2.2±3.87
PSU/1%wt.Ag	6.5±2.93	4.2±1.17	0±0.0	0.2±0.25
PSU/1.5%wt.Ag	7.4±0.67	3.2±2.95	0±0.0	0.6±0.98
PSU/2%wt.Ag	3.3±3.57	0.2±0.40	0±0.0	0±0.0
Cytotoxicity CT (%) - for human fibroblasts (CRL-7422 cell line, ATCC)				
	Undiluted extracts		**Fourfold diluted extracts**	
	after 24h	**after 48h**	**after 24h**	**after 48h**
PSU	51.3±23.8	19.0±7.65	21.6±5.01	7.3±6.86
PSU/1%wt.Ag	33.7±13.0	18.6±7.3	27.7±19.8	8.2±10.3
PSU/1.5%wt.Ag	20.3±8.78	31.1±20.0	27.6±5.86	12.2±4.3
PSU/2%wt.Ag	21.3±4.50	7.6±4.64	16.7±6.42	2.5±4.33

In comparison with osteoblasts, tests conducted with the use of human fibroblasts demonstrated higher cytotoxicity of the examined PSU and its composites with nanosilver. After a 24-h incubation of tested materials with fibroblasts, the percentage of the live cells was lower than in the case of the tests performed after 48 h of incubation. Such behavior suggests that, despite the toxic effect, the process of cell multiplication continues. In the case of diluted extracts, a significant decrease of cytotoxicity was observed. The cytotoxic effect of the materials on fibroblasts and the related lower vitality of these cells can be a beneficial phenomenon, if the implant material is intended for the regeneration or replacement of the bone tissue, as one should aim at the activation of the osteoblasts (osteogenic cells) and a simultaneous inhibition of the fibroblasts' activity.

Fibroblasts are the first cells to anchor on the implant's surface during the healing process. Their excessive adhesion and proliferation can lead to encapsulation, as a result of the overgrowth of the undesirable fibrous tissue, which, at the same time, can diminish the effectiveness of the implant's integration with the bone. The observed higher cytotoxicity of the tested PSU and its composites (PSU/Ag) for fibroblasts is probably connected with the porosity of the materials' surface. Introducing silver nanopowder with the particle size below 100 nm makes the surface porous and resembling the natural porosity of the bone [12]. The porosity of the implant's surface in the nanometric scale is thus highly desired in order to improve the adhesion and proliferation of osteoblasts [13], whereas, at the same time, it can cause a weakening of the same functions in the case of fibroblasts.

The results of the determination of cytotoxicity of PSU and its composites (PSU/Ag) for human osteoblasts and human fibroblasts by the LDH test are compiled in Table 3. For the composites with the polysulfone matrix with the highest content of nano-silver, the cytotoxicity of non-diluted extracts did not exceed 3% after 24 h and 7% after 48 h of their incubation with osteoblasts. In the case of diluted extracts, the values did not exceed 5% and 8%, after 24 and 48 h, respectively. Pure PSU demonstrated no cytotoxic effect, which is verified by the literature data on its biocompatibility. In the case of fibroblasts, the LDH test, similarly to the MTT test, showed a higher cytotoxicity of the examined materials.

TABLE 3 The lactate dehydrogenase (LDH) activity in supernatants of human osteoblasts and human fibroblasts cultures, expressed as cytotoxic effect (CT,%) after incubation with extracts of tested materials (PSU and PSU/Ag composites).

Cytotoxicity CT (%) – for human osteoblasts (HTB-85 cell line, ATCC)				
	Undiluted extracts		**Fourfold diluted extracts**	
	after 24h	**after 48h**	**after 24h**	**after 48h**
PSU	0.0±0.0	0.0±0.0	0.0±0.0	0.0±0.0
PSU/1%wt.Ag	0.3±0.58	0.4±0.63	1.2±2.02	0.0±0.0
PSU/1.5%wt.Ag	0.6±1.10	0.0±0.06	1.0±1.20	1.0±1.25
PSU/2%wt.Ag	1.6±1.41	2.9±3.38	2.3±3.10	5.2±2.75

TABLE 3 *(Continued)*

Cytotoxicity CT (%) - for human fibroblasts (CRL-7422 cell line, ATCC)				
	Undiluted extracts		**Fourfold diluted extracts**	
	after 24h	**after 48h**	**after 24h**	**after 48h**
PSU	1.2±0.87	4.1±2.08	0.0±0.0	0.0±0.0
PSU/1%wt.Ag	2.8±3.12	4.6±2.10	0.5±0.55	0.7±1.15
PSU/1.5%wt.Ag	1.5±1.27	5.2±1.47	1.2±1.15	2.0±2.36
PSU/2%wt.Ag	4.8±3.48	6.0±2.08	0.6±1.03	4.9±1.55

The cytotoxicity of nanoparticles depends on the applied cell line. The same type of nanomaterial can have a different cytotoxic effect for different cell lines. The size of the nanoparticles influences not only the cytotoxicity of the material, but also the possibility of the particles' penetration inside the cell. The smaller the nanoparticle can easier and faster to penetrate into the cell (nucleus) and determine the lower cytotoxicity. Both the antibacterial and the cytotoxic properties of nanosilver depend on the amount, size and shape of the applied nanoparticles and the amount of the released silver ions.

The results of the ICP-MS analysis made it possible to determine the total amount of silver which penetrated from the composite materials into the distilled water after a 1-, 4-, 6-, 12- and 24-wk incubation (Fig. 2). For the reference samples in the form of pure polysulfone, no release of silver ions was recorded and thus the presented diagram does not include reference.

On the basis of obtained results, it was observed that the content of the Ag^+ silver ions increase together with the increase of the weight fraction of the modifier phase in the composite. The content of determined silver ions rise with the time of the materials immersion in the water environment and also with the increasing content of the nano-addition present in the polymer matrix. The highest amount of silver ions was determined after 6 mo of incubation for the composites containing 2% weight fraction of nano-powder. For the PSU/2%wt.Ag material, the Ag^+ ion content after 24 incubation weeks increased nearly two and a half times (2.75 ×) as referred to the same samples determined after a week. One can thus state that the amount of the released silver ions correlates with the bactericidal efficiency. The higher content of the determined ions, higher the probability of bacteria elimination. The amount of the released ions is also affected by the homogeneous distribution of the nanoparticles in the composite.

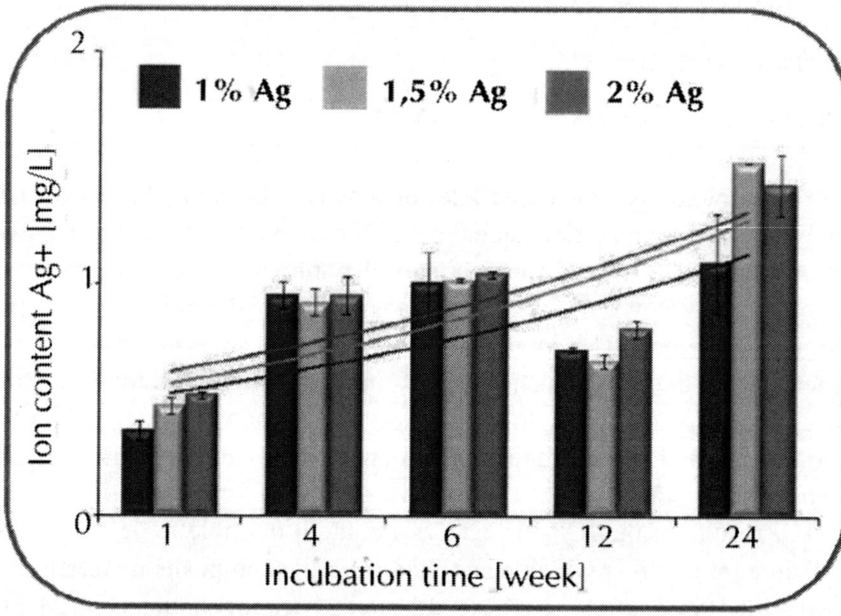

FIGURE 2 The content of Ag^+ ions in the aqueous extracts during the 24 wk of incubation of the polysulfone's composites (PSU/Ag).

Performed microscopic tests aimed at the illustration of the distribution of silver and its homogenization in the polymer matrix, or a possible effect of the production technique on the composites' homogeneity. Figure 3 presents images of the PSU/2%wt.Ag composite together with the distribution maps of the basic elements and the existing silver, both at the surface (Fig. 3A) and inside the composite (Fig. 3B).

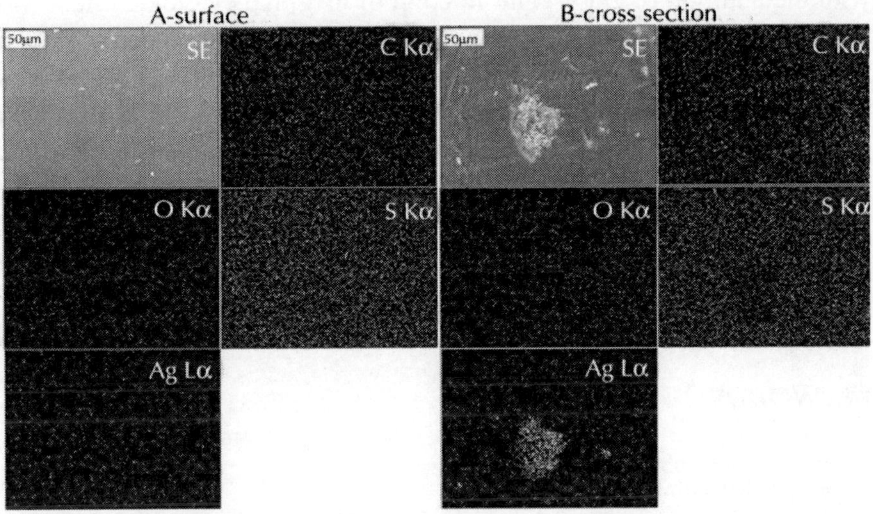

FIGURE 3 SEM image and element distribution maps for the PSU/2% wt.Ag, mag.500×, A) surface, B) cross section.

The composite's surface is smooth, and the silver distribution is uniform. So far, the results of performed analysis for the fractures demonstrated that, despite the homogeneous surface inside the composites, there are areas including agglomerates of silver nanopowder. A small surface development (5 m^2/g) and the size of the nAg below 80 nm favor the formation of inhomogeneous areas. In the composites containing higher amount of the modifier, the probability of the existence of agglomerates is much higher, which can influence not only the biological but also the mechanical properties of the composites.

24.7 CONCLUSION

Within the frames of this chapter, polymer materials and polymer-metallic composites (PSU/Ag) were obtained and examined. On the basis of the performed tests, it was stated that the polymer composites with a polysulfone matrix modified with nano-silver have a bactericidal function in contact with Gram-positive (*S. aureus*) and Gram-negative (*E. coli*) bacteria. The highest antibacterial efficacy was recorded for the composites with the addition of 2% weight fraction of nAg. No cytotoxic effect was observed for the *in vitro* tested polymer composites in contact with osteoblasts and a slight cytotoxicity was stated for fibroblasts. Performed ICP-MS analysis made it possible to conclude that nanosilver (as silver ions) is released into the water environment, and its amount results from the technique of the material (polymer) modification and the amount of the applied modifier. The performed microstructural observations demonstrated that the technique and the amount of the introduced nanosilver significantly affect the composite's homogeneity.

KEYWORDS

- **Bactericidal**
- ***Escherichia coli***
- **Polysulfone matrix**
- ***Staphyloccocus aureus***

REFERENCES

1. Gong, P.; Li H.; He, X.; Wang, K.; Hu, J.; Tan, W.; Zhang, S.; Yang, X. Preparation and antibacterial activity of Fe_3O_4 and Ag nanoparticles. *Nanotechnology,* **2007**, *18*, 604–611.
2. Marambio-Jones, C.; Hoek, E. M. V. A review of the antibacterial effects of silver nanomaterials and potential implications for human health and the environment. *J. Nanopart Res.,* **2010**, *12*, 1531–1551.
3. Borczyk, R.; Jaremczuk, B.; Puchała, P. Ocena preparatu NanoCare Silver Plus – nanoera w stomatologii (cz. I), In *Kwartalny Niezbędnik Edukacyjno-Informatyczny dla Stomatologów,* **2007**, *64*, 64–72.
4. Maia, L.; Wanga, D.; Zhangb, S.; Xiea, Y.; Huangc, C.; Zhang, Z. Synthesis and bactericidal ability of Ag/TiO2 composite films deposited on titanium plate, *Appl. Surf. Sci.,* **2010**, *257*, 974–978.
5. Suwanprateeb, J.; Thammarakcharoen, F.; Wasoontararat, K., et al. Preparation and characterization of nanosized silver phosphate loaded hydroxyapatite by single step co-conversion process, *Materials Science and Engineering* C., **2012**, *32*, 2122–2128.
6. Szczepański, M.; Kamianowska, M.; Skrzydlewska, E. Wpływ hiperglikemii na procesy oksydacyjno-redukcyjne w komórkach śródbłonka ludzkiej żyły pępowinowej. *Pol. Merk. Lek.,* **2007**, *23*, 136, 246.

7. Puzanowska-Tarasiewicz, H.; Kuźmicka, L.; Tarasiewicz, M. Wpływ reaktywnych form azotu i tlenu na organizm człowieka. *Pol. Merk. Lek.*, **2009**, *27*, 162, 496.
8. Xiaoyi, X.; Qingbiao, Y.; Yongzhi, W., et al. Biodegradable electrospun poly(L-lactide) fibers containing antibacterial silver nanoparticles; *Eur. Polym. J.,* **2006**, *42*, 2081–2087.
9. Cell proliferation kit (MTT). Instruction manual. Version 3. Roche Applied Science, Germany.
10. Mosmann, T. Rapid colorimetric assai for cellular growth and survival. Application to proliferation and cytotoxity assays. *J. Immunol. Methods.,* **1983**, *65*, 55–63.
11. Cytotoxicity detection kit (LDH). Instruction manual 5th Version. Roche Applied Science; Germany, 2004.
12. Cohen, A.; Liu-Synder, P.; Storey, D.; Webster, T. J. Decreased fibroblast and increased osteoblast functions on ionic plasma deposited nanostructured Ti coatings. *Nanoscale Res Lett.,* **2007**, *2*, 385–390.
13. Sato, M.; Sambito, M. A.; Aslani, A.; Kalkhoran, N. M.; Slamovich, E. B.; Webster T. J. Increased osteoblast functions on undoped and yttrium-doped nanocrystalline hydroxyapatite coatings on titanium. *Biomaterials,* **2006**, *27*, 2358–2369.

CHAPTER 25

THE TECHNIQUE OF HARVESTING THE CANCELLOUS BONE FROM PROXIMAL TIBIA FOR DIFFERENT APPLICATIONS IN MAXILLOFACIAL SURGERY

P. MALARA

CONTENTS

25.1 INTRODUCTION

The purpose of the chapter is to describe the operative technique of cancellous bone harvesting from proximal tibia for filling the bone cavities in the orofacial region on the basis of literature review and own experience. The chapter includes step-by-step description of the operative technique, postoperative care and early rehabilitation. Possible complications at the donor site, such as fractures due to the weakening of the tibia, donor site morbidity, the necessary period of postoperative hospital stay and mean volume of cancellous bone obtained are also described.

Own experience has shown that in majority of cases no early complications occur after the application of the technique of cancellous bone harvesting from proximal tibia by medial approach and that the harvested bone is a good material for filling the bone cavities in the orofacial region. Postoperative hospital stay is not necessary after the procedure. The volume of cancellous bone obtained ranges from 8 to 21 cc. The moderate pain experienced by the patients just after surgery allows for early ambulation.

The conclusion of this chapter based on the own experience to date and the literature review is going to support the statement that cancellous bone harvesting from proximal tibia by medial approach is a relatively complication-free surgical procedure.

A wide spectrum of diseases regarding the head and neck is within the scope of interest of maxillofacial surgeons and includes infectious diseases, cancers and traumas, as well as congenital and acquired malformations. An important part of this specialty is treating diseases of the oral cavity and jawbones, especially including the removal of impacted teeth and orthognathic treatment.

Bone cavities that can occur after resections of numerous changes in the orofacial region are still a serious problem in maxillofacial surgery. The problems that maxillofacial surgeons face in every-day practice include weakening of bone structures, which depends on the size of the bone cavity. Larger cavities left without any augmentative treatment may result in pathological bone fractures, aesthetical disorders and long-term functional complications. Another problem in this type of surgical treatment is shortage of soft-tissues at the site of the resection, which may cause further problems with tight closure of a bone cavity.

Tight closure of the postoperative cavity creates empty spaces in the bone. There are two scenarios that an empty space in the bone can undergo, depending mainly on its size. Regardless the resection site, post-operative empty spaces fill in with extravasated blood coming from the adjacent tissue. It is a beneficial process in terms of regeneration of the bone. The clot that forms from extravasated blood can lead to healing of the bone cavity as a result of a cascade of biochemical and cellular processes. It is caused by the fact that first of all it is overgrown by fibrous tissue which then goes through the process of mineralization, that results in filling the bone cavity with newly created, full-fledged bone tissue.

The presented process takes place according to the scenario above only in case of relatively small bone cavities. In case of larger bone cavities, exceeding 4 cm in the largest dimension, there is a risk of different complications in healing which first of all result from the retraction of the clot, which appears in the closed post-operative cavity. The problems may include the occurrence of defective bone-like tissue, which does not fulfill histological and strength criteria.

Intraoral approach that is often used in maxillofacial surgery is inevitably connected to superinfection of the surgical wound, as the oral cavity does not constitute a sterile environment. Infection of a hematoma in the bone cavity with bacterial flora of the oral cavity may result in its suppuration. The appearance of suppuration complications, which within the facial part of the skeleton might progressively spread to adjacent spaces, in many cases makes regenerative processes impossible, requires long and expensive treatment; it may also result in the occurrence of life-threatening complications.

The problem of empty spaces formation can be approached by direct filling of the bone cavities at the time of surgery. Maxillofacial surgeons can choose between autogenous bone grafts, xenogenic or alloplastic materials. Xenogenic along with alloplastic materials posses only the inorganic phase. They consist mainly from hydroxyapatite or calcium triphosphates. The advantage of alloplastic materials is the lack of necessity to operate on the donor site that results in the reduction of discomfort of a patient who underwent such a procedure. After the implantation into the cavity those materials form a kind of scaffold for infiltrating new blood vessels and connective tissue fibers. At the later time they also constitute a substrate of inorganic phase in the process of mineralization of the bone matrix. The lack of any organic phase is also a great disadvantage of those materials, for they mainly have osteoconductive properties with no osteogenic and osteoinductive properties. This results in the slower pace of healing of the bone.

This is why maxillofacial surgeons often use autologous bone for filling the bone cavities. Autografts are claimed to be a golden standard of reconstructive surgery, as together with the graft that consists of organic and inorganic parts, a large pool of osteogenic cells is also moved. As opposed to soft tissue grafts, inorganic phase of bone grafts creates a scaffold which functions as a bridge for the cells coming from the rim of the recipient site. Thanks to these processes the graft is colonized by cells, which undertaking their functions, allow the survival of the graft [1]. Considering the fact that autogenous grafts are performed within the same body, the problem of potential rejection of the graft on immunological basis does not exist. While it is a significant problem in case of heterogeneous and xenogenic grafts [2].

There are two types of bone tissue in the human organism. Cortical bone thanks to the presence of Haversian channels shows good osteoconductive properties. Thanks to its mechanical properties it can be used in cases when recreation of tridimensional cavities within the facial part of the skeleton is required. As opposed to cortical bone, cancellous bone is extremely rich in osteogenic cells. Living osteoblasts of cancellous bone may survive even for a few hours from the time of harvesting of the tissue; early revascularization in closed cavities takes place after 48 hr. The disadvantage of the cancellous bone grafts is their small mechanical endurance. It is also connected with the lack of possibility to use them in case of tridimensional reconstructions [3].

The process of healing of the grafted bone is connected with resorption and remodeling of the bone graft. The resorption degree and its pace depend on numerous factors, including the size of the grafted bone, its quality, as well as the quality of the donor site and the method of attachment of the graft in the recipient site [4]. The fixation method of the graft in the recipient site is of great importance. On one hand, the attachments should ensure stable mounting of the grafted bone. Even the slightest movements of the graft on its base significantly increase the rate of resorption. On the other hand stability of the graft should be ensured using the smallest possible number of binding materials. The smaller amount of binding material, the slower the rate of the graft resorption can be clinically observed. Moreover, the smaller volume of grafted bone tissue, the slower the rate of graft resorption.

The purpose of this chapter is to present the surgical technique of cancellous bone autograft from the proximal tibia to fill the bone cavities in the orofacial region.

25.2 DESCRIPTION OF THE SURGICAL TECHNIQUE

The tibia is a long bone. At the proximal end of tibia there are two condyles: medial and lateral. Between the condyles there is an intercondylar eminence limited on the sides by two intercondylar protuberences – medial and lateral. Both condyles are surrounded by vertically falling margo infraglenoidalis, below which tibial tuberosity is located at the front [5]. Tibial tuberosity is palpable, and defining its location is essential to avoid damaging the articular surface of a knee joint during the procedure. The patellar tendon, under which small branches of the upper and lower medial popliteal artery run, is attached to the proximal part of the tibial tuberosity. The tibialis anterior muscle is located on the lateral surface of the tibia below the preparation line which is necessary for proper access to the proximal base [6].

The procedure of harvesting cancellous bone from the proximal tibia can be carried out under general anesthesia or analgo-sedation. The skin around the knee and proximal part of the shin should be washed twice with colorful disinfectants and the surgical field should be protected with sterile wrappings. Medial access is possible by the skin incision 2–3 cm long led 2 cm below and 2 cm medially from the anterior tibial tuberosity (Fig. 1).

The area of incision to the level of periosteum is injected with 2 cc of 2% lignocaine with noradrenaline (1:80,000). The incision is performed with blade no.10 through all the layers to the periosteum. Then the periosteum is exfoliated with a raspator in a way that allows free access to the medial surface of the proximal base. Soft tissues are retracted with Langenbeck's hooks.

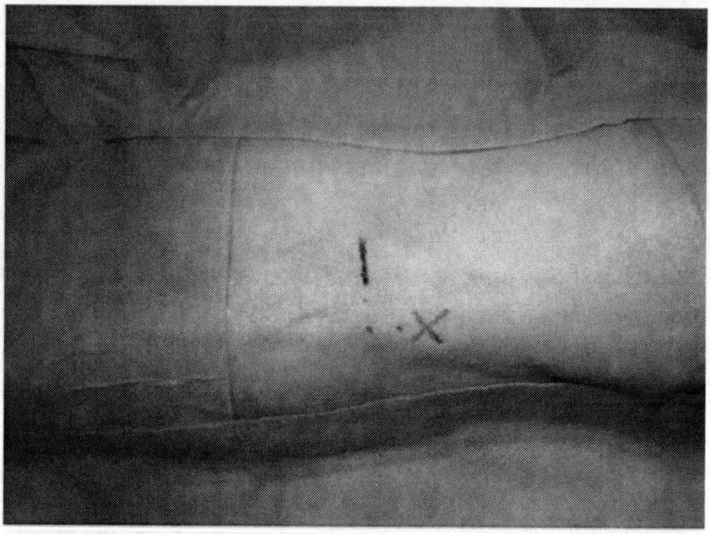

FIGURE 1 Through-and-through 2–3 cm long incision is run 2 cm below and 2 cm medial to anterior tibialis tuberosity.

The opening in the cortical plate of the tibia is performed using a trephine 8 mm in diameter mounted on a surgical handpiece at 120 revolutions per minute with significant cooling by sterile saline solution (Fig. 2).

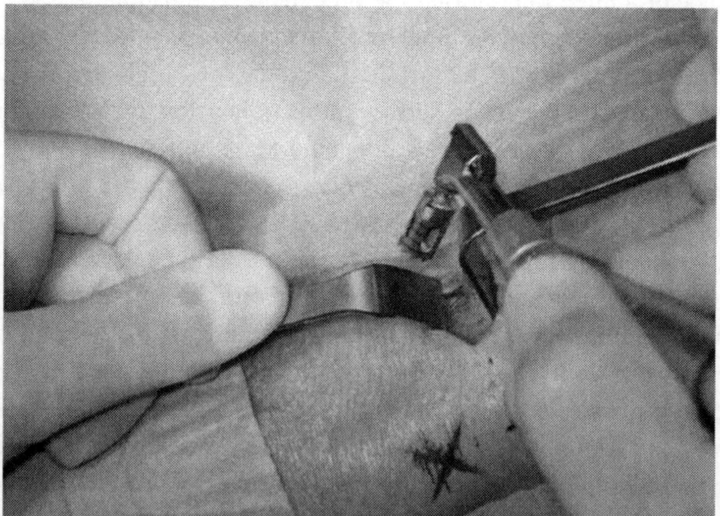

FIGURE 2 An opening in the cortical bone of tibia is made with a trephine 8 mm in diameter mounted on a contrangle surgical handpiece.

Then, through the opening straight and lateral bone curettes of various sizes are inserted, thanks to which mobilization of subsequent layers of cancellous bone can be performed. The tools are not used to extract cancellous bone, but to separate cancellous bone from compact bone. A bone collector mounted on a surgical suction is used to extract cancellous bone (Fig. 3).

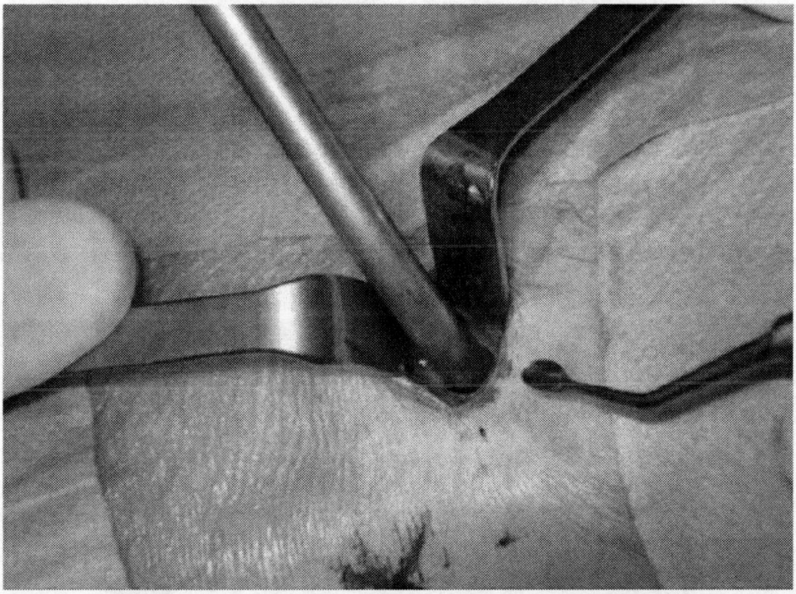

FIGURE 3 Cancellous bone from tibia is captured and removed with a bone collector connected to suction.

In order to easily isolate fragments of cancellous bone extensive flushing with saline is performed through the opening. The procedure is performed until the moment when it is not possible to extract larger amounts of cancellous bone from the proximal base with manual tools. Then a final flushing is performed with sterile saline solution, which is suctioned from the inside of the proximal base of the tibia using suction. The wound is closed in three layers. The periosteum is sutured using 3–0 absorbable sutures, subcutaneous tissue is sutured with 4–0 absorbable sutures. Skin is sutured with continous intradermal suture using 5–0 nylon thread (Fig. 4). A sterile lint dressing is applied directly on the wound.

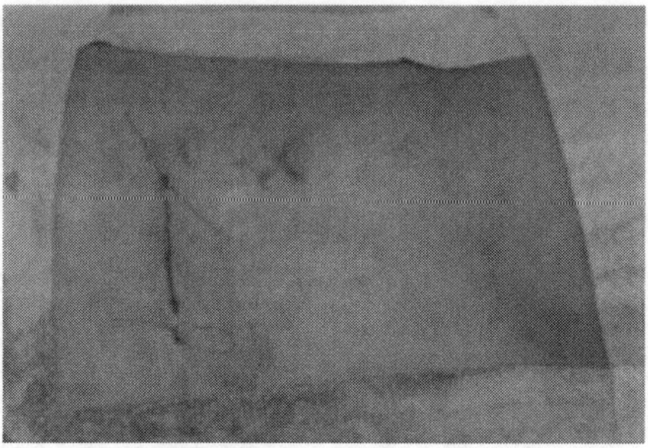

FIGURE 4 A skin wound is closed with intradermal nylon sutures.

25.3 PRE-SURGICAL PROCEDURES AND POSTOPERATIVE CARE

All the patients qualified for the treatment based on autogenous grafts from proximal tibia to have an X-ray of the knee joints in the antero-posterior and lateral views taken (Figs. 5 and 6).

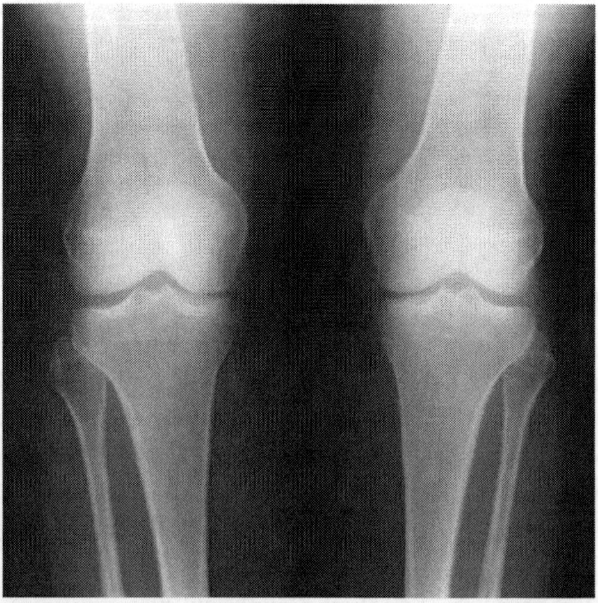

FIGURE 5 The A–P view of knee joints.

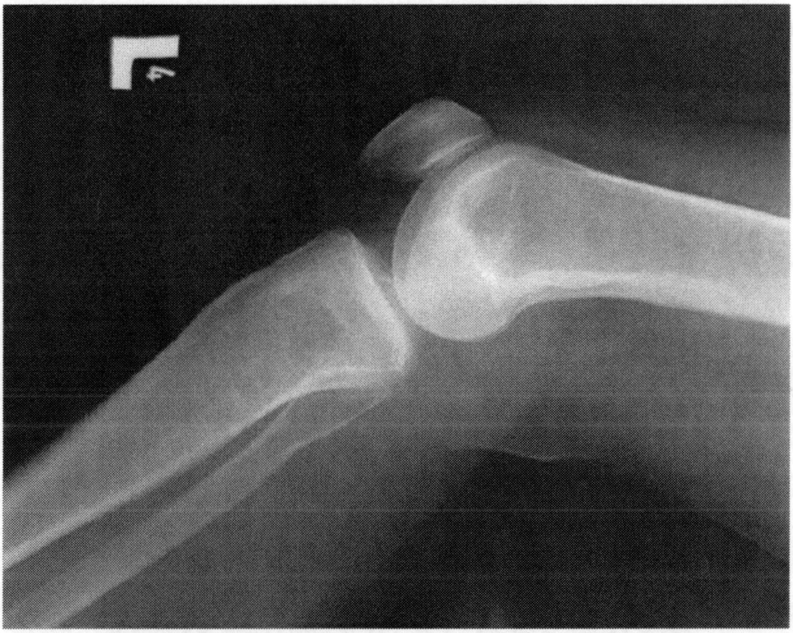

FIGURE 6 A lateral view of a knee joint.

It is essential to consider data from the interview selecting a limb to harvest cancellous bone (lack of previous fractures), as well as the results of radiological examination (lack of recognized changes of radiological image of the proximal tibia). In the situation when both limbs can be good donor sites, a patient's preferences should be taken into account.

Directly after the surgery the shin from the foot to the knee is wrapped in an elastic bandage. An elastic pressure dressing is held on the shin for 7 days. Antibiotics and non-steroid anti-inflammatory drugs are administered to patients, taking into consideration their general health and the extent of the surgery on the donor site. At afternoon hours on the day of the surgery patients are encouraged to walk. However, avoiding direct pressure on the operated limb, jumps, running the stairs, climbing a ladder, etc., are recommended for 3 mo after surgery. Control knee radiographs are taken after the surgery in order to confirm the accuracy of the opening in the cortical plate and to exclude possible fractures and infractions of the cortical plate (Fig. 7).

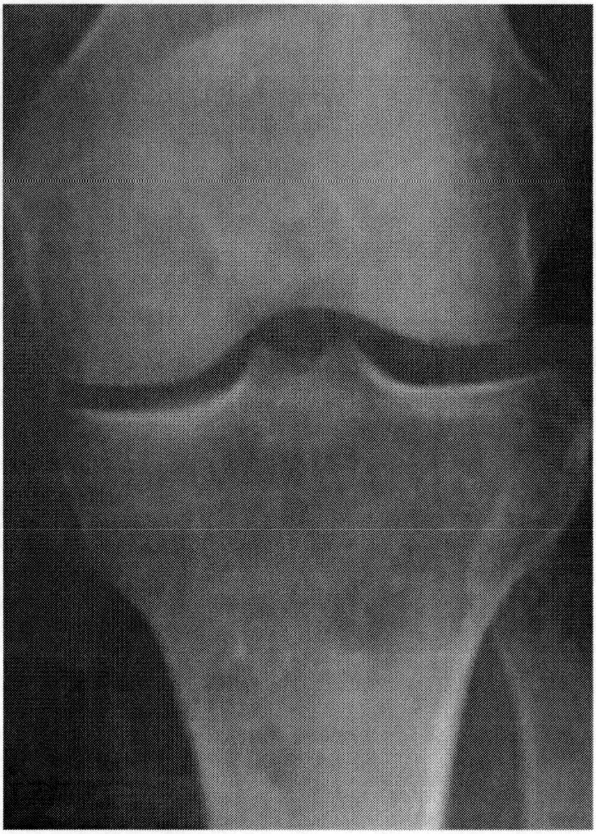

FIGURE 7 The P–A view of a knee joint taken after the harvesting procedure.

Sutures are removed between the 7th and 10th day after surgery. Hospitalization period depends on the extent of the surgery intraorally and the healing of the recipient site. Considering the donor site, patients do not require post-operative hospitalization and they can be released home on the day of the surgery.

25.4 DISCUSSION

In literature numerous donor sites for applications in maxillofacial surgery are offered, including calvarium, symphysis of the mandible, ribs, iliac crest and tibia [7]. It reflects the necessity of obtaining the autogenous bone grafts in order to fill bone cavities of the jaw and mandible. These treatments include reconstructive surgeries of clefts, post-cancer defects and post-cystectomy defects. Bone blocks, which consist exclusively of cortical bone or cortical bone and cancellous bone, are used when it is necessary to achieve good mechanical properties of the graft. Plates of cortical bone undergo slower resorption than cancellous bone, however, they include fewer cellular elements, and the process of osteogenesis shows slower dynamics [8]. Contrary, when cancellous bone is used in reconstructive surgery, the transfer of a significant pool of living pluripotent cells is achieved, allowing osteogenesis, osteoinduction and osteoconduction [9]. Therefore in case of bone cavities with the geometry, which allows filling with, *inlay* techniques, surgeons prefer cancellous bone grafts. The majority of post-cystectomy cavities in mandible belong to the category of such bone cavities.

Choosing a donor site of the graft, expected amount and quality of harvested bone tissue should be taken into consideration, as well as the smallest possible post-operative discomfort for patients, difficulties in walking during the post-operative run, the length of necessary hospitalization period, possibility of early and late complications, as well as skills and preferences of the operator [7]. The area of the proximal base of the tibia has been used as a donor site in orthopedic surgery for many years [10, 11]. Since then harvesting surgical technique has been undergoing constant development and its aim is the smallest possible loss of cortical bone plate of the tibia [12, 13] and saving the zone of growth in children [14]. A little later attention was brought to the tibia as a donor site of cancellous bone for the needs of cranio-maxillofacial surgery, especially in the case of palate clefts and filling osteotomy gaps [15].

Surgical access to cancellous bone in the proximal base of the tibia can be obtained from the lateral surface of the bone with vertical skin incision running along the medial edge of anterior tibialis muscle below the tibial tuberosity [11] and via medial access through a skin incision running 2 cm below and 2 cm medially from tibial tuberosity [5]. The author of this chapter prefers medial access due to a lower number of described complications of this procedure. The medial surface of the tibia is located in this area directly under the skin. Access to the bone is obtained by an incision run through all layers – skin, poorly developed subcutaneous tissue and periosteum. This way the risk of damaging muscles and larger blood vessels or nerves is reduced.

Average volume of cancellous bone obtained by different authors ranges from 10–42 cc [5, 6, 11]. My own experiences indicate that in operated patients it was possible to obtain slightly smaller volumes of cancellous bone – from 8 to 21 cc. It should be pointed out that while planning the surgical treatment using cancellous bone graft it must be considered that the amount of harvested bone from one limb is too small, which is not possible to foresee pre-operatively. Therefore it is recommended to obtain a patient's consent to harvest bone from both tibias before the surgery [15, 16].

In literature a very small percentage of early complications in the donor site (from 0 to 1.9%) draws attention; the complications include mostly impaired cutaneous wound healing, bleeding, severe postoperative pain and difficulty in walking manifested as limping [10, 17]. In the group of patients operated by the author no early or late complications in the donor site were observed. There are a few reports regarding a possibility of fractures of the tibia in patients in whom cancellous bone of the proximal base was harvested [15, 18]. Hughes and Revington [15] report an occurrence of this complication in 2 of the 75 operated patients (2.7%). Even though we have not observed this complication in our own material, it is possible that its occurrence is more frequent than it is shown by scientific reports, as many of such fractures can heal themselves without the need for surgical intervention, as suggested by Thor et al. [18].

Relatively short surgery time, short skin incision resulting in a small scar, a possibility to walk on the first day after surgery and short hospitalization period undoubtedly belong to the advantages of cancellous bone graft from the base of the proximal tibia [12, 14–16, 19]. All patients operated by the author are encouraged to walk in a few hours after the procedure. The majority of the patients do not require any help in walking (crutches or walking stick) on the day of the surgery. However patients are recommended to avoid excessive pressure on the operated limb, such as running, jumping or climbing a ladder for 3 mo. Contact sports are also discouraged in this period. This is also recommended by other authors [15, 16]. It should be pointed out that considering the process of wound healing in the donor site, patients could be released home on the first post-operative day. It should be highlighted that in the donor site a good aesthetical effect in the form of a small linear scar visible on the shin is obtained.

25.5 CONCLUSION

Cancellous bone grafting from the proximal tibia approached via medial access is a relatively safe surgical technique that can be recommended when significant volumes of autologous bone are needed to fill the recipient site. Advantages of this technique include relatively short surgery time, limited skin incision resulting in an aesthetically acceptable scar, the ambulation possibility on the first day after surgery and short hospitalization period. All these features make this technique an attractive alternative for different donor sites. This surgical technique significantly contributes to the reduction of complications in treating large bone cavities and to significant acceleration of bone tissue regeneration at the recipient site.

KEYWORDS

- **Alloplastic**
- **Autogenous bone grafts**
- **Haversian channels**
- **Proximal tibia**
- **Superinfection**
- **Xenogenic**

REFERENCES

1. Moy, P. Clinical experience with osseous site development using autogenous bone, bone substitutes and membrane barriers. Oral Maxillofac. Surg. Clin. North Am. **2001**, 13, 493–509.
2. Block, M. S.; Kent, J. N. Sinus augmentation for dental implants: The use of autogenous bone. J. Oral Maxillofac. Surg. **1997**, 5, 1281-1286.

3. Citardi, M. J.; Friedman C. D. Nonvascularized autogenous bone grafts in the treatment of the resorbed maxillary anterior alveolar ridge: Rationale and approach. Implant Dent. **1998**, 7, 169–176.
4. Fonseca, R. J.; Clark, P. J.; Burkes, E. J.; Baker, R. D. Revascularization and healing of onlay particulate autologous bone grafts in primates. J. Oral Maxillofac. Surg. **1980**, 38, 572–577.
5. Herford, A. S.; Brett, J. K.; Audia, F.; Becktor, J. Medial approach for tibial bone graft: Anatomic study and clinical technique. J. Oral Maxillofac. Surg. **2003**, 60, 358–363.
6. Kalaaji, A.; Lilja, J.; Elander, A.; Friede, H. Tibia as donor site for alveolar bone grafting in patients with cleft lip and palate: long-term experience. Scand. J. Plast. Reconstr. Surg. Hand Surg. **2001**, 35, 35–42.
7. Rawashdeh, M. A.; Telfah, H. Secondary alveolar bone grafting: the dilemma of donor site selection and morbidity. J. Oral Maxillofac. Surg. **2008**, 46, 665–670.
8. Ozaki, W.; Buchman, S. R. Volume maintenance of onlay bone grafts in the craniofacial skeleton: micro-architecture versus embryologic origin. Plast. Reconstr. Surg. **1998**, 102, 291–299.
9. Silva, R. V.; Camili, J. A.; Bertran, C. A.; Moreira, N. H. The use of hydroxyapatite and autogenous cancellous bone grafts to repair bone cavities in rats. Int. J. Oral Maxillofac. Surg. **2005**, 34, 178–184.
10. O'Keefe, R. M.; Reimer, B. L.; Butterfield, S. L. Harvesting of autogenous cancellous bone graft from the proximal tibial methaphysis: a review of 230 cases. *J. Orthop. Trauma* **1991**, *5*, 469–474.
11. Alt, V.; Nawab, A.; Seligson, D. Bone grafting from the proximal tibia. *Trauma* **1999**, *47*, 555–557.
12. Ilankovan, V.; Stronczek, M.; Telfer, M.; Peterson, L. J.; Stassen, L. F. A.; Ward-Booth, P. A prospective study of trephined bone grafts of the tibial shaft and iliac crest. *Br. J. Oral Maxillofac. Surg.* **1998**, *36*, 434–439.
13. Hashemi, H. M. Oblique use of a trephine bur for the harvesting of tibial bone grafts. *Br. J. Oral Maxillofac. Surg.* **2008**, *46*, 690–691.
14. Belsy, W.; Ward-Booth, P. Technique for harvesting tibial cancellous bone modified for use in children. *Br. J. Oral Maxillofac. Surg.* **1999**, *37*, 129–133.
15. Hughes, C. W.; Revington, P. J. The proximal tibia donor site in cleft alveolar bone grafting: experience of 75 consecutive cases. *J. Cranio Maxillofac. Surg.* **2002**, *30*, 12–16.
16. Chen, Y. C.; Chen, C. H.; Chen, P. L.; Huang, I. Y.; Shen, Y. S.; Chen, C. M. Donor site morbidity after harvesting of proximal tibia bone. *Head Neck* **2006**, *28*, 496–500.
17. Catone, G. A.; Reimer, B. L.; McNeir, D.; Ray, R. Tibial autogenous cancellous bone as an alternative donor site in maxillofacial surgery: a preliminary report. *J. Oral Maxillofac Surg.* **1992**, *50*, 1258–1263.
18. Thor, A.; Farzad, P.; Larsson, S. Fracture of the tibia: Complication of bone grafting to the anterior maxilla. *Br. J. Oral Maxillofac. Surg.*, **2006**, *44*, 46–48.
19. Walker, T. M. W.; Modayil, P. C.; Cascarini, L.; Williams, L.; Duncan, S. M.; Ward-Booth, P. Retrospective review of donor site complications after harvest of cancellous bone from the anteriomedial tibia. *Br. J. Oral Maxillofac. Surg.*, **2009**, *47*, 20–22.

CHAPTER 26

EXPERIMENTAL ADHESIVE BIOMATERIAL IN THE DEVELOPMENT OF RESTORATIVE CONCEPT TOWARDS THE BIOMIMICRIC DENTISTRY

T. KUPKA, R. ORLICKI, and G. E. ZAIKOV

CONTENTS

26.1 INTRODUCTION

The prosthesis is replacing the missing part of human body with an artificial part. The aim of this chapter is to present the project of clinical experiment using an adhesive biomaterial type WMGIC in the development of dental materials. Normal results of standard non-clinical studies of physicochemical (net solidification time, compressive strength, acidic erosion using spray technique, $C_{0.70}$ opacity, the content of acid soluble arsenic and lead) and biological (in vitro cytotoxicity study in direct contact) type authorized to perform standard clinical experiment using studied material in a minimally invasive direct adhesive restoration occlusal cavities of moderate size in permanent teeth, which was completed with the calculations and statistical analysis at the Medical University of Silesia in Katowice. Starting from teeth of *Sahelanthropus thadensis* with enamel erosion, estimated at 7 million years, Etruscan bridges 2,600 years old, through Celsus' *plumbum* 2,000 years old, the Mayan implantations of pieces of shell and obsidian to dental alveolus after extractions 1,412 years ago, the prototype of Guillemeau's porcelain teeth 460 years before, to the contemporary materials, tissue engineering and hard tissue formation, the development of civilization is a dynamic and accelerating process where the studied material found its place as part of the progress of dental materials science and evolution of restorative concept towards the biomimicric dentistry.

As a result of tooth hard tissue diseases, there is a partial or total loss of a tooth. The prosthesis is replacing the missing part of human body with an artificial part. Restorative dentistry deals with the anatomical restoration of partially or completely damaged tooth organ and/or surrounding tissue using the dental materials and techniques, either both approved or experimented [1].

The oldest Brunet's archaeological artifact (2) on the teeth speaks of enamel and dentin erosion and dental arch in the shape of the letter "U" in hominid *Sahelanthropus thandensis* estimated at 7 million years old. The first traces of drilling in human teeth aged 9,000 years were found in Mehrgarh settlement. The first mention of dentistry comes from the time of the great ancient civilizations, written with the Sumerian pictograph and cuneiform script, aged 6,000 years. For over 50 centuries we observe intense development of concept, materials and methods applied to reconstruct dental organ cavities. An concept of restoration of a missing teeth not only for aesthetic but also for functional reasons appeared 4,500 years ago in Cairo, the Egyptian Old Kingdom; a homologous molar with a gold ligature tied to an existing molar was applied to try to restore chewing function. 2,800 years ago Hesy-Re applied dental adhesive to "plug holes in the tooth," as a mechanical analgesic barrier, yet without a concept of restoration of the missing hard tissue, consisting of: Nubian soil, powdered malachite and turpentine oil; we can say that it was a prototype of contemporary ceramic-polymer composite. In turn, Celsus Cornelius wrote 2,100 years ago about analgesic plugging of dental cavities using *plumbum* according to the recipe by Andromachus, simply being a "cement" of lead and canvas; after 1,900 years Fox in London wrote about *plomber* of lead. Kung Su in the Tang Dynasty begins to think restoratively; 1,359 years ago he applied "silver cake" – the prototype of the modern silver amalgam, as a material *de facto* to the immediate restoration of the missing organ dental hard tissues, not just to "plug holes" against pain stimuli. First attempts of restoration using semi-direct method is found in the Maya peoples in Mesoamerica, who 1,312 years ago manually cut semi-precious and precious stones for ritual refills/dental jewelry, adjusted them into effect in their incisors and stuck with rich composition mineral cement confirmed by spectral analysis; they also applied plastic direct restorations of powdered iron pyrite with unidentifiable matrix; after 840 years the Incas made gold refills in the incisors to distinguish the hierarchy. A 1,127 years ago, Rasis proposed filling cavities using a paste made of, among others, mastic, alum and honey. Aztecs in Mesoamerica 1,112 years ago applied filling cavities with a powder of snail shells, sea salt and herbs. An important phase in the development of restorative concepts was using 499 years ago by Da Vigo, for d'Arcoli's example, a gold foil to direct restorations, but after the surgical preparation (old Greek name of surgery, *cheirourgia*, means "handicraft") of the "corrosion," as he called carious tissue. A 460 years ago, Guillemeau gave the prototype of mineral paste, future dental porcelain. 312 years ago in Wroclaw Purmann proposed making handmade wax model of the missing tooth and then cutting from positive from tusk; however, Pfaff introduced the factual basis for indirect restorations in Berlin 256 years ago using plasticized wax impression, plaster model, wax pattern/greenbite. This was a major milestone in the indirect restorative dentistry. A 300 years ago, Boetger applied cut-to-size porcelain applied to reconstruct cavities in anterior teeth. A 223 years ago, Hunter said about the possibilities of regenerative brain tissue and put the basis for dentistry based on scientific proof. A 242 years ago, Darcet applied low melting alloys for direct restorations: bismuth:lead:tin 8:5:3, obtained independently about 100 years ago by Newton and applied directly in the mouth 194 years ago by Regnart, who poured them melted directly to the cavities in mouth. Just two years later in Paris Maury after Delabarre raised an important voice searching for the use of safe materials, mainly gold and platinum alloys instead of those dangerous low melting alloys. A 233 years ago Czerwiakowski wrote "About the artificial teeth." A 225 years ago in Edinburgh arose the concept of invasive retention preparation for the gold non-adhesive restorations and 216 years ago Hirsch(feld) applied a stone paste, a type of cement; he suggested the retention shape (entrance diameter smaller than the diameter of the bottom of the filling) for non-adhesive restorations. A 214 years ago, Dubois-Foucou applied modified composition and firing temperature of porcelain dentures, porcelain teeth in three shades (white and blue, white and gray, white and yellow) by adding metal oxides. Biomimicric concept in imitating colors and shades of teeth began to germinate, which was strongly developed 206 years ago by Fonzi, who introduced 26 shades of ceramics. A 192 years ago Delabarre made the first step to obtain artificial gums – Allen's "continuing gums." A 206 years ago Fox proposed maximally invasive preparations and restorations for the adjacent teeth, which 197 years ago resulted in strong reaction by Delabarre and his smallest possible intervention; it is him that we owe the minimally invasive approach, as well as the use of materials safe for the human body, which was, in turn, applied 192 years ago by Maury, to whom we owe the semi-transparent porcelain teeth, and two layers of porcelain coated lacquer. A 186 years ago in Bremen Koecker after Da Vigo applied total removal of carious tissue, but to the border of healthy tissue; there the concept of non-invasive preparation appeared. In turn, 178 years ago Linderer applied face restorations; he adjusted pieces of walrus tusks, mounted in cylindrically prepared cavities by pressing or hammering or using quicklime with turpentine oil or fish glue (colloid proteins of hydrolyzed collagen from animal tissues; gr. *colla* means "glue") as a cement; he also introduced the concept of bearing surface. A 175 years ago Murphy applied cut-to-size glass as a material for restorations of cavities in anterior teeth. Two characters, Cameron and Evans, 172 years ago, laid the groundwork for articulators, occlusion and articulation, precisely clarified, defined and implemented by Bonwill 148 years ago, as the balanced contact occlusion in the stomatognathic system. Linderer junior, who 161 years ago in Berlin applied biomaterial composed of ground enamel of carnivores with phosphoric acid, is the one to whom we owe the beginnings of research for biomimicric materials. He was to have many followers. A 157 years ago in Paris Sorel introduced phosphate cement – the first synthetic dental cement. Meanwhile, 155 years ago in Baltimore Arthur recommended precise restoration of anatomical shapes, points of contact in direct restorations with gold foil. This was a very important step in the development of restorative concept in medical dentistry – anatomical restoration of missing tissues. A 154 years ago in Dresden Rostanigs reported obtaining the "artificial bone." Today, we can say it was only a zinc-oxide-phosphate cement; but the biomimicric concept in research for substitute materials imitating tooth tissues began to bloom. A 138 years ago in Great Britain Fletcher introduced, as he said, an "artificial dentin", which truly was only a zinc-sulphate cement; however, four years later he applied the first biomaterial for direct restorations, imitating hard tooth tissues, "artificial dentin, artificial enamel" – a silicon cement by Steenbock's recipe. At the same time the concept of multilayer restoration also appeared: Driscol proposed direct restorations of the plastic phosphate cement and amalgam as a durable and sealed layer, which was the prelude to McLean's sandwich technique developed 40 years ago. A 134 years ago White and Ash suggested ready porcelain cylinders for indirect restorations (inlays) with a set of burs for cutting (after the Maya peoples, 1312). A 124 years ago Land burnt indirect restorations / inlays and porcelain jacket crowns on platinum foil, for the first time with subgingival preparation, adjusted in the cavity. A 118 years ago Foster and Alexander applied indirect restorations cast in laboratory; a wax impression of the cavity, covered with gold foil, immersed in plaster; gold solder flowed through the hole into the model. A 87 years ago Bauer obtained methacrylate dental/technical material (liquid monomer and powdered polymer). Important from the point of view of biomimicric dentistry was introduced 77 years ago in Switzerland by Hagger the free radical polymerization in adhesive dental materials and the first synthetic ceramic-polymer composite of RC family for Paffenberger's direct restorations aged 74 years. A 67 years ago Masuhara and Fisher introduced epoxy polymers in direct restorative materials. The beginnings of CAD/CAM at DP Technology are dated for 67 years, which first results in the indirect restorative dentistry appeared 37 years ago. A 62 years ago Buonocore microinvasively etched enamel, which resulted in obtaining the mechanical retention of adhesive polymer for the ceramic-polymer composite. A 55 years ago Bowen patented the first ceramic-polymer composite, and 53 years ago Smith introduced zinc-polycarboxylic cement [3].

A very important step in the development of restorative materials in the context of safety, minimal invasiveness and biomimicry was the first actually self-adhesive GIC biomaterial called ASPA, achieved 52 years ago by a team of McLean, Wilson, Kent, Crisp. A 39 years ago in London Dart applied light cured ceramic-polymer composite, which gave the operator practically full restorative freedom in time. A 32 years ago the first connecting factors to the dentine tissue appeared. In the last 30 years various RC and GIC hybrids were obtained; many new materials appeared: RMGIC's and ormoceres (29 years ago), anhydrous GIC's (28 years ago), PMRC's (23 years ago), giomeres and ceromeres (9 years ago) and glass-carbomers (6 years ago).

26.2 MATERIALS AND METHODS

Technological project of the chemical compound for the purpose of experimental glass-ionomer cement called GJS2/W was established in the Department of Bioceramics of the Institute of Ceramics and Building Materials in Warsaw; is a fast-setting acid-base water-activated cement in the form of powder and liquid, classified as a class IIa implanted medical product for long-term use/of a solid contact, intended to be placed in the dental organ. Because of the way of the contact with human body/environment of the oral cavity it can be an externally contacting product (single-layer self-restoration) or implant (in the multilayer/closed sandwich approach). For the purposes of this chapter manufacturer provided a kit, which included: powder free from foreign inclusions, with an evenly distributed pigment, color A2, Series 0412, liquid in the form of distilled water (PN-EN ISO 3696:1999/Ap1:2004; ENISO 3696:1995), Series 0412, conditioner, Series II 04 12 UZD, protective lacquer, Series 04 12 LKR, measuring cup, scoop and blocks of waxed paper. The powder component contained opaque glass of granulation < 45 μm (in Na_2O-BaO-SrO-SiO_2-Al_2O_3-P_2O_5-F oxide system), PAA and TA anhydrides, as well as pigments, which were homogenized in a high-speed mixer (Henschel Kassel, Germany). PAA was obtained in solid form by freeze-drying of its aqueous solution (Sigma Aldrich, USA). Powdered components were dyed using pigments approved for use in the manufacturing of medical products. Composition of powder was enriched using TA additive, commonly applied in the composition of the GIC. Fluid for manual making of the powder was distilled water. PAA aqueous solution of mass concentration 25% was the component of conditioner characterized at the temperature 23.5°C with a density of 1.09 g/ml, refractive index of 1.37, a clear appearance and the absence of sediment fibrosis.

The results of non-clinical studies (net solidification time, compressive strength, acidic erosion using spray technique, $C_{0.70}$ opacity, the content of acid soluble arsenic and lead (Institute of Ceramics and Building Materials in Warsaw: PN-EN ISO 9917-1:2005; EN 29917:1994, EN ISO 9917-1:2003) and in vitro cytotoxicity study in direct contact (Department of Experimental Surgery and Biomaterials Research of the Medical University in Wroclaw and the Laboratory of Virology of the Institute of Immunology and Experimental Therapy of the Polish Academy of Sciences in Wroclaw: PN-EN ISO 10993-5:2001; EN ISO 10993:1999) authorized the project to conduct a clinical experiment (PN-EN ISO 14155-1:2006; PN-EN ISO 14155-2:2004; EN ISO 14155-1:2003; EN ISO 14155-1:2003; ISO/TS 11405:2003), which was performed by the Experimental Odontology Research Group in the Medical University of Silesia. On 8th March 2005 the Commission of Bioethics of the Medical University of Silesia in Katowice issued a favorable opinion on the draft of medical experiment (NN-6501-237). A 20 participants of both sexes – 9 women and 11 men, aged 8–39 years – were randomly assigned to the Clinical Trial. Their average age was 15.5 years, the group of youths (under 16 years of age) counted 16 people, which was 80% of whole; while for the group of adults (over 16 years of age) were qualified 4 persons. Each participant of the trial was assigned with a unique identification code to ensure confidentiality. Selection of carious cavities was carried out for occlusal cavities of moderate size in permanent teeth. General and dental interview, as well as extraoral and intraoral examinations (after interviews) was carried out prior to the application. Minimally invasive rotary preparation of pathological carious tissue, making the studied material and restoration with after treatments was performed in accordance with the standards of dentistry and recommendations of the manufacturer. After 20 seconds of conditioning, dentin was gently washed with water and dried with an oil free air. Cement was placed in the cavity in a single volume in less than 90 seconds after the start of mixing. Final restorations were carried out not less than 120 seconds after the completion of mixing. Then the surface contacting externally with the oral cavity was coated with a protective lacquer, which was dried for 10–15 seconds using a gentle stream of dry air. Studied material was made and applied by one researcher. A total of 54 direct restorations were carried out; in most cases (49 applications – 92.6%) Class I cavities according to Black's classification were restored on the occlusal surface in 26 molars and 23 premolars (Class 1.2 according to Mount and Hume); representatively single Class I cavities according to Black within foramen cecum were restored (Class 1.2 according to Mount and Hume); Class II proximal in molar (Class 2.2 according to Mount and Hume), Class III mesial in canine (Class 2.2 according to Mount and Hume) and Class V atrial in premolar (Class 3.2 according to Mount and Hume). In clinical observation the following parameters were applied: USPHS (I – color, II – smoothness and gloss, III – anatomical shape, IV – discoloration, V – marginal seal, VI – secondary caries). Scale developed by Ryge was optimized with the modification, detailing the valuation at 0.1 point. The two clusters with compartments of clinical adequacy of the studied material were separated: the acceptable cluster (excellent < 0.5; 0.5 < very good < 1.0; 1.0 < good < 1.5) and the unacceptable cluster (1.5 < satisfactory < 2.0; 2.0 < unsatisfactory < 3.0); satisfactory grade pointed to the need for a partial correction, and unsatisfactory grade qualified for the total re-restoration. Next clinical observation periods were planned, the first one during the application visit, next ones in 12th, 24th, 48th and 96th week after application, during which initially were carried out dental interviews (oversensitivity or pain and its character). Teeth with cement applied were subjected to standard procedures of cleaning, drying, isolating from the access of saliva and the viability study of pulpal tissue was performed. Preliminary evaluation of restorations of the studied GJS2/W material was carried out visually, intraorally, using a dental mirror and a dental probe. Detailed clinical analysis was carried out visually on the basis of archived computer images from high-resolution monitor, such as photos and videosequences obtained using the intraoral camera Vista Cam (Durr Dental GmbH & Co. KG. KG, Bietigheim-Bissin, Germany) with a high performance and unique macro option (6 LEDs with focusing len, light-sensitive element in the W" Color Interline Transfer CCD, the number of image pixels in PAL 512 582), integrated with the computer dedicated to the purpose of this clinical trial: Windows XP environment (Microsoft, Washington, USA), AV TV Capture Card (Sapphire, Shatin, NT, Hong Kong). Individual restorations were subjected to both primary direct and secondary detailed virtual clinical observation using CorelDRAW Graphics Suite v. 12 programs (Corel Corporation, Ottawa, Ontario, Canada) by the Chief Researcher/Author. Each study participant had guaranteed a friendly environment of an intimate room while maintaining punctuality, courtesy and the right aura, also by playing relaxing music. Intraoral images from the clinical phase were exported electronically, providing the possibility of interactive online medical consultation. Clinical observation results representing the quality characteristics were expressed in an ordinal (ranked) scale. Descriptive characteristics of these features were shown in the form of positional descriptive statistics, such as the median, quartiles, upper and lower extreme values (minimum and maximum). The distribution of numbers of clinical evaluations for each characteristic was presented as percentage (number) of the frequency of each rank. Statistical analysis of clinical evaluation was made using the tests referring to the numbers. The use of some of these tests was subject to fulfilling the assumptions about the number of compared groups (Chi^2 test) or the need to use a dichotomous scale (McNemar's Chi^2 test, Cochran's Q test); therefore, an appropriate, maximally positive sorting of results was carried out; the sorting type was to create one perfect subgroup, which included the cases evaluated with excellent mark of 0, while the second subgroup included cases with marks different from 0. Variability of the individual restoration parameters in subsequent clinical observation periods was comparatively evaluated analyzing the results of the two adjacent research periods using McNemar's Chi^2 test. Correlation between the features within the same research period was evaluated using the same test. The results obtained in the subsequent weeks of observation were compared to evaluations of the application phase (McNemar's Chi^2 tests). If the number of incompatible pairs was less than 10, a method of estimating the exact probabilities for binary variables (exact binomial study) was applied. More than two dependant attempts were compared using Cochran's Q test. Studying the differences in the evaluation of two parameters based on the juxtaposition of incompatible pairs (McNemar's Chi^2 test) and the frequency of excellent marks (Chi^2 independence test or Fisher's exact test, if any expected number was less than 5). Statistical evaluation of correlations was also carried out (Chi^2 independence test or Fisher's exact test) between parameters, and the strength of this correlation was characterized by the Cramer convergence factor of 9 for dichotomous variables. In case of the absence of even a single pair of inconsistent evaluations, this factor takes too inflated value, so in this case percentage of according evaluations was given. Independent attempts for the upper and lower jaw and premolars and molars for each characteristic from the same research period were compared using Chi^2 independence test or Fisher's exact test. For all statistical tests, in all tests performed, the level of significance a = 0.05 (Statistica v. 6.0 PL) was taken.

26.3 RESULTS

All 20 participants were subjected to clinical observation in the application phase. A 54 direct restoration of the studied material GJS2/W were evaluated. In the evaluation of parameters I, II and III, All made restorations done in the application phase achieved acceptable marks, as in the case of parameters IV, V and VI, which allowed the classification of the studied material to an acceptable cluster. A 18 participants (90%) reported on clinical observation at 12th week. Restorations in 8 women (40%) and 10 men (50%) were evaluated. The average age was 16.05 years. A 52 restorations (49 Class I cavities, 1 Class II cavity, 1 Class III cavity and 1 Class V cavity according to Black), 96.3%. In evaluation of the parameters I and II characterizing surface – gloss, smoothness and color were considered. All made restorations achieved marks within an acceptable cluster. Parameters defining the anatomical shape, discoloration, marginal seal and the occurrence of secondary caries due to the marks achieved allowed the classification of the studied material to the acceptable category. In the interview at 12th week prior to clinical observation, none of the study participants reported tooth pain, provoked or spontaneous. A 17 participants (85%) reported on clinical observation at 24th week. Restorations in 8 women (40%) and 9 men (45%) were evaluated. The average age was 16.5 years. Total of 49 restorations (90.7%) were subjected to clinical observation: 46 Class I cavities, 1 Class II cavity, 1 Class III cavity and 1 Class V cavity according to Black), 96.3%. In evaluation of the parameters I and II characterizing surface – gloss, smoothness and color of the direct restorations of individual classes of cavities according to Black were considered. All made restorations achieved acceptable marks and were classified as excellent, very good and good ones. Evaluated parameters of anatomical shape, discoloration, marginal fissure and secondary caries, due to achieved marks within an acceptable cluster, allowed the classification of most of them to the categories of excellent and very good, within their material groups. In one case (lower molar – Class I occlusal according to Black) restoration has been classified to the total re-reconstruction (unsatisfactory mark for a parameter specifying the anatomical shape and marginal seal). In one case restoration needed partial correction (satisfactory mark for a parameter specifying the marginal seal in lower molar – Class I occlusal according to Black). Cavities have been re-reconstructed. In the interview at 24th week prior to clinical observation, none of the study participants reported pain, provoked or spontaneous, in the teeth restored using the studied material GJS2/W. A 20 participants (100%) reported on clinical observation at 48th week. Restorations in 9 women (45%) and 11 men (55%) were evaluated. A 54 direct restorations (51 Class I cavities, 1 Class II cavity, 1 Class III cavity and 1 Class V cavity according to Black), representing 100% of All made restorations made using glass-polyalkenecarboxylic cement GJS2/W. In evaluation of the parameters I and II gloss, smoothness and surface color in all classes of cavities according to Black were considered. All made restorations achieved acceptable marks. Parameters defining the anatomical shape, the occurrence of discolorations, marginal microleakage and secondary caries, thanks to marks achieved in most cases within an acceptable cluster, allowed the classification of the studied material to the category of very good, within their material groups. In two cases restoration has been classified to the total re-reconstruction (unsatisfactory marks for a parameter specifying the anatomical shape – lower molar, Class I occlusal according to Black, and for a parameter specifying the marginal seal – lower molar, Class I occlusal according to Black). In one case restoration has been classified to the partial correction (satisfactory mark for a parameter specifying the anatomical shape and marginal seal – Class I occlusal according to Black in lower molar). Restorations have been re-reconstructed. In the interview at 48th week prior to clinical observation, none of the study participants reported pain, provoked or spontaneous, in the teeth restored using the studied material. 20 participants (100%) reported on clinical observation at 96th week. Restorations in 9 women (45%) and 11 men (55%) were evaluated. 54 direct restorations of cavities of hard structures of permanent teeth (51 Class I cavities, 1 Class II cavity, 1 Class III cavity and 1 Class V cavity according to Black), representing 100% of All made restorations made using GJS2/W cement. In evaluation of the parameters I and II characterizing the surface, gloss, smoothness and color in all classes of cavities according to Black were considered. All made restorations achieved marks within an acceptable cluster Parameters defining the anatomical shape, the occurrence of discolorations, marginal microleakage and secondary caries, thanks to marks achieved in most cases within an acceptable cluster, allowed the classification of the studied material to the category of very good. In the interview at 24th week prior to clinical observation, none of the study participants reported pain, provoked or spontaneous, in the teeth restored using the studied material.

26.4 RESULTS AND DISCUSSION

The results of clinical observation on the parameter I obtained at 12th week differed significantly ($p = 0.00001$) from the results of the application phase. In 25% of cases, excellent mark on color did not change after 12 weeks. This level of the number of marks consistent with the results of application phase, evaluated as excellent, remained at a constant level, with the exception of decline at 24th week. Particular attention should be paid to the improvement of the color evaluations after the first 12 weeks; restorations, which have been excellent at 12th week of observation, even though this evaluation in the application phase varied from excellent (very good or good), accounted for 51.9% of all restorations. After another 12 weeks, the percentage increased to 63.3% and remained at a similar level at 48th week (63.0%) and 96th week (59.3%). Taking into account the results from adjacent weeks of clinical observation, it is clear that the largest percentage of restorations, among which an improvement of the color occurred, falls on the 12th week. In the subsequent weeks, a gradual relative decrease in the percentage of restorations with improved color was evaluated; starting from 24th week the following values were respectively obtained: 14.4%, 8.2% and 1.9%. Improvement of the color evaluations can be commented on taking into account the percentage of cases of pairs, which achieved excellent marks in adjacent weeks of clinical observation. From 25% of excellent marks at 12th week in relation to the application phase, this percentage increased to the value of 68.8% for 12th and 24th week, 79.6% for 24th and 48th week and 81.5% for 48th and 96th week. Statistical analysis of the improvement of parameter I confirmed that a highly significant ($p = 0.00001$) progression within this parameter occurred in the first 12 weeks in relation to the application of the studied material. The subsequent periods of observation in relation to the previous ones revealed no statistically significant changes ($p = 0.3428$, $p = 0.6831$, $p = 3711$). Statistical evaluation of the entire 96-week period of clinical observation gave a highly significant result ($p = 0.00001$). From a small percentage (27.1%) of direct restorations made using the studied material GJS2/W, classified as excellent in terms of color in the application phase, high percentages of the maximal marks of the acceptable cluster in the subsequent weeks of observation were obtained; starting from 12th week the following values were respectively obtained: 75.0%, 83.3%, 87.5% and 81.2% at 96th week. Statistical analysis of the changes of the evaluations of smoothness and gloss in the subsequent weeks of clinical observation in relation to the application phase revealed no statistically significant result. Similarly, comparing the results of inconsistent marks in adjacent weeks of observation has shown that changes in this parameter were insignificant.

Overall evaluation of the entire study period revealed that changes in the parameter II within 96 weeks of clinical observation were not statistically significant ($p = 0.2911$). The first 12 weeks gave no significant changes in the parameter III ($p = 1.0000$). Comparing the results of the 24th week to the results of the application phase revealed a trend toward significant difference of inconsistent pairs ($p = 0.0625$), 48th week revealed a statistically significant change in the same correlation ($p = 0.0078$), as well as the situation at 96th week, which was confirmed by the lack of statistically significant changes ($p = 0.4795$) between 48th and 96th week. Influence of time on significance of changes of the evaluation of the anatomical shape in relation to the application phase was the result of a gradual increase in the number of pairs of inconsistent marks (excellent in the application phase and different from perfect in subsequent periods of clinical observation (I, RI)); from the initial 3.9% at 12th week, by 10.2% at 24th week, to 14.8% at weeks 48th and 96th. Analysis of changes of the evaluation of the anatomical shape in the subsequent weeks of clinical observation revealed no such process, showing statistical insignificance. Overall evaluation of the entire study period confirmed the variability of the parameter III on the level of statistical significance at $p = 0.0002$. In the subsequent weeks a gradual decrease of the percentage of the maximal marks was observed – from 87.5% in the application phase to 70.8% at 96th week. Influence of time on the change of parameter IV was revealed at 24th week; the difference of inconsistent results compared to the application phase proved to be statistically significant ($p = 0.0156$). The subsequent weeks revealed the increasing percentage of the evaluations of restorations various from perfect that in the application phase achieved excellent marks, which indicates deterioration. It was 46.3% of cases in 96th week. Statistical analysis of the changes of weeks 48th and 96th in relation to the application phase indicated a highly significant difference in the number of pairs of inconsistent marks, respectively, $p = 0.0001$ and $p = 0.00001$. Statistical breakthrough of the evaluations of parameter IV for the results of adjacent weeks occurred at 48th week. Change of the evaluation of this parameter in relation to the previous one, of the 24th week, turned out to be highly significant ($p = 0.0056$). Comparing the results of weeks 96th and 48th reveals a statistically significant difference ($p = 0.0233$). Analyzing the percentage of perfect results in all individual weeks of clinical observation revealed a gradual decrease in the number of restorations having such evaluation. Change of the percentage of excellent marks of this parameter in the entire period of clinical observation gave a highly significant result ($p = 0.00001$). During application phase a marginal seal achieved excellent marks in 98.1% (88.5% + 9.6%) cases; after 12 weeks, 9.6% of restorations revealed deterioration of marginal seal. This change revealed a trend toward statistical significance ($p = 0.0736$). Statistically significant ($p = 0.0078$) change of parameter V in relation to the application phase was observed at 24th week; deterioration of seal was found in 16.3% of cases. After another 24 weeks, or in 48th week of observation, level of seal remained at a similar level. The percentage of restorations having marks consistent or inconsistent with their status in the application phase was similar to that at 24th week of clinical observation. At 96th week the percentage of failures compared to the perfect initial seal decreased to 11.1%, but still was statistically significant ($p = 0.0313$). Analysis of the evaluations of parameter V allowed to notice a gradual reduction of the percentage of cases with excellent marks from 97.9% in the application phase to 81.3% in 48th week, and a slight increase to 85.4% in 96th week. Process of change in the number of restorations with excellent marks turned out to be statistically significant ($p = 0.0017$). It should be evaluated that comparing marginal seal in two adjacent weeks of clinical observation revealed no statistical significance. Among 100% of excellent marks in the application phase symptoms of secondary caries at 12th week of clinical observation were noticed only in 3.9% of cases. This percentage increased slightly at 24th week, to the value of 10.2%, but this change allowed to notice only the existence of trend ($p = 0.0625$) towards the statistical significance. In the subsequent weeks of observation (weeks 48th and 96th) the situation has improved to the extent that – as noticed – the difference in the percentage of pairs inconsistent with the evaluation in the application phase was statistically insignificant ($p = 0.2500$ and $p = 0.1250$). Comparing pairs of inconsistent marks achieved in adjacent weeks of observation allowed concluding that the changes of the evaluation of parameter VI in the subsequent weeks of observation were statistically insignificant. Statistical evaluation of changes in the entire study period gave a result very close to statistical significance ($p = 0.0552$). During 24 weeks decrease in the percentage of cases with excellent marks was observed, from 100% in the application phase to 89.6% at 24th week, which was followed by the improvement and the percentage of excellent marks increased to 93.8% at 48th week and 91.7% at 96th week (Table 1).

TABLE 1 Percentage (Number) of excellent marks in each week of the period of clinical observation for each of the studied parameters. Sample size 48; *p* – *p*-value of Cochran's Q test.

Week of observation	Parameters according to Ryge scale					
	Color	Smoothness and gloss	Shape	Discoloration	Aperture	Secondary caries
0	27.1 (13)	64.6 (31)	87.5 (42)	100.0 (48)	97.9 (47)	100.0 (48)
12	75.0 (36)	75.0 (36)	85.4 (41)	91.7 (44)	87.5 (42)	95.8 (46)
24	83.3 (40)	70.8 (34)	77.1 (37)	85.4 (41)	81.3 (39)	89.6 (43)
48	87.5 (42)	72.9 (35)	70.8 (34)	62.5 (30)	81.3 (39)	93.8 (45)
96	81.3 (39)	62.5 (30)	70.8 (34)	54.2 (26)	85.4 (26)	91.7 (44)
p	0.00001	0.2911	0.0002	0.00001	0.0017	0.0552

Modern methods of restoration of tooth defects (partial or total loss) could be grouped into three circles: natural (regeneration – including the remineralization of enamel epithelium, cloning, transplantation), foreign (non-organic: metals, non-metals (ceramics, carbon, etc.), organic (polymers), and their composites) and biomimicric (synthetic hard tissues).

Philosophy of direct application of foreign substitutes of tooth color is based on two pillars, namely RC and GIC, and their various qualitative and quantitative combinations, variations, hybridizations. RC's provide strong arguments for their potentially superior abilities of accurate imitation of the tooth hard tooth structures. On the other hand, pejorative resonance still accumulates in the complicated and ineffective over time, especially from the practical point of view, the adhesive technology, and their lack of anticariogenic activity. GIC's are characterized on the one hand by the phenomenon of bioactivity – an emission of fluorine, which is an advantage of action against recurrent caries and inherent adhesion to tooth substrate, resulting in dentin integration and biochemical binding, but unfortunately also insufficient aesthetics and questionable universal physicomechanical properties. The third millennium – the age of the information society, artificial intelligence, temporal travels, will definitely be the time of intense technological evolution of our civilization. Paraphrasing NHF Wilson [4], one shall hope that progress in other fields of science (aeronautics, microelectronics, biotechnology, cybernetics) – transfer of technology – will be invaluable contribution to the positive qualification and modernization of dentistry.

Due to the unsatisfactory strength values of GIC's and hence limited clinical application for the restorations of Classes III and V cavities according to Black, for many years modifications of the chemical composition of the glass filler were carried out, such as by incorporation of phase dispersive glass containing enhanced crystallites, such as alumina (Al_2O_3), rutile (TiO_2), brazilite (ZrO_2) or tielite (Al_2TiO_5), glasses reinforced with aluminum, silicon or carbon fibres, HAp whiskers and grain [$Ca_{10}(PO_4)_6(OH)_2$] in the form of aggregates consisting of primary hexagonal crystals (0.3–0.5 μm × 2–3 μm) or minerals such as borax $Na_2[B_4O_5(OH)_4] \cdot 8\ H_2O$ [5, 6]. Improvement of biocompatibility and strength properties of GIC was attempted to achieve by doping with up to 12% of the particles of synthetic HAp/ZrO_2. [7] The incorporation of the nanoparticles significantly reduced CS during the first 24 hr and also shortened worktime. Only higher, 10% share of fraction of $BaSO_4$ shortened net solidification time. According to the authors, the use of opaque nanoparticles, e.g., $BaSO_{4,}$ can positively influence the translucency of GIC's and their aesthetic values [8]. According to Randal and Wilson [9], meta-analysis carried out for the years from 1970 to 1996 for clinical studies on secondary caries, CGIC's and RMGIC's applied for direct restorations revealed that the size of the cavity, randomization of the treatment methods and control groups are the basis for achieving important results in a clinical study evaluating the therapeutic effectiveness. Results of such studies are rarely publicized. In turn, lack of clear and binding confirmations on medicinal influence on secondary caries in articles may be associated with a variety of GIC, different in terms of bacteriostatic effect, emission of fluorine, etc. Lasaffre et al. [10] found that in a number of clinical studies of the oral cavity, if several researchers take part in them, differences arise in their individual evaluations, which in turn has a significant influence on the final, overall result of the study. Several authors in clinical studies of dental materials ensured the participation of two independent observers and, in the case of non-compliance, the verification of an independent expert [11]. Research project involving one observer is one of choice in clinical studies [12]. Wilson [4] concludes that carrying out of the comparative clinical studies of various GIC's is difficult due to the patient factor. Researchers in clinical studies applied usually USPHS criteria, using Ryge rating scale (Romeo, Sierra, Tango, Victor) or its modification (Oscar – excellent; Alfa – very good; Bravo – visible cavities, but restoration does not require replacement; Charlie – clinical failure of restoration requiring immediate supply; Delta – massive damage of restoration (mobility or fallout) requiring immediate supply). Most clinical studies on GIC's had mass, quantitative dimension [13] Meta-analysis by Frencken et al. [14] revealed that WMGIC's with enhanced strength parameters introduced to the market of dental materials were alternative in restorations of occlusal cavities on one surface in permanent teeth.

Clinical experiment was based on a representative selection of Class I occlusal and one of each Class I cavity within foramen cecum, Class II cavity, Class III cavity and Class V cavity, according to Black, which is a classic sign of application predestination of GIC. Direct clinical study was based on standard USPHS parameters, according to 4-point (0–3) Ryge scale, while indirect/virtual analysis in own modification using CorelDRAW Graphics 12, aimed at its optimizing and detailing. Concerns articulated by other dental researchers on number and calibration of the studying persons influenced the decision on the two-level direct-indirect method of one observer. The relatively low number of patients and trials resulted from the assumption to carry out the project with high quality and precision of the restoration and frequency [15, 16]. Planned periods of clinical observation, reached 96th week of study and the results obtained appear to correspond with the literature data. Studied material GJS2/W, polydispersive in the structure, characterized by granulation <45 μm, well-thought-out fusion, distribution and reactivity of strontium and barium oxides, sodium with acidic polymer lyophilisate PAA/TA, as well as average CS of 160 MPa and a fast coagulation, found its qualitative and quantitative basis for experimental clinical use in direct restorations of occlusal cavities of moderate size in permanent teeth. Medical experiment project included the treatment of tooth hard tissues with 25% aqueous solution of PAA for 20 sec, followed by gentle rinsing and drying. According to Yiu et al. [17] GIC's, due to the sorption of water from the moist dentine may lead to local postoperative hypersensitivity; is likely to create a silicon-rich phases in the air voids present in the tissue associated with the GIC, and all this can be due to movements of the water in the intermediate/gel layer. Studied material GJS2/W in no case applied hypersensitivity. Non-mass nature of the study was reflected in the conditions of the study site itself, in its intimacy, maintaining punctuality, courtesy and the right aura that is creating a climate of good practice and clinical study, also by playing relaxing music. Results statistically confirm that GJS2/W could be successfully applied in restoration of occlusal cavities of moderate size 2 by Mount and Hume in permanent dentition.

Starting from teeth of *Sahelanthropus thadensis* with enamel erosion, estimated at 7 million years, the settlement of Mehrgahr with traces of drilling teeth aged 9,000 years; 4,500 years old dentures of ligature wire allogeneic teeth in Giza, 2,600 years old Etruscan cosmetic bridges with human or calf teeth, combined using metal ligature, soldering and rings, through 2,000 years old Celsus' *plumbum* – plugging of cavities using canvas and lead, 1,412 years old first Mayan implantations of pieces of shell and obsidian to dental alveolus after extractions, 460 years old prototype of Guillemeau's porcelain teeth, to the contemporary intraosseal ceramic implantations and CAD/CAM restorations, regenerations using extracellular matrix, resorbable scaffolds, banked cryopreservations of stem cells, creation of autogenous tooth, creation of tooth hard tissue – the development of civilization is a dynamic and accelerating process, where the studied material found its place as part of the progress of dental materials science and evolution of restorative concept towards the biomimicric dentistry.

KEYWORDS

- **Continuing gums**
- **Guillemeau's porcelain teeth**
- **Mehrgarh settlement**
- **Occlusal cavities**
- ***Plumbum***
- ***Sahelanthropus thadensis***
- **Silver cake**

REFERENCES

1. Kupka, T. W. WMGIC experimental adhesive-type biomaterial on the road of the development towards the concept of restorative dentistry Biomimicry. In *13th International Scientific Conference "Dental Engineering Biomaterials."* Ustroń; 1–2.06.2012; Abstr. 2.5; 46–47.
2. Brunet, M.; Guy, F.; Pilbeam, D.; Mackaye, H. T.; Likius, A.; Ahounta, D.; Beauvilain, A.; Blondel, C.; Bocherens, H.; Boisserie, J. R A new hominid from the Upper Miocene of Chad, Central Africa. *Nature* **2002**, *418 (6894)*, 145–151.
3. Hoffman-Axthelm, W. *History of Dentistry*. Chicago, 1981.
4. Wilson, N. H. F. Direct adhesive materials: current perceptions and evidence – future solutions. *J. Dent.* **2001**, *29*, 307–316.
5. Lucas, M. E.; Arita, K.; Nishino, M. Toughness, bonding and fluoride-release properties of hydroxyapatite-added glass ionomer cement. *Biomaterials,* **2003**, *24*, 3787–3794.
6. Nagaraja, U. P.; Kishore, G. Glass Ionomer Cement – the Different Generations. *Trends Biomater Artif Organs* **2005**, *18*, 158–165.
7. Gu, Y. W.; Yap, A. U.; Cheang, P.; Khor, K. A. Effect of incorporation of HA/ZrO_2 into glass ionomer cement (GIC). *Biomaterials* **2005**, *26*, 713–720.
8. Prentice, L. H.; Tyas, M. J.; Burrow, M. F. The effect of ytterbium fluoride and barium sulphate nanoparticles on the reactivity and strength of a glass ionomer cement. *Dent. Mater.* **2006**, *22*, 746–751.
9. Randal, R. C.; Wilson, N. H. Glass ionomer restoratives: a systematic review of a secondary caries effect. *J. Dent. Res.* **1999**; *78*, 628–637.
10. Lesaffre, E.; Mwalili, S. M.; Declerck, D. Analysis of caries experience taking interobserver bias and variability into account. *J. Dent. Res.* **2004**, *83*, 951–955.
11. Chinelatti, M. A.; Jan, W. V.; Pallesen, U. Clinical performance of a resin-modified glass ionomer and two polyacid-modified resin composites in cervical lesions restorations: 1-year follow-up. *J. Oral Rehabil.* **2004**, *31*, 251–257.

12. Księżarek, S. Clinical evaluation of a new Dentinmet silver amalgam. 1-year observation. *Ann. Acad. Med. Siles.* **2002**, *46*, 191–193.
13. Frencken, J. E.; Taifour, D.; van't Hof, M. A. Survival of ART and amalgam restorations in permanent teeth of children after 6.3 years. *J. Dent. Res.* **2006**, *85*, 622–626.
14. Frencken, J. E.; van't Hof, M. A.; van Amerongen, E. C.; Holmgren, C. J. Effectiveness of single-surface ART restorations in the permanent dentition: a meta-analysis. *J. Dent. Res.* **2004**, *83*, 120–123.
15. Hickel, R.; Peschke, A.; Tyas, M.; Mjör, I.; Bayne, S.; Peters, M.; Hiller, K. A,; Randall, R.; Vanherle, G.; Heintze, S. D. FDI World Dental Federation – clinical criteria for the evaluation of direct and indirect restorations. Update and clinical examples. *J. Adhes. Dent.* **2010**, *12*, 259–272.
16. Naik, A. D., Petersen, L. A. The neglected purpose of comparative-effectiveness research. *New England J. Med.* **2009**, *360 (19)*, 1929–1931.
17. Yiu, C. K. Y.; Tay1, F. R.; King, N. M.; Pashley, D. H.; Sidhu, S. K.; Neo, J. C. L.; Toledano, M.; Wong, S. L. Interaction of glass-ionomer cements with moist dentin. *J. Dent. Res.* **2004**, *83*, 283–289.

CHAPTER 27

MICROSTRUCTURE OF DENTAL CASTING ALLOY NI-CR-MO (RODENT)

A. KORNEVA, I. ORLICKA, K. SZTWIERTNIA, and G. E. ZAIKOV

CONTENTS

27.1 INTRODUCTION

Microstructure of the dental alloy Ni-Cr-Mo Rodent after cast in the vacuum-pressure furnace has been examined. Chemical analysis by using energy dispersive spectroscopy as well as crystallographic orientations topography by using backscattered electron diffraction in the scanning electron microscope was carried out. It was found that after cast microstructure consists of dendrites of solid-solution strengthened nickel–chromium matrix (γ phase). Inside the dendrites the precipitations of molybdenum silicides, eutectics γ–P and few particles of aluminum oxide were observed. It is observed that the rising interest for cobalt-chrome and chrome-nickel alloys since they became available for dentists, which used them for casting removable partial dentures. At present these alloys have replaced the use of type IV gold ones, and almost all prostheses skeletons are made from them. Furthermore, these materials are used as substitutes for gold alloy type III [1]. Chrome-nickel and chromium-cobalt alloys are also used in metal-ceramic prosthetic restorations.

New dental casting alloy Ni-Cr-Mo (Rodent) has been developed for permanent, porcelain faced dentures at the Meissner Higher School of Dental Engineering in Ustron. Its chemical composition fulfills the requirements the European standard ADA. However, both the chemical composition, the microstructure and hardness slightly differ from others dental alloys of the same type [2].

27.2 RESEARCH METODOLOGIES

The Rodent for testing was obtained by melting of the original alloys. At first a prototype was made of wax then the mold was prepared. Casting and annealing of the mold was performed in a vacuum/pressure furnace.

Examinations of the microstructure were carried out on an optical microscope Leica QWin and in the scanning electron microscope SEM XL30, Philips. In order to illustrate microstructures (the phase contrast and crystallographic orientation maps) backscattered electrons (BSE) and diffractions of them (EBSD) in SEM were used. For observations of surface morphology secondary electrons (SE) were used.

Microsections of samples were prepared by grinding with abrasive paper and then cloth polished using diamond pastes of decreasing grit (up to 0.25 μm). Electrolytic polishing and etching was carried out in the reagent composed of: 100 ml of perchloric acid HCrO4 and 900 ml of acetic acid CH3COOH, at the voltage of 20 and 13 V, respectively.

The chemical composition of the Rodent alloy was tested in XL30 SEM, Philips by EDS Link ISIS X-ray. The results of analysis are shown in Table 1. The share of alumina was calculated using appropriate software QWin Leica for optical microscope.

TABLE 1 The chemical composition of the Rodent alloy.

The content of elements, % wt.				
Ni	Cr	Mo	Si	Fe
62.7±0.2	24.3±0.1	10.2±0.1	1.7±0.1	1.1±0.1

X-ray phase analysis was carried out with help of PW1710 diffractometer (Philips), CoKα.

27.3 RESULTS AND DISCUSSION

Microstructure after casting of Rodent consists of grains/dendrites sizes up to a few hundred micrometers. Inside the grains are uniformly distributed second phase precipitations that originated during alloy solidification as a result of a superfusion on growing dendrites boundaries (Fig. 1a).

The matrix of alloy is a solution of chromium and molybdenum in nickel, crystallizes as FCC austenite (γ phase), Fig. 1b. EDS/SEM investigations have revealed that in the matrix are dispersed mainly precipitations of three other phases. These are eutectics γ-P (shown in Fig. 1b as the A), precipitations of molybdenum silicides (marked as B) and precipitations of alumina (marked as C). Very few particles of alumina are evenly distributed in the matrix, and their global share was estimated at approximately 0.1%. The chemical composition of the matrix and the main precipitates of the Rodent alloy are shown in the Table 2. In comparison with matrix phase P contains 2.5 times more Mo, while the precipitations of molybdenum silicide contain 5 times more Mo, 4.4 times more Si and 1.9 times less Ni, 1.5 times less Cr as well as 4 times less Fe.

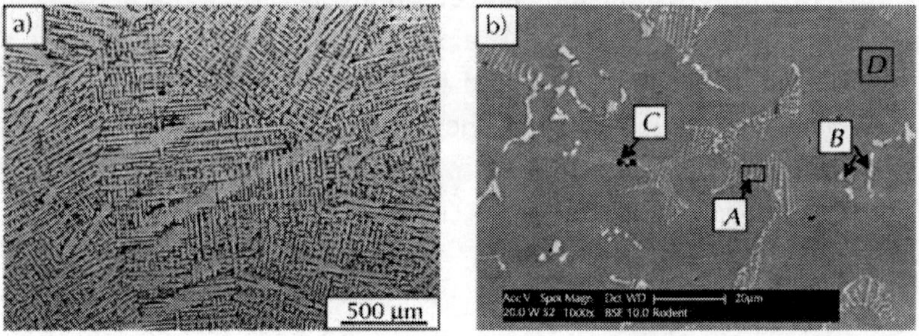

FIGURE 1 Microstructures after casting.

TABLE 2 The chemical composition of the matrix and the main precipitates of the Rodent alloy.

The place of rearseach	The content of elements, % wt.				
	Ni	Cr	Mo	Si	Fe
faza γ-P (place ***A***)	51.4±0.2	22.4±0.1	21.9±0.2	3.6±0.1	0.7±0.1
molybdenum silicides (place ***B***)	34.5±0.2	16.7±0.1	42.4±0.2	6.1±0.1	0.3±0.1
γ phase (matrix) (place ***D***)	64.4±0.2	24.5±0.1	8.5±0.1	1.4±0.1	1.2±0.1

Figures 2–4 show the microstructure images of the Rodent achieved under BSE, SE and as results of chemical microanalysis EDS/SEM within the precipitations. Obtained data confirm the presence of alumina particles (Fig. 3a, b), γ-P eutectics (Fig. 3c, e) and the precipitations of molybdenum silicide (Fig. 3e, f) in the matrix of γ phase (Fig. 3c, d).

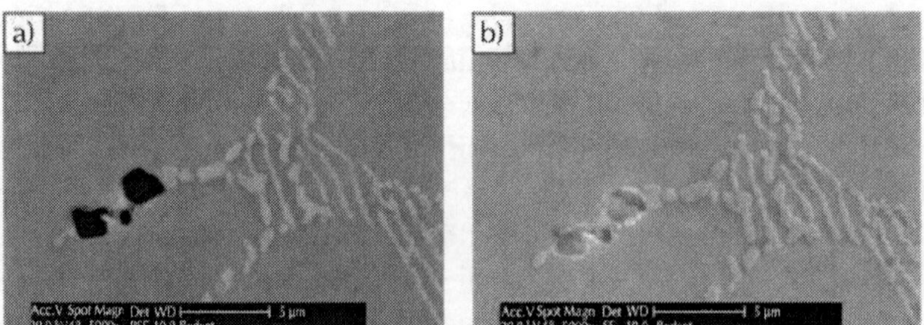

FIGURE 2 Microstructure images of the Rodent.

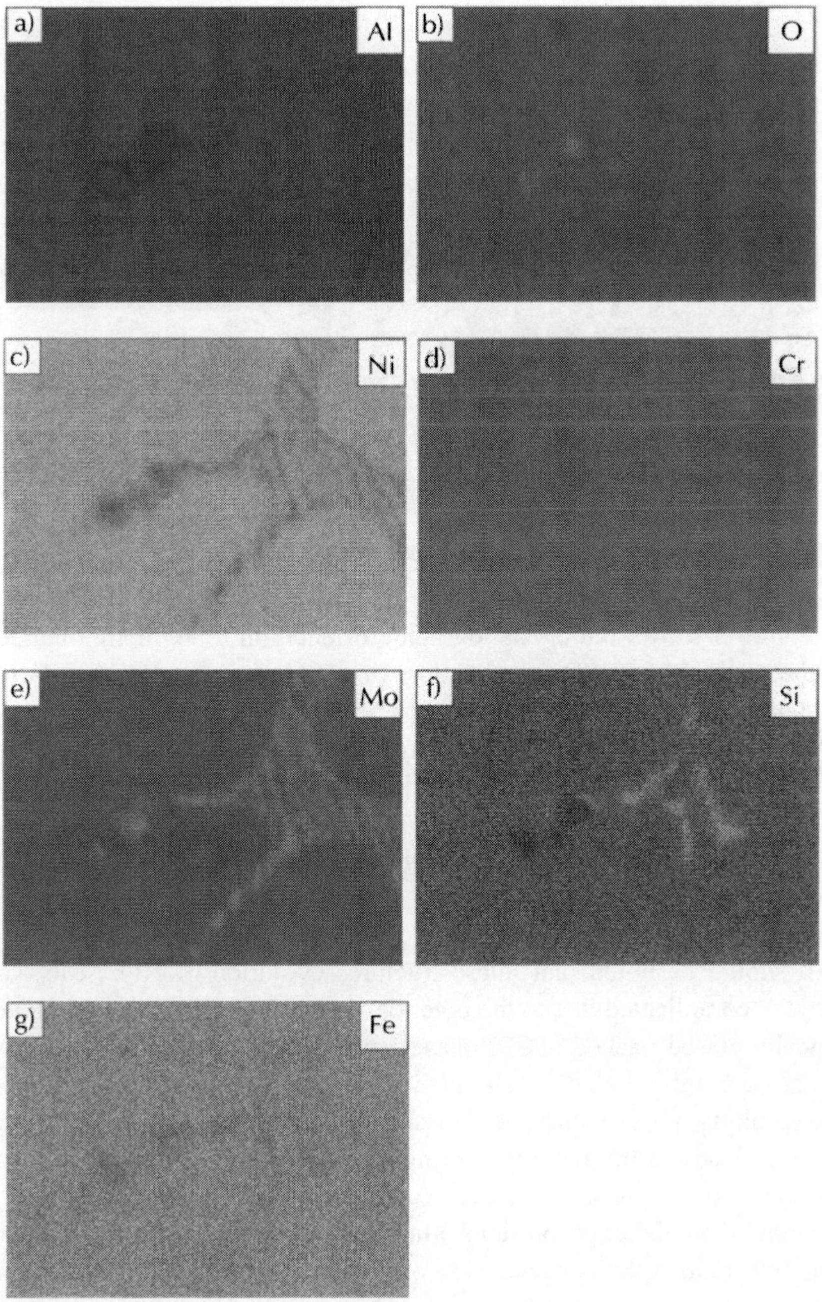

FIGURE 3 Microstructure images of the Rodent.

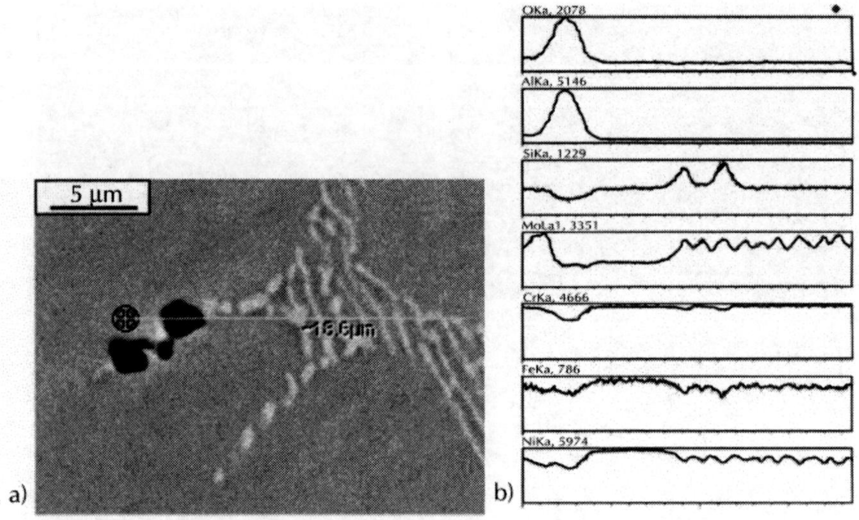

FIGURE 4 Microstructure images of the Rodent.

Figure 5 shows the crystallographic orientation topography obtained by the method of EBSD/SEM. Colors of the basing triangle identify the crystallographic direction, which is parallel to the normal to the sample surface. Changes of disorientation angle through boundary of two grains (line 1–1) as well as variation a disorientation angle inside the grain (line 2–2) are shown. Generally the grain/dendrites boundaries are characterized by rather high disorientation angles, when microstructure inside the grains is characterized by very small changes of orientation ($<1^{\circ}$), which indicates a lack of cell structure.

It should be noted that microstructures evolution of Ni-Cr-Mo type alloys is complicated due to the potential formation of intermetallic topologically closed packed (TCP) phases such as tetragonal σ, hexagonal μ and orthorhombic P that are stabilized by the presence of Mo [3]. Generally speaking, all of which are undesirable because their complex crystal structure leads to limited slip systems, which makes these phases brittle and can lead to reduced toughness and ductility if they are present in the material in too higher proportions. Furthermore, their high Cr and Mo contents will reduce the concentration of these components in the austenite matrix, reducing corrosion resistance. The μ and P phases are particularly enriched in Mo (chromium content remains like in the solution), and the σ phase is high in Cr [4].

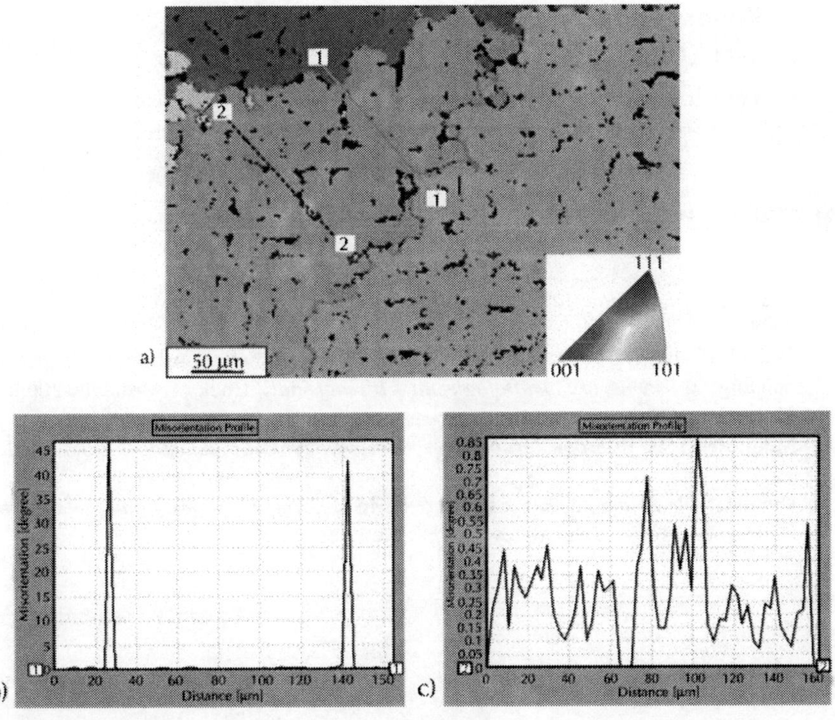

FIGURE 5 Crystallographic orientation topography.

27.4 CONCLUSION

Microstructure of Rodent after casting is composed of large grains/dendrites of the austenite matrix, which is the solid solution of nickel strengthened chromium and molybdenum. The matrix is additionally strengthened by dispersion of molybdenum silicide precipitates. A small number of carbide and aluminum oxide particles were also observed. The presence of eutectics γ-P rather weakens the structure of the alloy.

KEYWORDS

- **Nickel–chromium matrix**
- **QWin Leica**
- **Rodent alloy**
- **SEM XL30**
- **Topologically closed packed**

REFERENCES

1. Grain, R. G. In *Materiały Stomatologiczne*, Elsevier Urban and Partner: Wrocław, 2008.
2. Orlicki, R.; Pawlik, Ł. Casting alloy RODENT in prosthetic construction for porcelan bonding, In *Inżynieria Stomatologiczna – Biomateriały,* Book of Abstracts, 2009; 34–35.
3. Mikułowski, B. In *Stopy żaroodporne i żarowytrzymałe – nadstopy*, AGH: Kraków, 1997.
4. DuPont, J. N.; Lippold, J. C.; Kiser, S.D. In *Welding Metallurgy and Weldability of Nickel-Base Alloys*, John Wiley & Sons: New Jersey, 2009.

CHAPTER 28

DEVELOPING THE SCIENTIFIC BASIS FOR THE RATIONAL DESIGNING DEVICES GAS PURIFICATION

R. R. USMANOVA and G. E. ZAIKOV

CONTENTS

28.1 INTRODUCTION

The problem of cleaning gas emissions of fine impurities is one of the most pressing in the gas treatment and long highlighted by experimental and theoretical studies.

One of the most promising methods for increasing the efficiency of dust collection of fine particles is scrubbing. For this method are characterized by complex mass transfer processes in the course of interaction with the gas-dispersion flow scrubbing liquid droplets, resulting in the speed and the concentration of the phase determining gas cleaning.

Existing studies in this area show a strong sensitivity of the output characteristics of the regime and design of the device, indicating that a qualitatively different flow hydrodynamics at different values of routine-design parameters.

Thus, a systematic review of the effectiveness of hydrodynamics and vortex devices, receipt and compilation dependencies between regime-design parameters of the device, the establishment of effective and manufacturable design and development of their mass production for wide use in industrial practice is an urgent task.

28.2 DEVELOPMENT OF THE CONSTRUCTION BUBBLE-VORTEX APPARATUS WITH ADJUSTABLE BLADES

To optimize the bubble-vortex apparatus conducted experimental studies. The experiments were performed by a single method [1] of comparative tests on dust collectors bubble-vortex apparatus with a cylindrical chamber 0.6 m and a diameter of 0.2 and 0.4 m bubble-vortex machine with adjustable blades in accordance with Fig. 1 comprises a cylindrical chamber 1 inlet pipe 2. The cylindrical chamber 1 is three swirl gas flow, which is a four blades, curved sinusoidal curve. Adjusting the blades 2 is done by turning the eccentrics, sealed with a cylindrical chamber 1 through the spring washers and lock nuts. [2]

Swirler before the midpoint of the gas flow nozzle 4, and after a swirler located peripheral nozzle 5, which serves scrubbing liquid. Removal of dispersed particles produced by sludge overflow pipe 6 chip catcher 7.

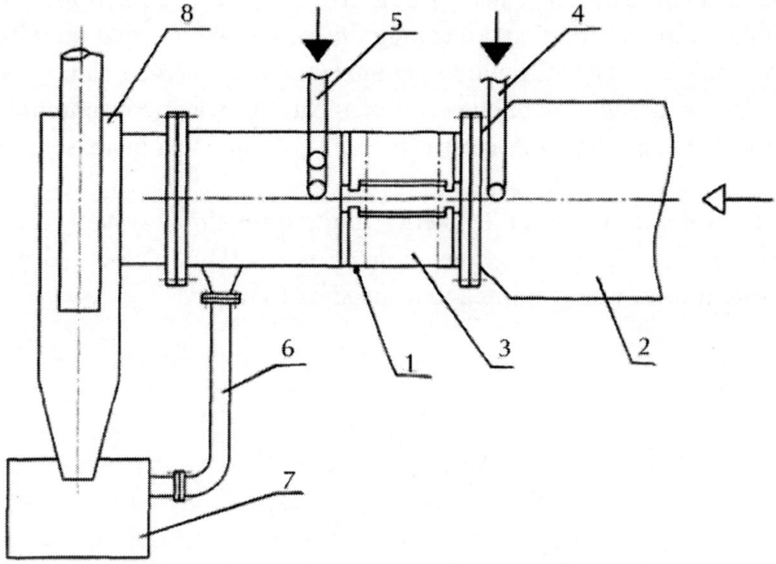

1 – cylindrical chamber, 2 – inlet pipe, 3 – swirl, 4 – central nozzle, 5 – peripheral nozzles, 6 – overflow pipe cuttings, 7 – chip catcher, 8 – cyclone.

FIGURE 1 Research facility.

Bubble-vortex machine with adjustable blades works as follows. The dusty gas is fed into a cylindrical chamber 1, where the swirl 3 with blades mounted in the radial grooves of the rod, deflects the flow and gives it a rotation. Under the action occurring at the same centrifugal forces dispersed particles are moved to the sides of the unit. For adjusting the blades at the entrance and exit of each blade 3 have two tabs. With their help, the blade is in contact with a pair of eccentrics. Eccentrics are turning vanes on the inlet and outlet sections of the cylindrical chamber 1 in different directions. With this blade 3 are installed in the position corresponding to the most efficient gas purification.

28.3 STUDY OF THE EFFECTIVENESS OF AIR CLEANING

Studies were conducted on a bubble-vortex apparatus with a cylindrical chamber with a diameter of 0.2 and 0.4 m as a model system were studied air and talcum powder with a particle size of $d = 2 \div 30$ microns. It identifies the total and fractional cleaning efficiency. We studied the effect on performance of cleaning mode parameters, which served as a - the total flow of air through the bubble-vortex apparatus, water consumption, spin factor K. Found that with increasing air flow rate is an increase cleaning, Fig. 2.

The optimal in terms of energy limits throughput system: the lower limit corresponds to the conventional cross-sectional velocity 5 m/s, the maximum flow rate is limited to a speed of 15 m/s.

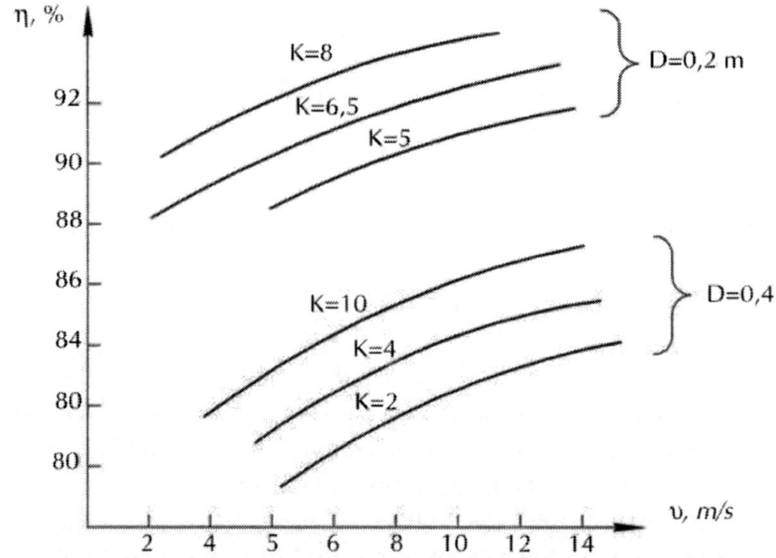

Dust: talc $d_m = 10mic; \sigma = 3,64; \rho_n = 2650\, ^{êg}\!/_{m^3}\; z = 5\, ^{g}\!/_{m^3}$;

FIGURE 2 Dependence on the effectiveness of cleaning the gas flow rate.

In bubble-vortex device at the minimum speed is reduced purity, with a maximum – a sharp increase hydraulic resistance.

It has also the effect of the coefficient K spin rotator on the effective dust collection: as K increases the degree of purification. Found a range of values K, at which the relatively high trapping efficiency ($K_{min} = 5$, $K_{max} = 8$). When $K = 1$, there is a significant reduction in treatment efficiency, with $K > 10$ performance remains almost constant, but the pressure drop in the apparatus to increase substantially. The influence of the diameter of the unit to the cleaning efficiency: with increasing diameter of collection efficiency is reduced, and the smaller the median diameter of the particles, the greater the fall cleaning efficiency. The method of calculation, one with the method of calculating cyclone [3], in which the total and fractional dust collection efficiency can be determined analytically:

$$\eta = 50 \cdot \left[1 + \Phi\left(x'\right)\right] \quad (1)$$

where

$$x' = \frac{\lg\left[\dfrac{d'_{50}}{d_{50} \cdot k \cdot 10^3 \cdot \sqrt{D \cdot \dfrac{\mu_g}{\rho_p} \cdot \vartheta_g}}\right]}{\sqrt{\sigma^2 + \lg^2\left(\dfrac{d_{50}}{d_{16}}\right)}},$$

$$\eta_f = \frac{1}{2} \cdot \left[1 + \Phi(x)\right]$$

where

$$x = \frac{\lg\left[\dfrac{d_+}{d_+ \cdot k \cdot 10^3 \cdot \sqrt{D \cdot \dfrac{\mu_g}{\rho_p} \cdot \vartheta_g}}\right]}{\sigma},$$

where, d'_{50} is the median distribution of dust particles entering the device, m; d_{50} is diameter particles captured with an efficiency of 50%, m; ϑ_g is a common velocity of the gas in the apparatus, m/s; μ_g is the dynamic viscosity of gas, Pa · m/s^2 ; ρ_p is particle density, kg/m^3 ; d_{16} is particle diameter at the entrance to the unit in which the mass of all the particles having a size less than 16% of the total dust mass, m; σ is value that characterizes the dispersion of the particles; κ is coefficient for this unit received $\kappa = 34,76$.

When preliminary calculations the crude treatment can be defined graphically, Fig. 3.

C is a function of the geometric parameters of the machine and can be calculated for vehicles designed by the known method [4]. Ψ is modified inertia parameter of the state of dust-gas systems.

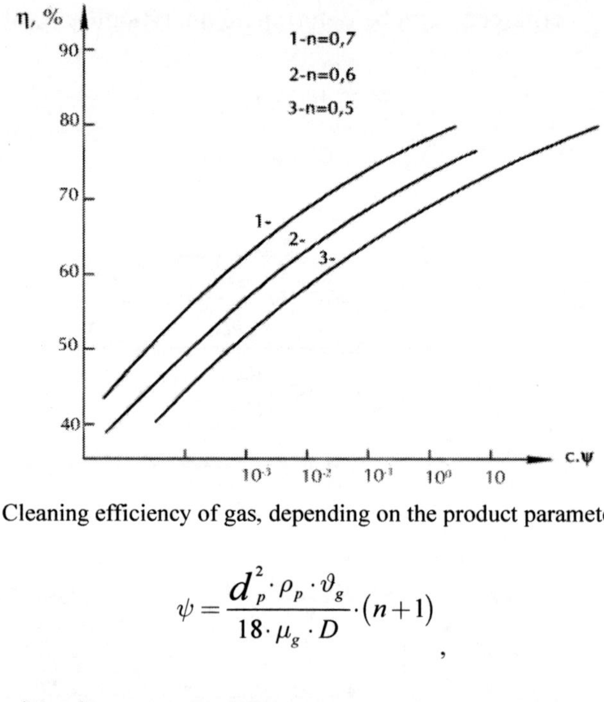

FIGURE 3 Cleaning efficiency of gas, depending on the product parameters $c \cdot \psi$

$$\psi = \frac{d_p^2 \cdot \rho_p \cdot \vartheta_g}{18 \cdot \mu_g \cdot D} \cdot (n+1),$$

Factor n is given by,

$$n = 1 - \left(1 - 0.0165 \cdot D^{0.15}\right) \cdot \left(\frac{T_g}{283}\right)^{0,3},$$

where, T_g is the absolute temperature of the gas, K; *D* is diameter apparatus, m.

28.4 STUDY OF HYDRODYNAMIC CHARACTERISTICS

Investigated the resistance bubble-vortex apparatus, depending on its classification regime-design parameters.

Found that the most efficient and cost effective mode of operation is at $K = 5–8$ [5].

The technique of pressure loss and specific energy consumption for dust control, which are determined by the formulas:

$$\Delta P = \xi \cdot \frac{\rho_g \cdot \vartheta^2}{2}, Ps; E = \frac{\Delta P}{3600}, \kappa Wt \cdot h / 1000 m^3$$

On this basis was constructed empirical mathematical model for calculating the coefficient of hydraulic resistance, including a formula to calculate ξ «dry» machine,

$$\xi_{dry} = \frac{1}{n}\left((R_m)^{2n} - 1\right) + \frac{\alpha}{K^2} \cdot \left(\frac{\vartheta_{out}}{\vartheta_{in}}\right)^2$$

empirical relationship for calculating the pressure drop in the gas transport liquid,

$$\xi_{tr} = 4 \cdot \left(\frac{Q}{G}\right)^{0,4} \cdot \sqrt{1 + \frac{1}{K^2}}$$

and ultimate dependence for calculating ξ irrigation apparatus,

$$\xi_{irr} = \frac{1}{n}\left((R_m)^{2n} - 1\right) + \alpha \frac{\varepsilon^2}{K^2} \cdot \left(1 + \frac{\rho_l}{\rho_g}\right) \cdot \left(\frac{\vartheta_{out}}{\vartheta_{in}}\right)^2 + 4\left(\frac{Q}{G}\right)^{0,4} \cdot \sqrt{1 + \frac{1}{K^2}} \quad (2)$$

where ξ is coefficient of hydraulic resistance; R_m is radius of the cylindrical chamber, m; $\rho_{n,}$ ρ_l is density of gas and liquid, kg/m³; υ_{in} υ_{ou} is velocity of gas at the inlet and outlet of the unit, m/s; *n, ε* is indicators vortex movement, *K*-factor twist swirler; *Q, G* is liquid and gas flow kg/m³; *α*-twist angle of the flow, °.

The resulting formula takes into account the presence of the dispersed phase and the partial loss of swirling flow.

The intensity of the twisting of the gas flow ratio was estimated geometric twist K_g.

$$K_g = \frac{32}{\pi^2} \cdot \frac{\vartheta_\varphi}{\vartheta_x} \cdot \frac{l}{D},$$

Because the value K_g does not match the real twist coefficient, taken the following relation:

$$K = 1.4 \cdot K_g^{0,72},$$

The experimental results are presented graphically dependencies of hydraulic resistance of regime-the design parameters, Figs. 4 and 5.

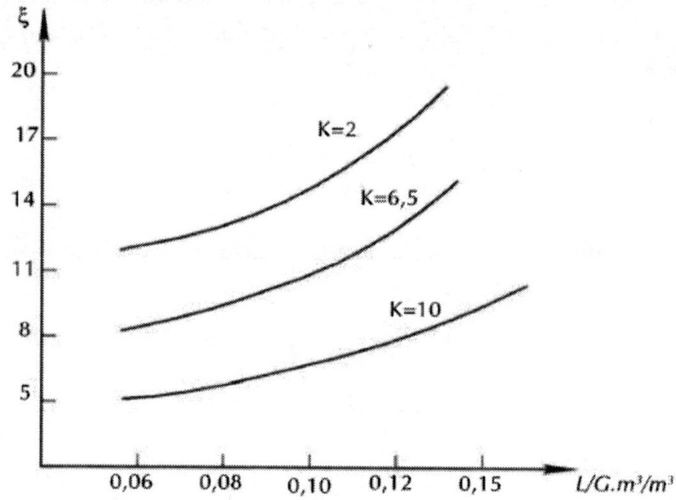

FIGURE 4 Dependence of ξ on the specific irrigation apparatus.

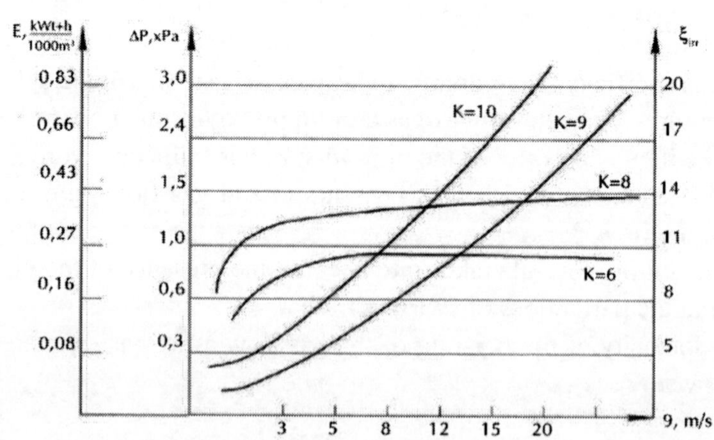

FIGURE 5 The dependence of the energy consumption for dust control and hydraulic resistance of the gas velocity in the machine.

28.5 CALCULATING THE COST OF DUST CONTROL

Using previously obtained correlations between the dust collection efficiency and pressure drop with regime – the design parameters of design procedure. The technique allows to select the unit with operational and design parameters under which it would provide the required process parameters at a minimum cost of cleaning.

The main components of the cost of treatment are worth the dust is not trapped unit (C_n), and the cost of energy in transmission through the apparatus of the gas flow (C_e).

The cost of treatment is determined by the formula:

$$C_0 = C_n + C_e,$$

Cost crude dust C_n decreases with increasing the efficiency of the device, with a decrease in the initial concentration of the dust, and with a decrease in the cost of the collected dust C_u.

$$C_n = (1-\eta)\cdot z_1 \cdot C_u,$$

The cost of electricity consumed for dust control, increases with increasing hydraulic resistance unit and is calculated by the formula:

$$\tilde{N}_e = \Delta P \cdot Q \cdot r \cdot C_{ue}$$

The full expression for calculating the cost of cleaning a gas cubic meter can be obtained using the Eq. (1) to calculate the efficiency and Eq. (2) to calculate the hydraulic resistance.

28.6 CONCLUSION

1. Developed a method of calculating the total and fractional dust collection efficiency, taking into account the geometric parameters of the device.
2. Developed a method for calculating the hydraulic resistance of a bubble-a vortex system, taking into account the structural parameters of the swirler and the presence of the dispersed phase.

3. The resulting formulas were the basis for the development of methods for calculating bubble-vortex apparatus. The developed method allows calculating the optimum geometry devices working optimally. Optimization criterion is the minimum cost of cleaning unit volume gas along with the required efficiency of dust collection.
4. The developed method can be used in the calculation and design of gas cleaning devices, as constituent relations define the relationship between the technological characteristics of the dust collectors and their geometrical and operational parameters.

KEYWORDS

- **Bubble-vortex apparatus**
- **Dry machine**
- **Dust-gas**
- **Hydraulic resistance**

REFERENCES

1. Adler, J. P.; Markov, E. V.; Granovsky, Y. In *Experimental Design in the Search for Optimal Conditions*, Nauka: Moscow, 1986; 279 p.
2. Usmanova, R. R.; Panov, A. K. Patent 2234358, RF bubble-vortex machine with adjustable blades, *Bull*, **2004**, 23.
3. Uzhov, V. N.; Valdberg, A. In *Preparation of Industrial Gases to Clean*, M.: Chemistry, 1975; 216 p.
4. Leith, D.; Licht, W. Symposium series, *Air*, 1971; 12 p.
5. Usmanova, R. R.; Panov, A. K.; Zaikov, G. E. Complex aerohydrodynamic research and the efficiency of arresting particles for barbotage – rotation, *J. Balkan Tribologi. Asso.*. **2006**, *3*, 368–373.

CHAPTER 29

SYNTHESIS OF BIOINORGANIC POLYMERS

V. KABLOV, D. KONDRUTSKY, A. D'AMORE, and O. ZINEEVA

CONTENTS

29.1 INTRODUCTION

All over the world there is a big interest to template synthesis as to a perspective method of reception new materials connections from more simple blocks more and more grows, reception which is complicated by traditional methods, and sometimes and at all it is impossible [1,2].

In this chapter, results of research template synthesis of bioinorganic polymers on ions of transitive metals are discussed. The chemical structure of bioinorganic polymers, sorption properties and thermal-oxidative degradation are investigated.

The urgency of work is commensurable with opportunities which are opened with use of the polymers received by template synthesis, as ecologically safe and economic sorption materials, and also as components fire-heat-shielding of coverings.

29.2 EXPERIMENTAL METHODS

In this method polymer food gelatin is used as the supplier of an ion of metal – various salts ($NiCl_2{\cdot}6H_2O$, $ZnSO_2{\cdot}7H_2O$, $CoCl_2{\cdot}6H_2O$), as the sewing agent – formalin, and as forming porous agent – a hydrocarbonate of sodium.

The gelatin preliminary dissolved in water, mixed up with a solution of salt of nickel, zinc or cobalt. To this mix, the solution of a hydrocarbonate of sodium was added. To the received complex of gelatin with salt and soda, formalin was slowly flowed.

Synthesis was spent in soft conditions, in the water environment at temperature 30–40°C. Reaction proceeds with high speed, in a minute after mixture of all reagents the product is formed. The received polymer represents a macromolecular complex on ions of nickel (BIP-Ni), zinc (BIP-Zn) or cobalt (BIP-Co).

The bioinorganic matrix of comparison on gelatin BP-M gelatin was received with use of the same reagents except for salt of metal.

For the study, sorption properties of polymers was received a bioinorganic polymer-matrix (BIP-M) by extraction template ion of nickel by 2% hydrochloric acid at intensive hashing in current of 20–30 min.

For the study, properties of the received bioinorganic polymers were done thermal-oxidative degradation by a method of pyrolysis at presence of oxygen of air at high-temperature electric furnace SNOL.

Samples of bioinorganic polymers were preliminary weighed and were located in crucibles with known weight. Thermal-oxidative degradation was spent up to 500°C – the temperature value necessary for decomposition of an organic basis of polymers.

A SPECORD-M82 was used to directly characterize the polypeptides. IR spectrum was performed for the product in form of thin suspension into vaseline.

For researches of sorption properties a series of solutions of chloride of nickel with $C_н(Ni^{2+})$ = 0.05 mol/l, 0.01 mol/l, and 0.005 mol/l and sulfate of copper with $C_н$ (Cu^{2+}) = 0.05 mol/l, 0.01 mol/l, and 0.005 mol/l were prepared. For sorption were used samples of water BIP-M and BP-M gelatin. Samples of polymeric matrixes in weight 5 g were located in plastic glasses and were filled in with the prepared solutions of chloride of nickel and sulfate of copper. Later 5 d contents glasses was filtered and washed out by the distilled water. The received polymer was left on drying for definition sorption abilities BIP-M and BP-M Gelatin and calculation to the SEC.

Differential-thermal analysis (DTA) of product and gelatin was carried out from 25 to 600°C using a derivatograph "MOM," Hungary. The mass of samples was 120 mg. Rate of heating was 10°C per min. Sensitivity DTA was 1/5°C.

29.3 RESULTS AND DISCUSSION

29.3.1 IR SPECTROMETRY OF PRODUCTS AND REAGENTS

IR spectral research BIP-Ni, BIP-M, gelatin and BP-M gelatin has shown, that in spectra of substances there are no valent fluctuations of amin-groups to which wave numbers in the field of 3,100–3,400/sm, but which are present at a spectrum of gelatin.

Frequencies of absorption of IR in areas 1,540–1,650/sm, 1,360–1,450/sm, 1,075–1,190/sm corresponding carboxy group, are displaced in case of BIP-Ni and BIP-M in area with greater wave numbers that it is possible to explain their participation in formation of coordination communications on ions of nickel.

Strong strips of absorption in areas 1,630–1,680, and 1,550/sm adequating to fluctuations aamide groups (Amide I and Amide II) also are displaced in case of BIP-Ni in more long-wave area that speaks processes intermolecular forming of complex.

In all substances the maintenance OH-groups, that shows in deformation fluctuations in the field of 850–960/sm.

In BIP-Ni, BIP-M and BP-M Gelatin the maintenance – CH_2 – groups which source is formaldehyde increases, on what specifies increase in intensity of absorption of IR in the field of 1,235–1,255/sm.

Comparison IR of spectra for polymers on ions Ni^{2+}, Zn^{2+} and Co^{2+} and on gelatin has shown matrixes of comparison, that each of three ions approximately to the same extent displaces waves carboxyl groups, amide groups in comparison with the polymer synthesized without participation of template ion that show occurring processes of complex forming.

29.3.2 THE PROBABLE MECHANISM OF REACTION TEMPLATE SYNTHESIS

Studied reaction represents reception of a complex of gelatin and salts of transitive metal and the subsequent interaction of formaldehyde with this complex. Reaction goes to some stages.

At the first stage there is an interaction of gelatin with salt of transitive metal with formation of a complex of gelatin with an ion of metal:

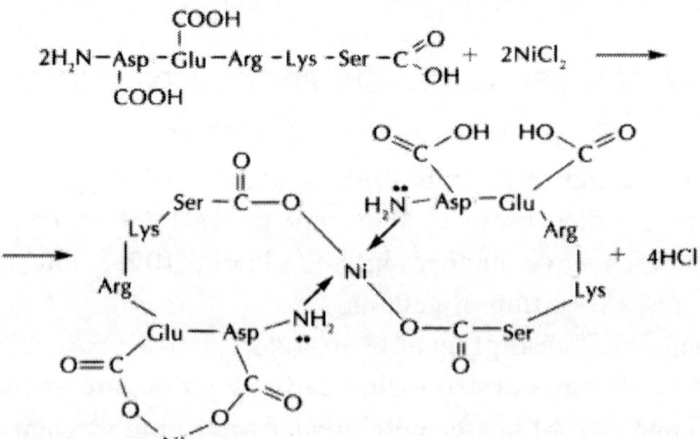

In the same time proceeds the reaction of forming porous with participation of a sodium hydrocarbonate.

At the second stage, there is an attack of atom of carbon groups of formaldehyde in free electronic pair atom of nitrogen of a trailer amino group of gelatin:

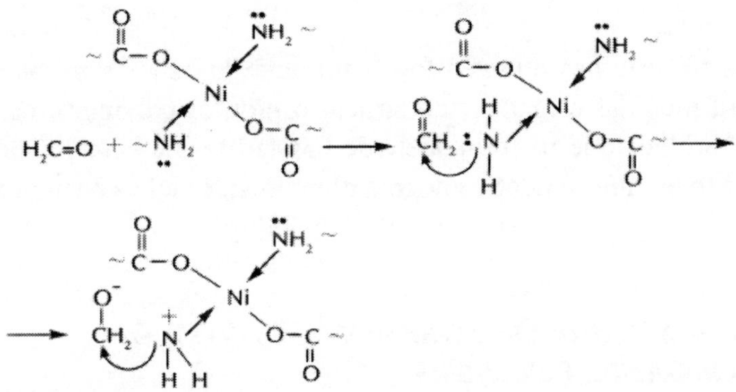

Formed intermediate complex attaches a free proton of water and turns in kation:

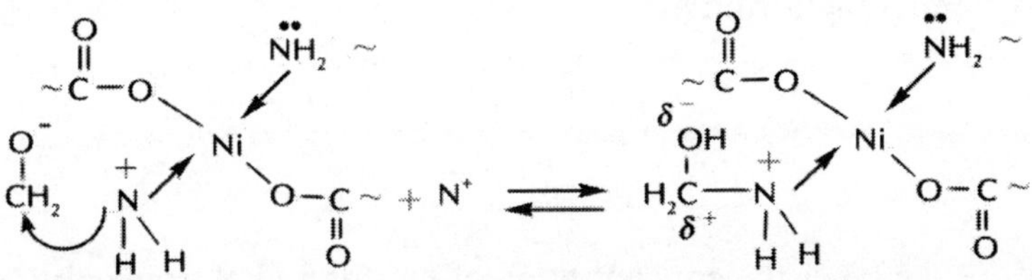

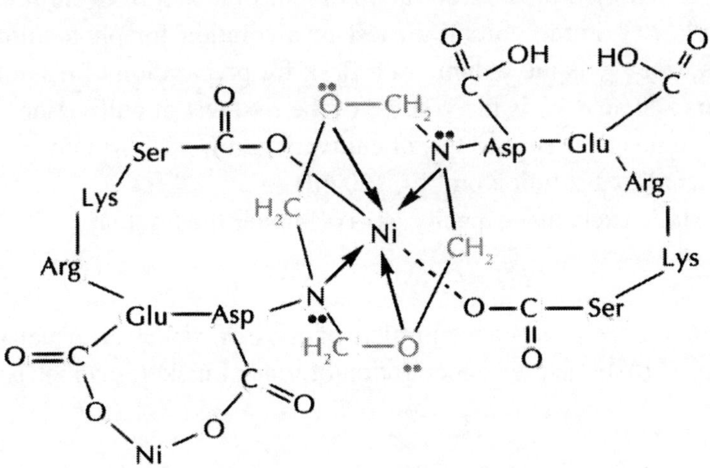

Molecules of polymer form the ordered structure around of an ion of metal, which is sewed by formaldehyde bridges.

29.3.3 RESEARCH OF STABILITY TO THERMAL-OXIDATIVE DEGRADATION

A series of tests has allowed to obtain data that shows introduction of an ion of metal as a matrix and structurization of polymer cross-section metilen bridges due to formaldehyde has led to increase in stability of polymer to thermal-oxidative degradation. Besides it has been proved that ions of nickel possess greater structuring ability.

29.3.4 RESEARCH OF SORPTION ACTIVITY OF BIOINORGANIC POLYMERS

Initial and final concentration of ions Ni^2 in investigated poly-measures pays off under the formula:

$$C(Ni) = \frac{C^{c.s.}(Ni^{2+}) \cdot V^{c.s.} \cdot V_{n2} \cdot V_{n1} \cdot V_{n0}}{V_{c1} \cdot \breve{n}_{samp}}, mol/l$$

where, $\tilde{N}^{\tilde{n}.s.}(Ni^{2+})$ is the concentration of ions of nickel from calibre schedule, mg/l; $V^{c.s.}$ is the volume of test of a solution for photocolorimeter analysis, ml; V_{n0} is the volume of a flask for preparation of a solution on sorption (500 ml); V_{n1} is the volume of the first test of cultivation, ml; V_{n2} is the volume of the second test of cultivation, ml; V_{c1} is volume of a flask with an analyzed solution on FEC (50 ml).

The static exchange capacity pays off under the Eq. (3):

$$SEC = C_F(Ni) - C_0(Ni), mg\text{-}eq/g$$

where, $C_F(Ni)-$ is the concentration of ions of nickel in a matrix after sorption; $C_0(Ni)-$ is the concentration of ions of nickel in a matrix before sorption.

$K_d = \dfrac{SEC}{\tilde{N}_s}$, The factor of distribution:

where, $C_s(Ni^{2+})$ is the concentration of sorption solution.

Results of experiments and calculations are presented in Tables 1–4.

TABLE 1 Static exchange capacity and factor of distribution for BIP-M in a solution of chloride of nickel.

No.	$C_0(Ni^{2+})$, mg-eq/g	$C_s(Ni^{2+})$, mol/l	$C_F(Ni^{2+})$, mg-eq/g	SEC (Ni^{2+}), mg-eq/g	% absorption (Ni^{2+})	$K_d(Ni^{2+})$
1		0.0518	0.54	0.47	98	9.1
2	0.07	0.0097	0.22	0.15	40	15.5
3		0.0053	0.15	0.08	27	15.1

TABLE 2 Static exchange capacity and factor of distribution for BP-M gelatin in a solution of chloride of nickel.

No.	$C_s(Ni^{2+})$, mol/l	SEC (Ni^{2+}), mg-eq/g	$K_d(Ni^{2+})$
1	0.0801	0.36	4.5
2	0.0102	0.14	13.7
3	0.0046	0.07	15.2

TABLE 3 Static exchange capacity and factor of distribution for BIP-M in a solution of sulfate of copper.

No.	$C^0_s(Cu^{2+})$, mol/l	$C^f_s(Cu^{2+})$, mol/l	SEC (Cu^{2+}), mg-eq/g	$K_d(Cu^{2+})$
1	0.0510	0.044	0.40	7.8
2	0.0255	0.0209	0.12	12.6
3	0.0116	0.0093	0.07	13.9

The static exchange capacity in case of sorption BP-M gelatin in a solution of sulfate of copper pays off under the Eq. (4):

$$SEC_{BIP-M}(Cu) = \frac{\tilde{N}_i^0(\tilde{N}\grave{e}) - \tilde{N}_i^F(\tilde{N}\grave{e})}{\breve{n}_{sorp.dry}}$$

TABLE 4 Static exchange capacity and factor of distribution for BP-M gelatin in a solution of sulfate of copper.

No.	$C^0_s(Cu^{2+})$, mol/l	$C^f_s(Cu^{2+})$, mol/l	SEC (Cu^{2+}), mg-eq/g	$K_d(Cu^{2+})$
1	0.0510	0.044	0.787	15.4
2	0.0255	0.0209	0.510	20.0
3	0.0116	0.0093	0.254	21.9

The generalized data on sorption investigated polymers in solutions of "related" and "another's" ions are presented in Fig. 1.

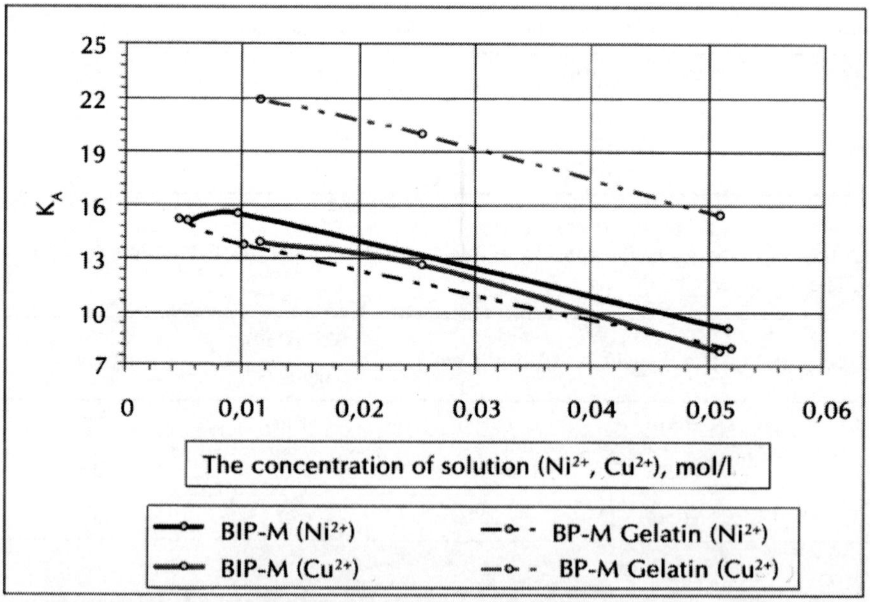

FIGURE 1 Dependence of factor of distribution on concentration of solutions Ni^{2+} и Cu^{2+} for BIP-M and BP-gelatin.

Thus, according to the purposes it is proved, that template synthesis leads to reception of the materials possessing selective properties in relation to template ion. The bioinorganic matrix of polymer (BIP-M) synthesized on ions of nickel in a series of tests proves as template sorbent possessing high absorbing selectivity to related ions and low to another's. So, factors of distribution for BIP-M in solutions with concentration 0.01 mol/l are K_d (Ni^{2+}) =15.5 and K_d (Cu^{2+}) =12.6.

It is proved, that the bioinorganic matrix of comparison on gelatin (BP-M Gelatin) possesses opposite selective properties. This matrix has been synthesized without template ions of metal, and it does not possess properties of template memory.

29.3.5 DIFFERENTIAL THERMAL ANALYSIS AND THERMO GRAVIMETRIC OF POLYMERS AND GELATIN

Results of research thermal degradation of BIP-Ni, BIP-Zn and BIP-Co and initial gelatin by method of differential thermal analysis have shown, that introduction of template ion raises stability of polymer to thermal-oxidative degradation.

During heating materials up to 100°C begins allocation in a gas phase adsorb waters with expenses energy.

Up to 300oC processes of structurization of materials proceed. There is an allocation flying vapor waters, low-molecular flying substances due to formation of amide. This process is displayed in the form of peaks on curves DTA for bioinorganic polymers. Gelatin is exposed significant destruction with allocation of flying substances that is displayed in the form of peak on curve DTA (Fig. 2).

After 250–300°C on curve sharp loss of weight by samples is clearly visible.

Thus, the received polymers have limits of thermostability 250–300°C. Thermostability of similar materials is caused by structuring influence of ions of metals, which interfere with allocation of flying substances.

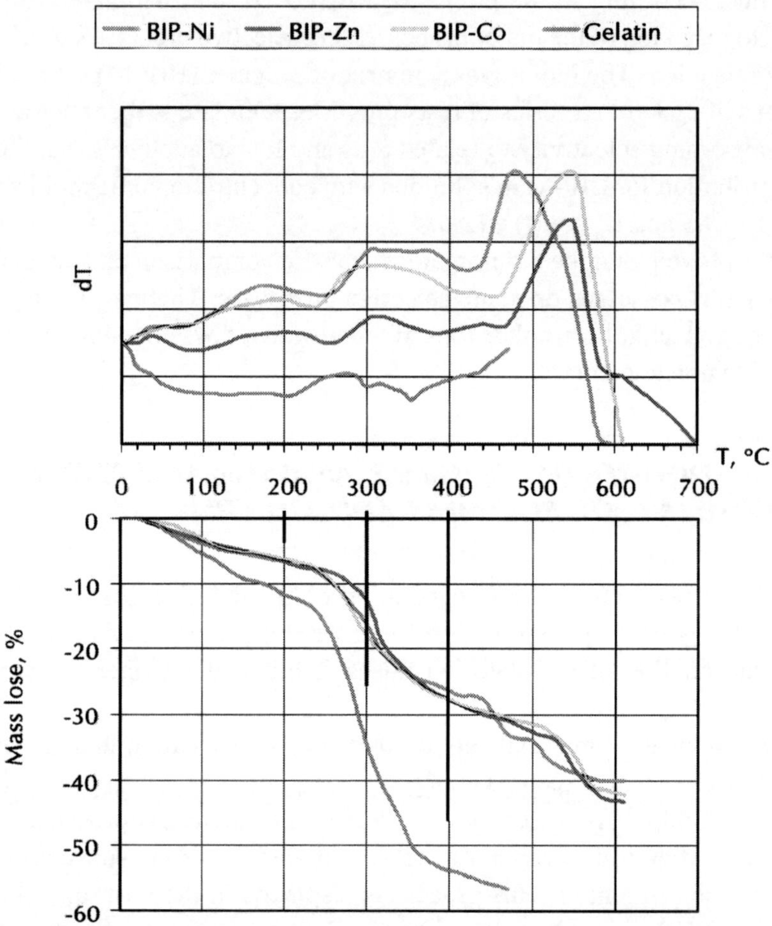

FIGURE 2 DTA and TG curves for BIP-Ni, BIP-Zn and BIP-Co and gelatin.

29.4 CONCLUSION

On the basis of the lead experiments and calculations is perspective using of the synthesized template bioinorganic polymers as ecologically safe and economic sorption materials in processes of water purification, selective extraction of ions of nickel of technological solutions or sewage, in hydrometallurgy, and also as components fire-heat-shielding of materials.

KEYWORDS

- **Fire-heat-shielding**
- **Photocolorimeter analysis**
- **Calibre schedule**
- **Hydrometallurgy**

REFERENCES

1. Pat.: CN101332999 (A) C01G3/02, Method for preparing Cu_2O or CuO hollow sub-microspheres with particle diameter controllable by water phase soft template method, Yun Fang [CN]; Jin Hu [CN]; Yueping Ren [CN]; Yongmei Xia [CN].
2. Altshuler, G. N.; Abramova, L. P.; Malyshenko, N. V.; Shkurenko, G. U.; Ostapova, E. V. Ion exchange activity of the crosslink calixarene polymer produced by template synthesis on base of Na^+, K^+ и Ba^{2+} matrixes, *Proc. Russian Acad.Sci. Chem. Ser.* **2005**, *8*, 1919.
3. Kablov, V. F.; Kondrutsky, D. A.; Sudnitsina, M. V. Synthesis and properties of phosphorus-containing polypeptides for effective ion-exchange materials of improved capacity, In *Progress in Monomers, Oligomers, Polymers, Composites and Nanocomposites*, Nova Science Publishers: New York 2009; pp. 113–123.
4. Kablov, V. F.; Kondrutsky, D. A.; Sudnitsina, M. V. The characteristics of phosphoprous-containing polypeptides synthesis and their properties, *Proceedings of Volgograd State Technical University*. Volgograd, **2008**, *5* (Ser. Chemistry and technology hetero-organic monomers and polymer materials), pp. 121–125.

CHAPTER 30

GENERATION OF STABLE MACRORADICALS IN LIGNIN ON EXPOSURE TO NITROGEN DIOXIDE

E. YA. DAVYDOV, I. S. GAPONOVA, A. D'AMORE, S. M. LOMAKIN, G. B. PARIISKII, T. V. POKHOLOK, and G. E. ZAIKOV

CONTENTS

30.1 INTRODUCTION

N.M. Emanuel Institute of Biochemical Physics (Russian Academy of Sciences) has several perfect installation of electron spin resonance (ESR). ESR-method is very effective and very sensitive for investigation of radical reaction in chemistry and biochemistry. This method was created in 1944 [1] by Prof. K.M. Zavoiskii (Academy of Sciences of USSR). V.V. Voevodskii, L.A. Blumenfeld, Ya.S. Lebedev, A.L. Buchachenko, E.G. Rozantsev, A.M. Wasserman, and A.L. Kovarskii did very much for application of this method for investigation of chemical and biochemical reactions [2–8].

The contributors of this chapter published the book in 2009 [4] about interaction of polymers with polluted atmosphere (nitrogen oxides). Their new paper is devoted for the problems of interaction between nitrogen dioxide and lignin. Mankind accumulated tremendous amount of lignin and did not find real ways for applications of it.

Therefore, any kind of research about lignin (applications) is very important for looking for the fields of utilization.

30.2 METHODS OF EXPERIMENTS

All information about application of ESR as well as optical spectroscopy for investigation of nitrogen dioxide reaction with polymers was published in 2005 [3].

30.3 EXPERIMENTAL RESULTS AND DISCUSSION

Detection of specific stable radicals in plants by EPR can be considered as a sensitive method of air pollution monitoring. The stable radicals are associated with products of free radical reactions initiated by various air pollutants, particularly nitrogen oxides.

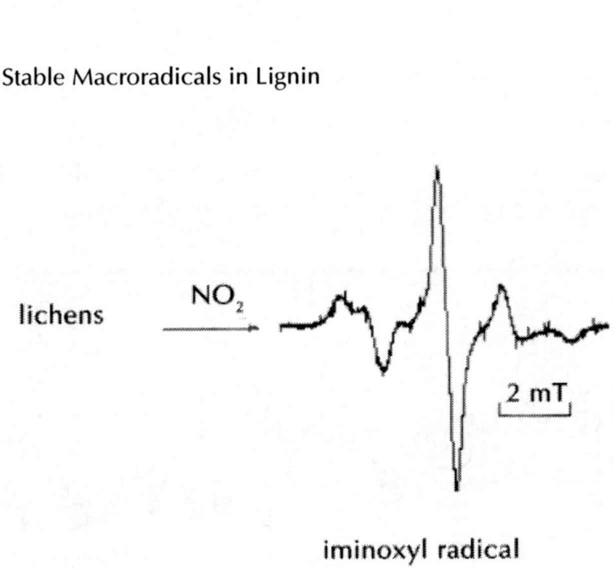

The nitrogen dioxide pre-treatment of wood pulp before oxygen bleaching in the process of paper production leads to appreciably decreased lignin contents (Fig. 1) [9, 10].

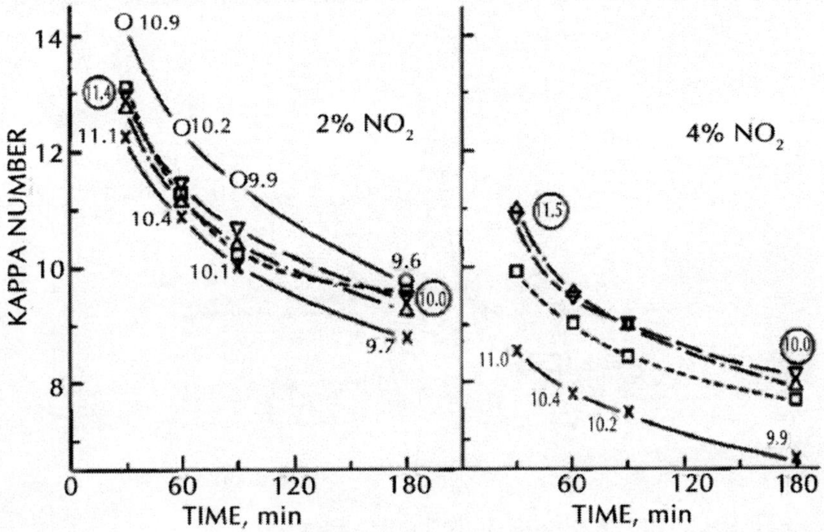

FIGURE 1 Dependence of kappa number on time for different conditions (2% NO_2 and 4% NO_2).

The impact of NO_2 on pines growing in cities (Vilnius, Kaunas) has been revealed. The reduction in NO_2 concentration in the atmosphere (1990–2006) determines an increase in pine radial increment. It is quite possible that this result is conditioned by interaction of NO_2 with reactive groups of lignin (Fig. 2) [11].

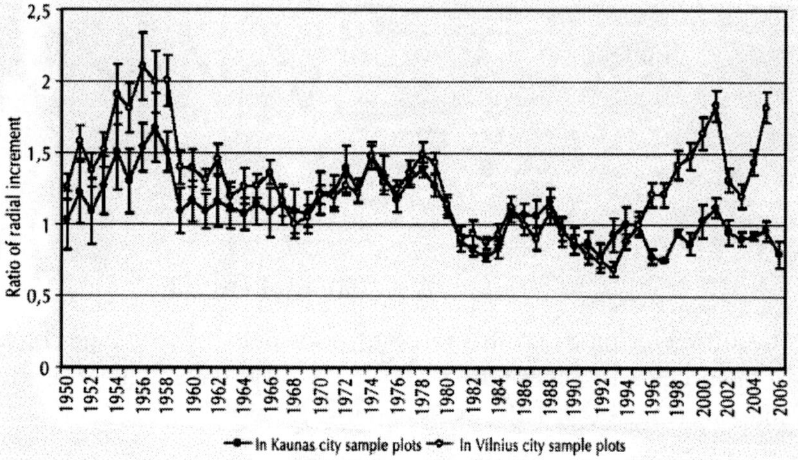

FIGURE 2 Dependence of ratio of radial increment on the year in Kaunas and Vilnius cities.

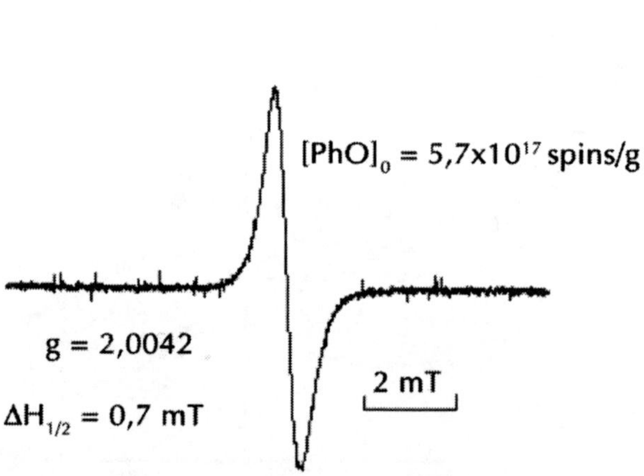

ESR spectrum of phenpxyl radicals detected in lignin before exposure to NO_2

Lignin is evidently reactive to nitrogen dioxide and that results in its destruction in NO_2 atmosphere [12]. The purpose of the present research was to examine the mechanism of primary radical reactions determining subsequent conversions of lignin units (Fig. 3).

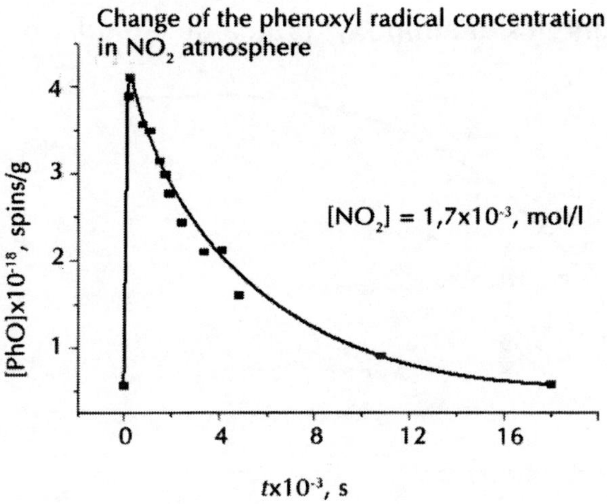

FIGURE 3 Dependence of phenoxyl radical concentration on time in nitrogen dioxide atmosphere.

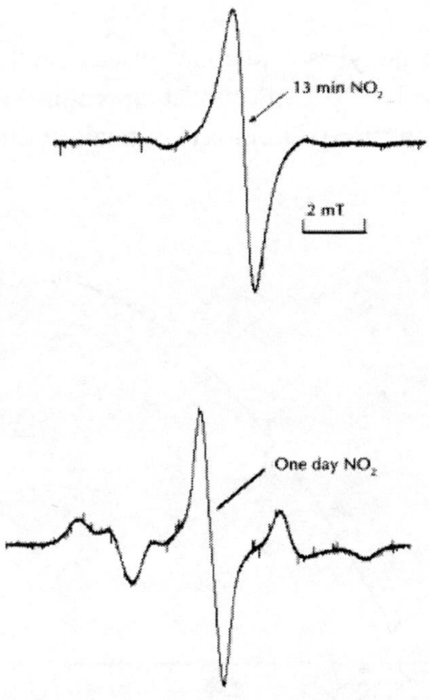

Decay of phenoxyl radicals is accompanied by appearance of iminoxyl radicals (Fig. 4).

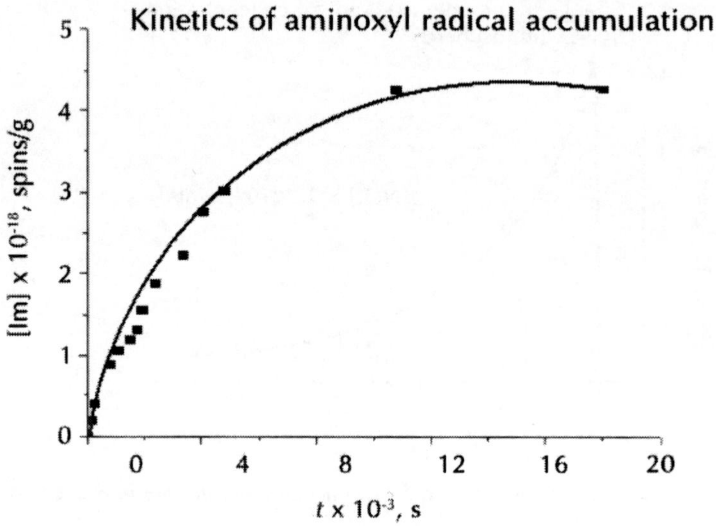

FIGURE 4 Accumulation of iminoxyl radicals in time.

Change in the iminoxyl radical concentration in the course of thermolysis of lignin exposed to NO_2 at room temperature. Figure 5 gives information about concentration of iminoxyl radicals at different temperatures.

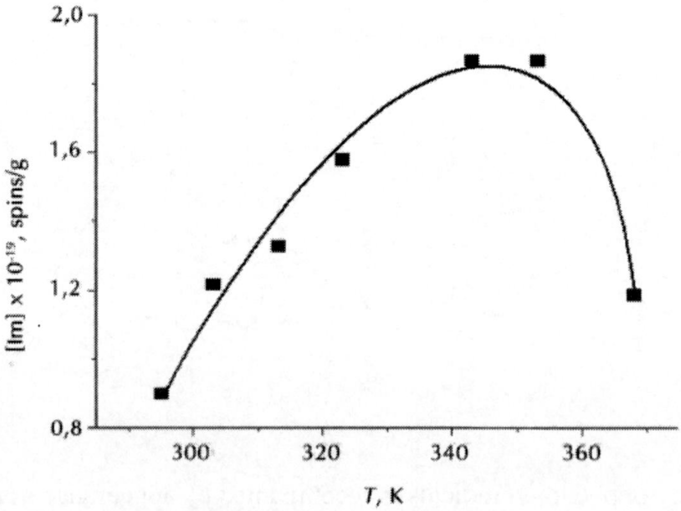

FIGURE 5 Concentration of iminoxyl radicals for different temperatures.

Thermolysis shifts the equilibrium decay of iminoxyl radicals by recombination with NO_2 to stable radicals.

$$\dot{\mathrm{Im}} + NO_2 \rightleftarrows \mathrm{Im}\,NO_2$$

Rates of iminoxyl radical accumulation and phenoxyl radical decay are approximately equal at various NO_2 concentrations (Fig. 6).

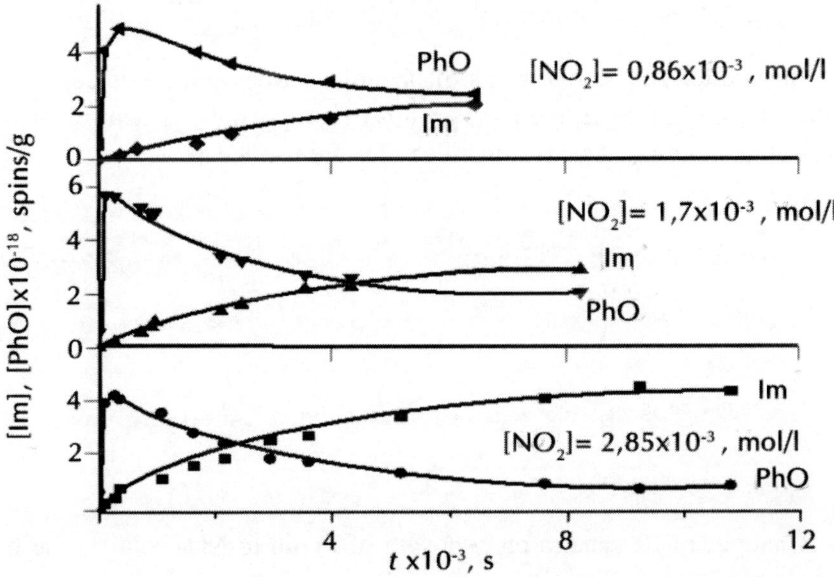

FIGURE 6 Dependence iminoxyl radical accumulation and phenoxyl radical decay on time.

The following scheme presents nitrosation mechanism of monomethoxyphenol groups of lignin with the formation of iminoxyl radicals. Initiators are dimers of NO_2 in the form of nitrosyl nitrate.

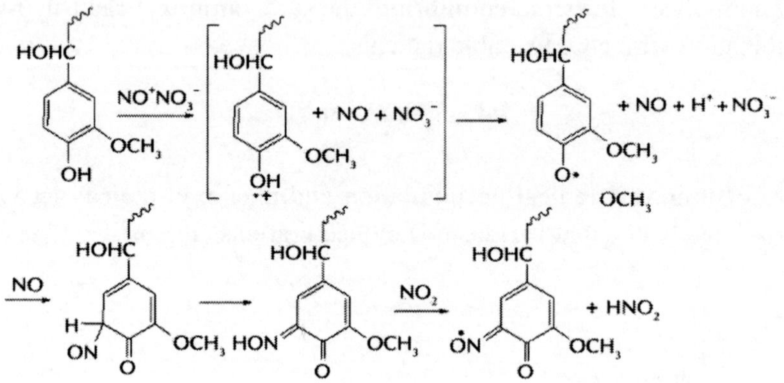

In parallel with conversions of phenols, nitrosyl nitrate is capable of oxidising multitude of hydroxyl groups of nonphenolic structures linking aryl rings in lignin. As a result, aldehydes are formed.

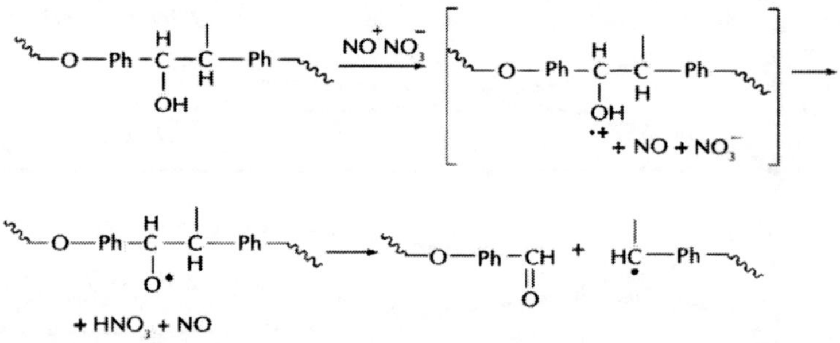

Changes of IR spectra on exposure of lignin to NO_2 confirm the presented mechanism (Fig. 7).

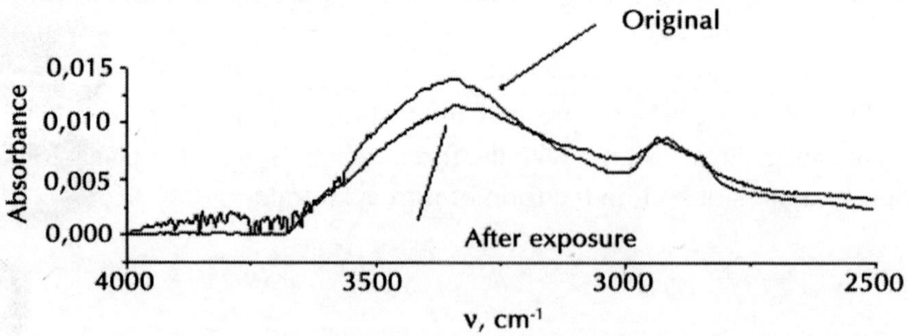

FIGURE 7 Infrared spectra of lignin before and after interaction with nitrogen dioxides.

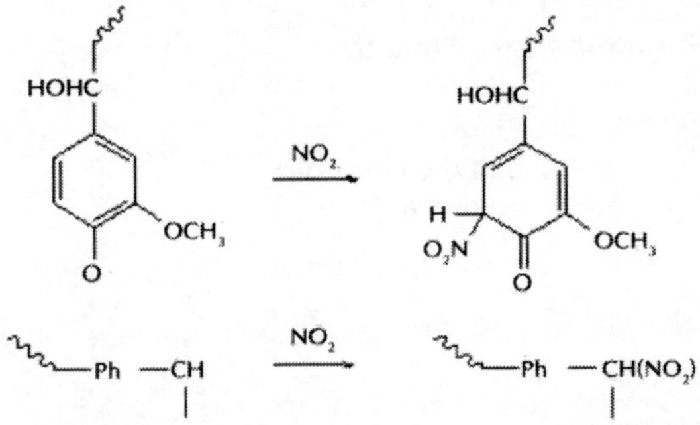

Aldehyde groups should be accumulated in the course of the lignin exposure to NO_2. The band of hydroxyl groups in lignin (3,300/cm) appreciably reduces for two days of NO_2 exposure. In addition, the decrease in intensity is observed for bands corresponding to stretching vibrations of C=C bonds of phenyl rings (1,512 and 1,450/cm). Moreover, the appearance of new intense bands at 1,558 and 1,337/cm belonging to asymmetric and symmetric stretching vibrations of N=O bonds in nitro groups takes place.

Aldehydes are precursors of acylaminoxyl radicals observed after prolonged exposure of lignin to NO_2.

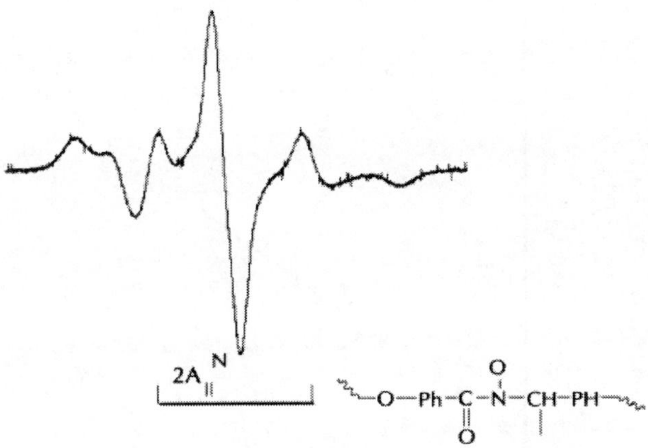

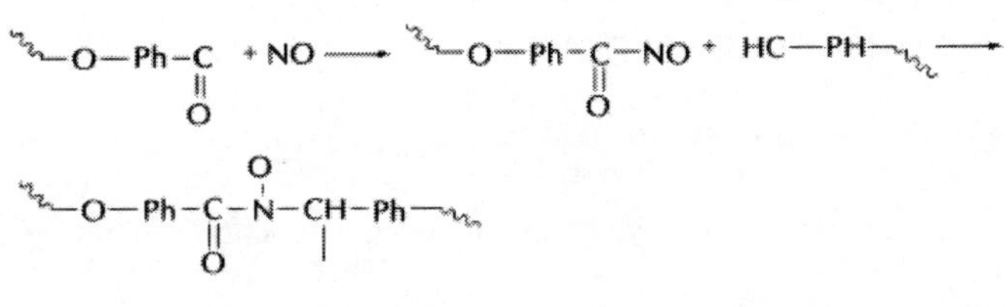

Dependencies of the iminoxyl radical yield in lignin on NO_2 concentrations is presented on the Figs. 8 and 9.

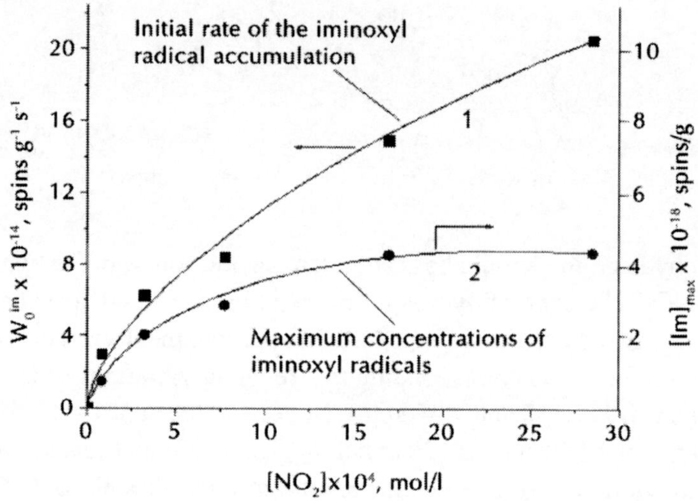

FIGURE 8 Dependence of accumulation rate of iminoxyl radicals on concentration of nitrogen dioxide.

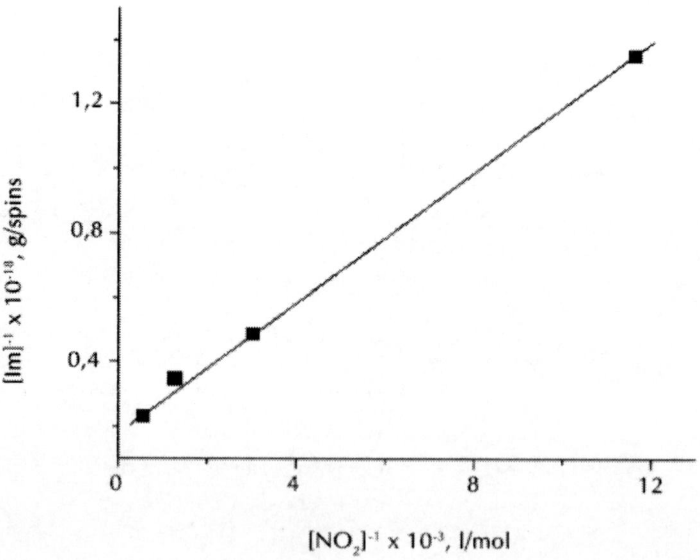

FIGURE 9 The iminoxyl radical yield in lignin as a function of nitrogen dioxide concentration.

$$[R_{im}]_{st} = \frac{[R_{im}]_{st}^{max}.b.[NO_2]}{1+b[NO_2]}$$

By ploting concentrations of R_{im} on steady-state levels ($[R_{im}]_{st}$) on NO_2 concentrations, the dependence reminding in form the Langmuir isotherm can be obtained. The adsorption of nitrogen dioxide on the lignin surface is the rate-determining factor for the iminoxyl radical generation.

30.4 CONCLUSION

1. Lignin is the very sensitive to the nitrogen dioxide action as evidenced by reasonably high concentrations of stable radicals, which can be accumulated in exposed samples (~2×10^{19} spins/g).
2. The generation of nitrogen containing radicals is connected with oxidative electron-transfer reactions initiated by dimers of NO_2 in phenol units and nonphenolic structures. The subsequent radical conversions include processes of degradation and modification (i.e., nitrosation and nitration) of lignin.
3. The formation of stable iminoxyl radicals in the presence of NO_2 is the typical process for lignin.

The detailed information about radical processes initiated by nitrogen dioxide in lignin may be found in Ref. [13].

KEYWORDS

- **Acylaminoxyl radicals**
- **Iminoxyl radical**
- **Langmuir isotherm**
- **Phenyl rings**

REFERENCES

1. Emanuel, N. M.; Zaikov, G. E.; Kritsman, V. A. In *Chemical Kinetics and Chain Reactions. Historical Aspects*, New York: Nova Science Publishers, 1992; 625 pp.
2. Burlakova, E. B.; Shilov, A. E.; Varfolomeev, S. D.; Zaikov, G. E. In *Chemical Kinetics*. VSP International Science Publishers: Leiden-Utrecht (The Netherlands), 2005; 682 pp.
3. Burlakova, E. B.; Shilov, A. E.; Varfolomeev, S. D.; Zaikov, G. E. In *Biological Kinetics*. VSP International Science Publishers: Leiden-Utrecht (The Netherlands), 2005; 356 pp.
4. Zaikov, G. E.; Davydov, E. Ya.; Pariiskii, G. B.; Gaponova, I. S.; Pokholok, T. V. In *Interaction of Polymers with Polluted Atmospheres. Nitrogen Oxides*, Smithers, Shawbury, Shrewsbury, Shropshire: UK 2009; 270 pp.
5. Zaikov, G. E. In *Chemical and Biochemical Physics. New Frontiers*, Nova Science Publishers: New York, 2006, 272 pp.
6. In *Resent Advances in Polymer Nanocomposites: Synthesis and Classification*, Thomas, S.; Zaikov, G. E.; Valsaraj, S. V.; Meera, A. P., Eds.; Brill Academic Publishers: Leiden-Boston (The Netherlands – USA), 2010; 436 pp.
7. Thomas, S.; Zaikov, G. E.; Valsaraj, S. V. In *Resent Advances in Polymer Nanocomposites*. VSP International Science Publishers: Leiden-Boston (The Netherlands – USA), 2009; 528 pp.
8. Zaikov, G. E.; Rakovsky, S. K. In *Ozonation of Organic and Polymeric Compounds*, Smithers, Shawbury, Shrewsbury, Shropshire: UK, 2009; 414 pp.
9. Engström T.; Samuelson O. Treatment of wood pulp with nitrogen dioxide before oxyge bleaching. *Polym. Bulletin.* **1981**, *4,* 219–223.
10. Bihani, B.; Sanuelson, O. Consumption of nitrogen oxides during pretreatment of wood with nitrogen dioxide and dioxide oxygen. *Wood Sci. Technol.* **1984**, *18*, 295–306.
11. Šimatonyte A.; Vendoviene J. Impact of sulfur and nitrogen dioxide concentration on radical increment dynamics of Scots pine (*Pinus Silvestris* L.) growing in cities. *Environ. Res. Eng. Manage.* **2009**, *48*, 25–34.
12. Dimmel, D. R.; Karim, M. R.; Savidakis, M. C.; Bozell, J. J. Pulping catalysts from lignin (5). Nitrogen dioxide oxidation of lignin models to benzoquinone. *J. Wood Chem. Techn.* **1995**, *16*, 169–189.
13. Gaponova, I. S.; Davydov, E. Ya.; Lomakin, S. M.; Pariiskii, G. B.; Pokholok, T. V. Features of stable radical generation in lignin on exposure to nitrogen dioxide. *Polym. Degr. Stability.* **2010**, *95*, 1177–1182.

INDEX

B

E

N

T

W

X

Y

Z